ZUMBERGE'S LABORATORY MANUAL FOR

Physical Geology

The McGraw·Hill Companies

ZUMBERGE'S LABORATORY MANUAL FOR PHYSICAL GEOLOGY, FIFTEENTH EDITION

Published by McGraw-Hill, a business unit of The McGraw-Hill Companies, Inc., 1221 Avenue of the Americas, New York, NY 10020.

This book is printed on acid-free paper.

1 2 3 4 5 6 7 8 9 0 DOW/DOW 1 0 9 8 7 6 5 4 3 2 1 0

ISBN 978-0-07-352415-3
MHID 0-07-352415-8

Vice President & Editor-in-Chief: *Marty Lange*
Vice President EDP/Central Publishing Services: *Kimberly Meriwether David*
Publisher: *Ryan Blankenship*
Senior Sponsoring Editor: *Debra Hash*
Senior Marketing Manager: *Lisa Nicks*
Senior Project Manager: *Jane Mohr*
Design Coordinator: *Brenda A. Rolwes*
Cover Designer: *Studio Montage, St. Louis, Missouri*
Senior Photo Research Coordinator: *Lori Hancock*
Cover Image: *Martin R. Selznick*
Buyer: *Sandy Ludovissy*
Media Project Manager: *Balaji Sundararaman*
Compositor: *S4Carlisle Publishing Services*
Typeface: *10/12 Times Roman*
Printer: *R. R. Donnelley*

Photos: 1.1, 1.3, 1.6-1.21, 1.26-1.34, 1.36-1.41, 1.48-1.55, 1.58-1.66: © Robert Rutford/James Carter, photographer
Page 1: © Martin Selznick and James Carter

The credits section for this book begins on page 283 and is considered an extension of the copyright page.

Some of the laboratory experiments included in this text may be hazardous if materials are handled improperly or if procedures are conducted incorrectly. Safety precautions are necessary when you are working with chemicals, glass test tubes, hot water baths, sharp instruments, and the like, or for any procedures that generally require caution. Your school may have set regulations regarding safety procedures that your instructor will explain to you. Should you have any problems with materials or procedures, please ask your instructor for help.

www.mhhe.com

ZUMBERGE'S LABORATORY MANUAL FOR

Physical Geology

FIFTEENTH EDITION

Robert H. Rutford
University of Texas, Dallas

James L. Carter
University of Texas, Dallas

Dedicated to the memory of

James H. Zumberge

whose dedication to his students and commitment to the geosciences

live on.

Jim was born and raised in Minneapolis, Minnesota, and received his university degrees from The University of Minnesota. He started his teaching career at the University of Michigan, and it was there that the first edition of this laboratory manual was written and published by Wm. C. Brown in 1951.

At both the national and international levels, Jim was active in polar science research and policy matters, serving on numerous government boards and as president of an international organization. He had a distinguished career in higher education administration, serving as president or chancellor of three universities.

He was a dedicated teacher with a love for geology. His enthusiasm for whatever he was involved in was contagious, and his students and colleagues were caught up in his view of the future. It is in recognition of his excellence that this manual carries his name.

CONTENTS

PART IV

Geologic Interpretation of Topographic Maps, Aerial Photographs, and Earth Satellite Images 105

PART V

Structural Geology 199

Plate Tectonics and Related Geologic Phenomena 247

NOTES TO USERS OF THIS MANUAL

In previous editions this section has been titled "Materials Needed by Students Using This Manual." Reviewers of the manual have suggested that we use this page to offer some advice and help as you begin your geology laboratory experience.

In the course of the laboratory exercises, you will be using a number of items that require some care. They are not "dangerous," but you should handle them with some caution until you are comfortable with their use. In the study of minerals, your laboratory instructor will provide you with a small bottle of dilute hydrochloric acid. Although this is a very weak acid, you should take care to keep it away from your eyes and from making contact with your clothes. In the event that you do spill, wash the affected area in water immediately and advise your instructor of the problem.

You also will be provided with a glass plate for hardness testing and a piece of unglazed porcelain to use as a streak plate. When using the glass plate and the porcelain streak plate, hold them flat on the table top, not in your hand.

Following is a list of materials that will be useful for completing the exercises in this manual. For example, the hand lens should be available for the identification of the minerals in rocks. Other materials will be required to complete the exercises dealing with maps or photos. Since the authors do not know which of the exercises you will be assigned, we urge you to ask your instructor about the exact materials that you will need during the semester.

1. 10x hand lens.
2. Scale ("ruler") in metric and English units.
3. Colored pencils (red, blue, and assorted other colors).
4. Felt-tip pens (1/8" × 1/4" tip), three assorted colors).
5. Several medium to medium-soft pencils (2H or No. 2).
6. Small magnifying glass (optional) for map reading.
7. 8 ½" × 11" tracing paper.
8. Eraser (art gum or equivalent).
9. Inexpensive pencil sharpener.
10. Inexpensive compass, for drawing circles.
11. Dividers (optional), for measuring distances on maps.

Finally, a word about units of measure. While the United States has "officially" adopted the metric system, the truth is that we continue to use both metric and English (U.S.) units of measure in this country. Therefore, we have made a conscious decision to use both in this manual in the hope that you will become more comfortable with the conversion of feet and miles into meters and kilometers and the reverse. Most of the maps used in the United States are in English units. On the inside front cover you will find a conversion table for reference as you make the conversions required in the various exercises.

PREFACE

The geologic sciences continue to undergo remarkable changes. Those changes that have endured over time have been incorporated into each edition of this manual since the first edition was published in 1951. Although the subject matter has changed and expanded in scope, the number of laboratory sessions in a given academic quarter or semester has not increased. Because the time available in a quarter or semester cannot be expanded without disrupting the class schedule for the entire college or university, the problem of too much material for too little time poses a dilemma for authors, instructors, and students.

Approach

On the assumption that the subject matter to be covered in any course is the prerogative of the instructor and not the authors, we have written a manual that contains more material than can be covered in a single laboratory course, thereby leaving the selection of individual exercises to the instructor. While we believe the overall scope of this manual is in keeping with the general subject material covered in a beginning laboratory course, we think the instructors should determine the specific exercises that are in keeping with their own ideas of how to organize and present subject material.

We are aware that some instructors will wish to introduce Plate Tectonics early in the course. Part VI, Plate Tectonics and Related Geologic Phenomena, is written so that it stands alone and can be used at any time. At The University of Texas at Dallas the Physical Geology lecture and laboratory courses are taught by two professors, one who opens the course with plate tectonics, the other who closes with it. In Parts IV and V the majority of the exercises also are self-contained so that instructors may use them in the order they prefer.

In addition to the variety of laboratory exercises offered, we also provide you with background material for each exercise. By allowing you to review the important concepts and geologic terms you will encounter in the laboratory, we hope to enhance your chance for a successful completion of the exercises. According to reviewers and users of past editions, the supplemental material provided is particularly useful in those instances when you do not routinely bring your textbooks to class or when you are not concurrently enrolled in the lecture class. To supplement this background material, we include a *Glossary* (pp. 000–000) as an integral part of the manual.

The fifteenth edition follows the same overall organization of past editions. You will note that we provide answer sheets for all of the exercises. At the request of many users, we have provided tearout worksheets for mineral and rock identification.

New to This Edition

This edition of Zumberge's *Laboratory Manual for Physical Geology* reflects the continued effort by the authors and publisher to improve the quality of the figures and tables. Consistent colors and symbols for the various rock types are used, the sharpness and detail of the photos has been increased, and efforts have been made to make this edition more user friendly.

This manual has also been updated in other ways, including:

Part I:

- Revision of exercise format, revision of some tables for clarity
- Addition of illustrations
- Revision of the Classification of Sedimentary Rocks
- Updated exercises

Part II:

- Considerable rewrite of text and addition of text materials
- Refinement of line drawings
- Additional materials in section on correlation and fossils
- Complete rewrite of Exercise 7 dealing with radiometric dating and geologic ages
- Updated exercises

Part III:

- Minor changes to both text and illustrations for clarity and consistency
- Updated exercises

Part IV:

- New and revised figures
- Replaced Raisz landform map of the contiguous U.S. with U.S.G.S. Digital Shaded Relief Map of the Coterminous U.S.
- Updated information and photo of South Cascade Glacier
- Updated information on Lake Michigan water levels

Part V:

- Revision of illustrations and text for clarity and consistency

Part VI:

- Revision of illustrations and text for clarity and consistency
- New figure on the interior of the earth (fig. 6.2)
- New materials on recent earthquakes in Haiti and Chile
- New figure on earthquake magnitude and energy (box 6.1, fig. 3)
- Expansion of section dealing with volcanic islands and hot spots
- Updated exercises

The cover of this edition is a photo of a most interesting area in Southwest Colorado. A description of Red Mountain is provided on the outside back cover.

Website

www.mhhe.com/zumberge15e
This text-specific site gives you the opportunity to further explore topics presented in the book using the Internet. Students will find flashcards, animations, additional photos, and all of the weblinks listed in the lab manual. Included in the password-protected Instructor's Edition is an Instructor's Manual and a list of slides that accompany the fourteenth edition.

Acknowledgements

We acknowledge with special thanks the contributions of the graduate teaching assistants who have assisted during the revisions of this manual. The advice and counsel of our faculty and staff colleagues at The University of Texas at Dallas are also greatly appreciated.

To those who reviewed this and past editions, we express our thanks and appreciation for their critical comments and suggestions for improvement. These include:

John R. Anderson, *Georgia Perimeter College*
Anne Argast, *Indiana University-Purdue University Fort Wayne*
Jamal M. Assad, *California State University, Bakersfield*
Abbed Babaei, *Cleveland State*
Lynne Beatty, *Johnson County Community College*
Polly A. Bouker, *Georgia Perimeter College*
Phyllis Camilleri, *Austin Peay State University*
Professor Roseann J. Carlson, *Tidewater Community College*
Beth A. Christensen, *Georgia State University*
James N. Connelly, *University of Texas—Austin*
John Dassinger, *Chandler-Gilbert Community College*
Linda Davis, *Northern Illinois University*
Rene A. De Hon, *University of Louisiana at Monroe*
Jack Deibert, *Austin Peay State University*
Janice J. Dependahl, *Santa Barbara City College*
Chris Dewey, *Mississippi State University*
Paul K. Doss, *University of Southern Indiana*
David Gaylord, *Washington State University*
Cathy A. Grace, *University of Mississippi*
Nathan L. Green, *University of Alabama*
Jeff Grover, *Cuesta College*
Daniel Habib, *Queens College*
Professor Vicki Harder, *Texas A & M University*
Timothy H. Heaton, *University of South Dakota*
Dr. Thomas E. Hendrix, *Grand Valley State University*
Stephen C. Hildreth, Jr., *University of South Dakota*
Robert B. Jorstad, *Eastern Illinois University*
Steve Kadel, *Glendale Community College*
Dr. Phillip R. Kemmerly, *Austin Peay State University*
Professor Ray Kenny, *Arizona State University*
Dr. Rudi H. Kiefer, *University of North Carolina, Wilmington*
David T. King, Jr., *Auburn University*
Gary L. Kinsland, *University of Louisiana—Lafayette*
M. John Kocurko, *Midwestern State University*
Mark Kulp, *University of New Orleans*
Ming-Kuo Lee, *Auburn University*
Kari Lavalli, *Boston University*
Neil Lundberg, *Florida State University—Emeritus*
Jerry F. Magloughlin, *Colorado State University*
Nasser M. Mansoor, *State University of New Jersey, Rutgers*
Glenn M. Mason, *Indiana State University*
Ryan Mathur, *Juniata College*
Joseph Meert, *University of Florida*
Linda D. Morse, *College of William and Mary*
John E. Mylroie, *Mississippi State University*
Jacob A. Napieralski, *University of Michigan—Dearborn*
Terry Naumann, *University of Alaska, Anchorage*
Max Neams, *Olivet Nazarene University*
Professor Anne Pasch, *University of Alaska, Anchorage*
Robert W. Pinker, *Johnson County Community College*
Dr. Mary Jo Richardson, *Texas A & M University*
David Steffy, *Jacksonville State University*
Professor Howard Stowell, *University of Alabama*
Lorraine W. Wolf, *Auburn University*
Aaron Yoshinobu, *Texas Tech University*

We especially acknowledge the contributions of Judy Taylor and John Craddock. Judy's assistance with a number of the figures was most greatly appreciated. John Craddock, a Professor of geology at Macalester College in St. Paul, Minnesota, is a user, reviewer, and contributor. His suggestions and input were most useful as we proceeded with this revision of the manual.

As authors we accept the full responsibility for any inadvertent errors that have crept into these pages, and we welcome comments from users if they discover such errors. We also hope that users will make suggestions to us that will assist us in the continued improvement of this manual in the future.

Finally, we extend our gratitude to the professional men and women of McGraw-Hill for their design of the format and expert help in transforming our manuscript into a final product.

About the Authors

Robert H. Rutford

Bob, Excellence in Education Foundation Chaired Professor in Geosciences Emeritus and President Emeritus of The University of Texas at Dallas, was born and raised in Minnesota where he attended the University of Minnesota, graduating with a B.A. in Geography in 1954.

He was commissioned a Second Lieutenant in the U.S. Army and spent a year in Greenland. He returned to the University of Minnesota as a graduate student in geography, completing his M.A. in 1963.

In 1959 he made the first of numerous trips to Antarctica. He changed his major to geology and received his Ph.D. in 1969. His dissertation dealt with the glacial geology and geomorphology of the Ellsworth Mountains, Antarctica.

His academic career began at the University of South Dakota, then at the University of Nebraska-Lincoln and following a period at NSF in Washington, D.C. as Director of the Office of Polar programs, he returned to Nebraska to serve as Vice Chancellor. In 1982 he became the president of the University of Texas at Dallas, a position he held for 12 years. In 1994 he returned to the faculty where he was an active teacher with classes in physical geology, global environments, glaciers, and deserts until 2007. He served as the U.S. Delegate to the International Scientific Committee on Antarctic Research and served as president of that organization from 1998 to 2002.

Bob became a co-author of this manual in 1979, and as the senior author since 1992, has attempted to continue Jim Zumberge's record of excellence.

James L. Carter

James, who became a co-author of this manual in 1995, spent his early years in southern Texas where he developed his love for the outdoors. He received his B.S. in Geological/Mining Engineering at Texas Western (now University of Texas at El Paso) in 1961 and a Ph.D. from Rice University in geochemistry in 1965. His dissertation dealt with the chemistry and mineralogy of the earth's upper mantle as revealed by mantle xenoliths in basalts. His research and teaching interests are broad and include lunar studies and extraterrestrial resources, ore deposit genesis, geochemical explorations, environmental issues, and Late Cretaceous dinosaurs. He has studied ore deposits in Central America, China, India, Mexico, Chile and Peru, as well as in the United States. James was a Principal Investigator on the characterization of lunar regolith samples returned to earth from the six Apollo missions and the Russian LUNA 20 unmanned mission. He also made the lunar regolith simulants JSC-1, JSC-1A, JSC-1A-C, JSC-1A-F, JSC-1A-VF and JSC-1A-VFR.

He taught a wide range of graduate and undergraduate courses at The University of Texas at Dallas from its inception in 1969 until 2007 when he retired. In 2003 he was the recipient of the AAPG (American Association of Petroleum Geologists) Distinguished Educator Award in the Southwest Section. He is a member of the American Association of Professional Geologists.

PART I

Earth Materials

Background

The materials that make up the crust of the earth fall into two broad categories: minerals and rocks. Minerals are elements or chemical compounds formed by a number of natural processes. Rocks are aggregates of minerals or organic substances that occur in many different architectural forms over the face of the earth, and they contain a significant part of the geologic history of the region where they occur. To identify them and understand their history, you must be able to identify the minerals that make up the rocks.

The goal of Part I is to introduce students of geology to the identification of minerals and rocks through the use of simplified identification methods and classification schemes. Students will be provided with samples of minerals and rocks in the laboratory. These samples are called **hand specimens.** Ordinarily their study does not require a microscope or any means of magnification because the naked or corrected eye is sufficient to perceive their diagnostic characteristics. A feature of a mineral or rock that can be distinguished without the aid of magnification is said to be **macroscopic** (also **megascopic**) in size. Conversely, a feature that can be identified only with the aid of magnifiers is said to be **microscopic** in size. The exercises that deal with the identification and classification of minerals and rocks in Part I are based only on macroscopic features.

Minerals

Definition

A **mineral** is a naturally occurring, crystalline, inorganic, homogeneous solid with a chemical composition that is either fixed or varies within certain fixed limits, and a characteristic internal structure manifested in its exterior form and physical properties.

Mineral Identification

Common minerals are identified or recognized by testing them for general or specific physical properties. For example, the common substance table salt is actually a mineral composed of sodium chloride (NaCl) and bears the mineral name halite. The salty taste of halite is distinctive and is sufficient for identifying and distinguishing it from other substances such as sugar (not a mineral) that have a similar appearance. Chemical composition alone is not sufficient to identify minerals. The mineral graphite and the mineral diamond are both composed of a single element, carbon (C), but their physical properties are very different.

The taste test applied to halite is restrictive because it is the only mineral with the taste of common table salt. Other minerals may have a specific taste different from that of halite. Other common minerals can be tested by visual inspection for the physical properties of **crystal form, cleavage,** or **color,** or by using simple tools such as a knife blade or glass plate to test for the physical property of hardness.

The first step in learning how to identify common minerals is to become acquainted with the various physical properties that individually or collectively characterize a mineral specimen.

Properties of Minerals

The physical properties of minerals are those that can be observed generally in all minerals. They include such common features as luster, color, hardness, cleavage, streak, and specific gravity. **Special properties** are those that are found in only a few minerals. These include magnetism, double refraction, taste, odor, feel, and chemical reaction with acid. In your work in the laboratory, use the hand specimens sparingly when applying tests for the various properties.

General Physical Properties

Luster

The appearance of a fresh mineral surface in reflected light is its **luster.** A mineral that looks like a metal is said to have a **metallic luster.** Minerals that are **nonmetallic** are described by one of the following adjectives: **vitreous** (having the luster of glass); **resinous** (having the luster of resin); **pearly; silky; dull** or **earthy** (not bright or shiny).

Your laboratory instructor will display examples of minerals that possess these various lusters.

Color

The color of a mineral is determined by examining a fresh surface in reflected light. Color and luster are not the same. Some minerals are clear and transparent; others are opaque. The variations in color of a mineral are called **varieties** of the mineral (fig. 1.1).

Color is not a diagnostic property for the majority of nonmetallic minerals. Some nonmetallic minerals, however, have a constant color, which can be used as a diagnostic property. Examples are malachite, which is green, and azurite, which is blue. Most minerals with a metallic luster

Figure 1.1

The specimens in this photograph are all varieties of quartz. The difference in colors is due to various impurities. Clockwise from left: smoky quartz, rock crystal, rose quartz, citrine, amethyst.

vary little in color, and color of a freshly broken surface is a diagnostic property.

Hardness

The **hardness** of a mineral is its resistance to abrasion (scratching). Hardness can be determined either by trying to scratch a mineral of unknown hardness with a substance of known hardness or by using the unknown mineral to scratch a substance of known hardness. Hardness is determined on a relative scale (linear scale) called the **Mohs scale of hardness,** which consists of 10 common minerals arranged in order of their increasing hardness (table 1.1). In the laboratory, convenient materials other than these 10 specific minerals may be used for determining hardness.

In contrast to the Mohs scale, the Vickers scale is an example of a nonlinear scale that points out the great difference in hardness between the various minerals (fig. 1.2). Nonlinear hardness scales are used to test the hardness of materials, for example, by measuring the volume of an indentation left in the surface of the material under a known pressure.

In this manual, a mineral that scratches glass will be considered "hard," and one that does not scratch glass will be considered "soft." *In making hardness tests on a glass plate, do not hold the glass in your hand; keep it firmly on the table top.* If you think that you made a scratch on the glass, try to rub the scratch off. What appears to be a scratch may only be some of the mineral that has rubbed off on the glass.

Table 1.1 Mineral Hardness According to the Mohs Scale (A) and Some Common Materials (B)

Hardness	A	B
1	Talc	
2	Gypsum	
2.5		Fingernail
3	Calcite	
3.5		Copper coin
4	Fluorite	
5	Apatite	
5–5.5		Knife blade
5.5		Glass plate
6	Feldspar	
6.5		Steel file, Streak plate
7	Quartz	
8	Topaz	
9	Corundum	
10	Diamond	

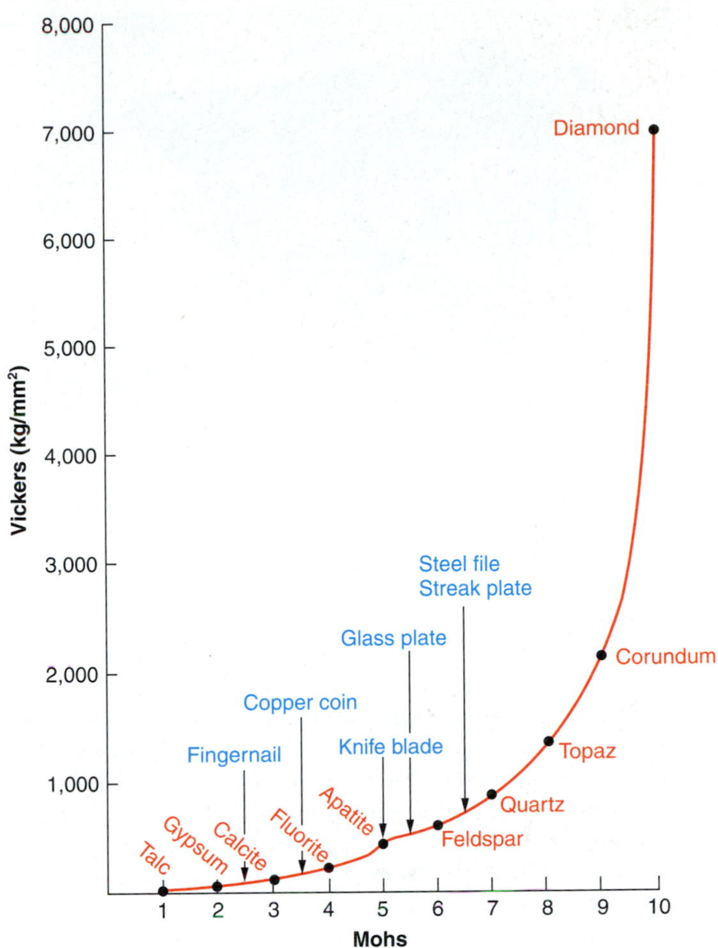

Figure 1.2

Mohs hardness scale plotted against Vickers indentation values (kg/mm^2).

Cleavage

Cleavage is the tendency of a mineral to break along definite planes of weakness that exist in the internal (atomic) structure of the mineral. Cleavage planes are related to the crystal system of the mineral and are always parallel to crystal faces or possible crystal faces. Cleavage may be conspicuous and is a characteristic physical property that is useful in mineral identification. It is almost impossible to break some minerals in such a way that cleavage planes do not develop. An example is calcite, with its rhombohedral cleavage.

Perfect cleavage describes cleavage planes with surfaces that are very smooth and flat and that reflect light much like a mirror. Other descriptors such as **good, fair,** and **poor** are used to describe cleavage surfaces that are less well defined. Some minerals exhibit excellent crystal faces but have no cleavage; quartz is such a mineral.

The cleavage surfaces of some minerals such as calcite, muscovite (fig. 1.3), halite, and fluorite are so well developed that they are easily detected. In others, the cleavage surfaces may be so discontinuous as to escape detection by casual inspection. Before deciding that a mineral has no cleavage,

Figure 1.3

Muscovite is a mineral with one direction of cleavage (basal cleavage).

turn it around in a strong light and observe whether there is some position in which the surface of the specimen reflects the light as if it were the reflecting surface of a dull mirror. If so, the mineral has cleavage, but the cleavage surface consists of several discontinuous parallel planes minutely separated, and rather than perfect cleavage, it has good, fair, or poor cleavage. As will be noted in the discussion of crystal form, it is important to differentiate between cleavage surfaces and crystal faces (the actual breaking of a mineral crystal may be useful in making this differentiation).

In assigning the number of cleavage planes to a specimen, do not make the mistake of calling two parallel planes bounding the opposite sides of a specimen two cleavage planes. In this case the specimen has two cleavage surfaces but only one plane of cleavage, that is, one direction of cleavage (fig. 1.3 and fig. 1.4 example 1). Halite has cubic cleavage, thus six sides, but only three planes of cleavage because the six sides are made up of three parallel pairs of cleavage surfaces.

The angle at which two cleavage planes intersect is diagnostic. This angle can be determined by inspection. In most cases, you will need to know whether the angle is 90 degrees, almost 90 degrees, or more or less than 90 degrees. The cleavage relationships that you will encounter during the course of your study of common minerals are tabulated for convenience in figure 1.4.

Parting

Minerals may exhibit the characteristic of **parting,** sometimes called false cleavage. Parting occurs along planes of weakness in the mineral, but usually the planes are more widely separated and often are due to twinning deformation or inclusions. Parting is not present in all specimens of a given mineral.

Fracture

Some minerals have no cleavage but show **fracture** that forms a surface with no relationship to the internal structure of the mineral; that is, the break occurs in a direction other than a cleavage plane. The broken surface may exhibit **conchoidal** fracture, where the fractured surface is curved and smooth with fine, concentric ridges (see fig. 1.9). The mineral asbestos (crysotile) is characterized by **fibrous** fracture. Other descriptive terms often used to describe fracture types include **hackly, uneven** (rough), **even** (smooth), and **earthy** (dull but smooth fracture surfaces common in soft mineral aggregates such as kaolinite).

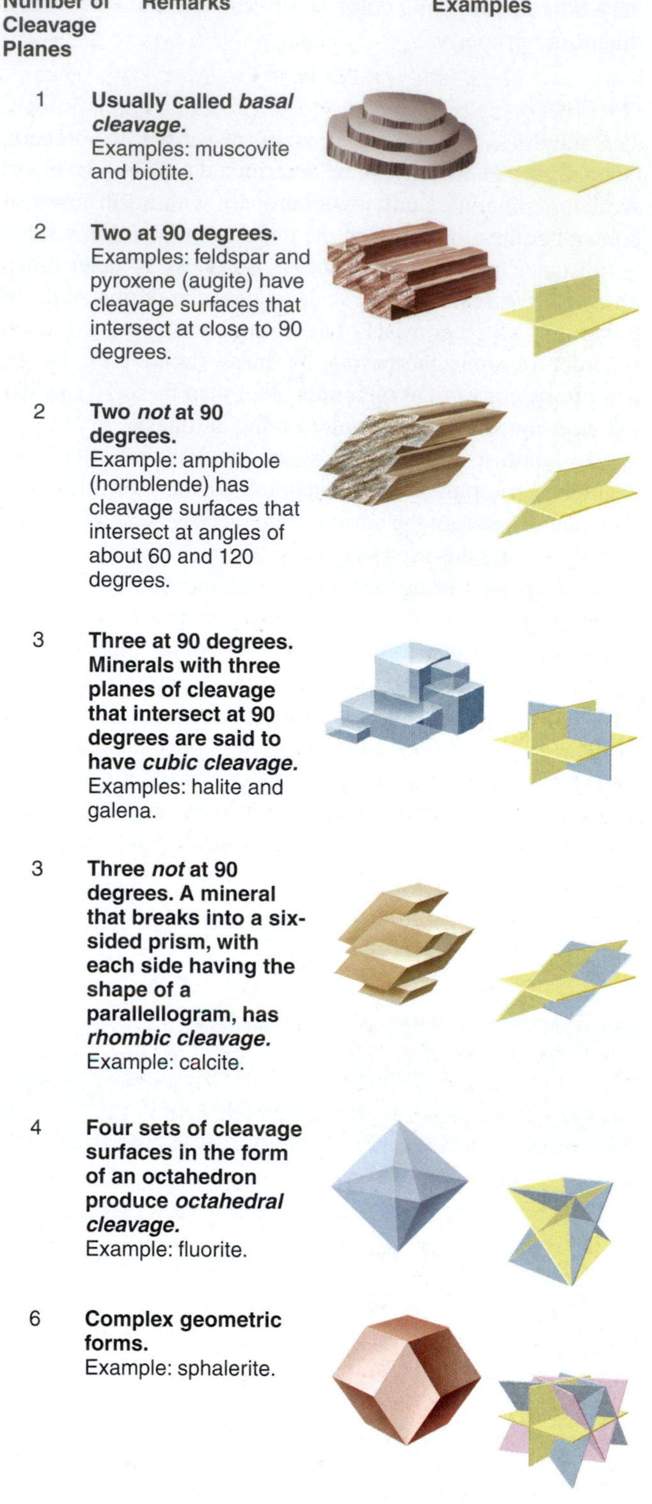

Number of Cleavage Planes	Remarks	Examples
1	**Usually called *basal* cleavage.** Examples: muscovite and biotite.	
2	**Two at 90 degrees.** Examples: feldspar and pyroxene (augite) have cleavage surfaces that intersect at close to 90 degrees.	
2	**Two *not* at 90 degrees.** Example: amphibole (hornblende) has cleavage surfaces that intersect at angles of about 60 and 120 degrees.	
3	**Three at 90 degrees. Minerals with three planes of cleavage that intersect at 90 degrees are said to have *cubic cleavage.*** Examples: halite and galena.	
3	**Three *not* at 90 degrees. A mineral that breaks into a six-sided prism, with each side having the shape of a parallellogram, has *rhombic cleavage.*** Example: calcite.	
4	**Four sets of cleavage surfaces in the form of an octahedron produce *octahedral cleavage.*** Example: fluorite.	
6	**Complex geometric forms.** Example: sphalerite.	

Figure 1.4

Descriptive notes on cleavage planes.

Streak

The color of a mineral's powder is its **streak.** The streak is determined by rubbing the hand specimen on a piece of unglazed porcelain **(streak plate).** Some minerals have a streak that is the same as the color of the hand specimen; others have a streak that differs in color from the hand specimen. The streak of minerals with a metallic luster is especially diagnostic.

Specific Gravity

The **specific gravity** (G) of a mineral is a number that represents the ratio of the mineral's weight to the weight of an equal volume of water. In contrast to density, defined as mass per unit volume, specific gravity is a dimensionless number. The higher the specific gravity, the greater the density of a mineral.

For purposes of estimating the specific gravity of the minerals in the laboratory, it is sufficient to utilize a simple **heft** test; that is, to compare the relative specific gravity of one sample by lifting the sample in question in one hand while lifting a sample of a known specific gravity of similar size in the other hand. For example, compare a sample of graphite (G = 2.2) in one hand with a sample of galena (G = 7.6) in the other hand. *Take care to compare samples of similar size.* This allows you to determine the relative specific gravity of minerals. When hefted, minerals such as graphite (G = 2.2) and gypsum (G = 2.3) are relatively "light," quartz (G = 2.65) and calcite (G = 2.7) are "average," whereas corundum (G = 4.0), magnetite (G = 5.2), and galena (G = 7.6) are "heavy."

Diaphaneity

The ability of a mineral to transmit light is its **diaphaneity.** If a mineral transmits light freely so that an object viewed through it is clearly outlined, the mineral is said to be **transparent.** If light passes through the mineral but the object viewed is not clearly outlined, the mineral is **translucent.** Some minerals are transparent in thin slices and translucent in thicker sections. If a mineral allows no light to pass through it, even in the thinnest slices, it is said to be **opaque.**

Tenacity

Tenacity is an index of a mineral's resistance to being broken or bent. It is not to be confused with hardness. Some of the terms used to describe tenacity are

Brittle—The mineral shatters when struck with a hammer or dropped on a hard surface.

Elastic—The mineral bends without breaking and returns to the original shape when stress is released.

Flexible—The mineral bends without breaking but does not return to its original shape when the stress is released.

Crystal Form

A **crystal** is a solid bounded by surfaces (crystal faces) that reflect the internal (atomic) structure of the mineral. **Crystal form** refers to the assemblage of crystal faces that constitute the exterior surface of the crystal. **Crystal symmetry** is the geometric relationship between the faces.

Symmetry in a crystal is determined by completing a few geometric operations. For example, a cube has six faces, each at right angles to the adjacent faces. A planar surface that divides the cube into portions such that the faces on one side of the plane are mirror images of the faces on the other side of the plane is called a **plane of symmetry.** A cube has nine such planes of symmetry. In the same way, imagine a line (axis) connecting the center of one face on a cube with the center of the face opposite it (see cubic model at the top left of fig. 1.5). Rotation of the cube about this axis will show that during a complete rotation a crystal face identical with the first face observed will appear in the same position four times. This is a fourfold **axis of symmetry.** Rotation of the cube around an axis connecting opposite corners will show that three times during a complete rotation an identical face appears, thus a **threefold axis of symmetry.**

The same mineral always shows the same angular relations between crystal faces, a relationship known as the **law of constancy of interfacial angles.** The symmetric relationship of crystal faces, related to the constancy of interfacial angles, is the basis for the recognition of the six crystal systems by crystallographers, and all crystalline substances crystallize in one of the six crystal systems (fig. 1.5). Some common substances, such as glass, are often described as crystalline, but in reality they are **amorphous**—they have solidified with no fixed or regular internal atomic structure.

The six crystal systems can be recognized by the symmetry they display. Figure 1.5 summarizes the basic elements of symmetry for each system and shows some examples of the **crystal habit** (the crystal form commonly taken by a given mineral) of some minerals you may see in the laboratory or a museum.

Perfect crystals usually form under special conditions in which there is open space for them to grow unrestricted during crystallization. In nature, they are the exception rather than the rule. Crystals are more commonly distorted, and their external form (crystal habit) is not perfectly developed. Regardless of the external form, the internal arrangement of the atoms within the crystals is fixed.

Many of the hand specimens you see in the laboratory will be made up of many minute crystals, so few crystal faces, or none, can be seen, and the specimen will appear granular. Other hand specimens may be fragments of larger crystals, so only one or two imperfect crystal faces can be recognized. Although perfect crystals are rare, most student laboratory collections contain some reasonably good crystals of quartz, calcite, gypsum, fluorite, and pyrite.

A word of caution: Cleavage fragments of minerals such as halite, calcite, and galena are often mistaken for crystals. This is because their cleavage fragments have the same geometric form as the crystal.

Two or more crystals of some minerals may be grown together in such a way that the individual parts are related through their internal structures. The external form that

CRYSTAL SYSTEM	CHARACTERISTICS	EXAMPLES*
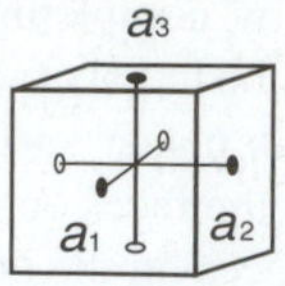 CUBIC (ISOMETRIC)	Three mutually perpendicular axes, all of the same length ($a_1 = a_2 = a_3$). Fourfold axis of symmetry around a_1, a_2, and a_3.	 Halite (cube) Pyrite Fluorite Galena; Magnetite (octahedron); Pyrite; Fluorite (twinned)
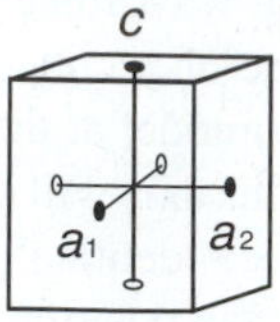 TETRAGONAL	Three mutually perpendicular axes, two of the same length ($a_1 = a_2$) and a third (c) of a length not equal to the other two. Fourfold axis of symmetry around c.	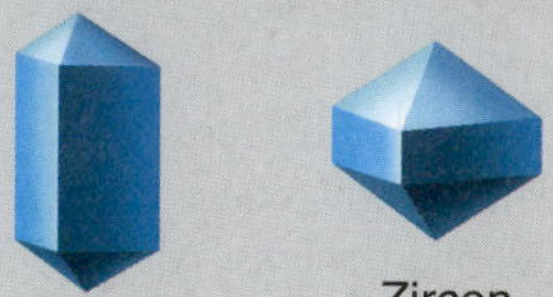 Zircon; Zircon
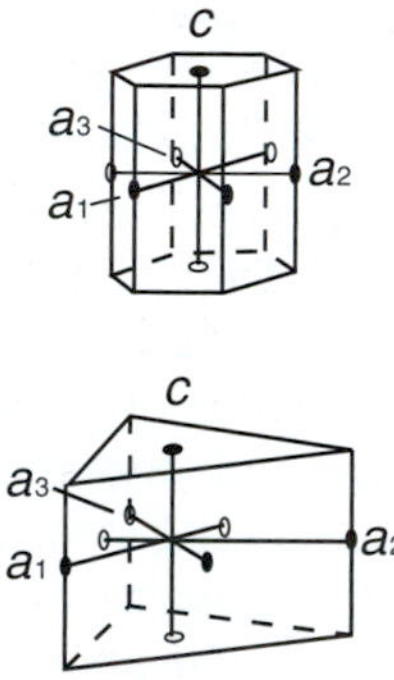 HEXAGONAL	Three horizontal axes of the same length ($a_1 = a_2 = a_3$) and intersecting at 120 degrees. The fourth axis (c) is perpendicular to the other three. Sixfold or threefold axis of symmetry around c.	 Apatite; Apatite; Quartz; Corundum; 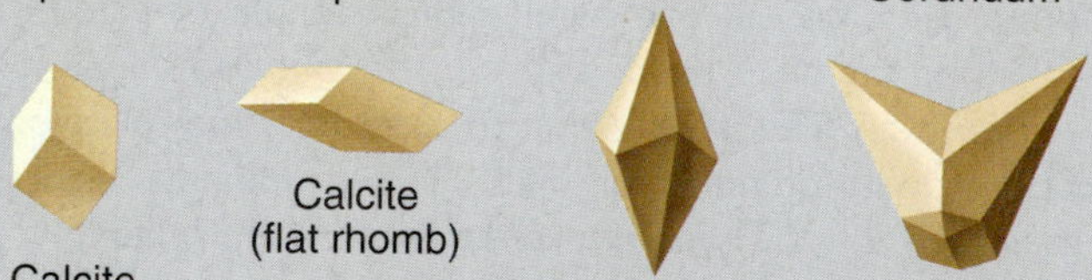 Calcite (steep rhomb); Calcite (flat rhomb); Calcite (scalenohedron); Calcite (twinned)
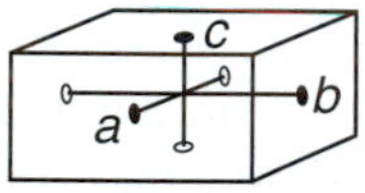 ORTHORHOMBIC	Three mutually perpendicular axes of different length. ($a \neq b \neq c$). Twofold axis of symmetry around a, b, and c.	 Topaz; Staurolite** (twinned)
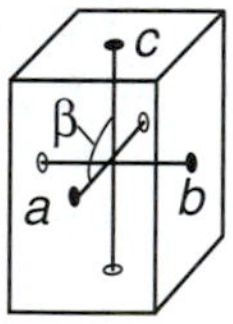 MONOCLINIC	Two mutually perpendicular axes (b and c) of any length. A third axis (a) at an oblique angle (β) to the plane of the other two. Twofold axis of symmetry around b.	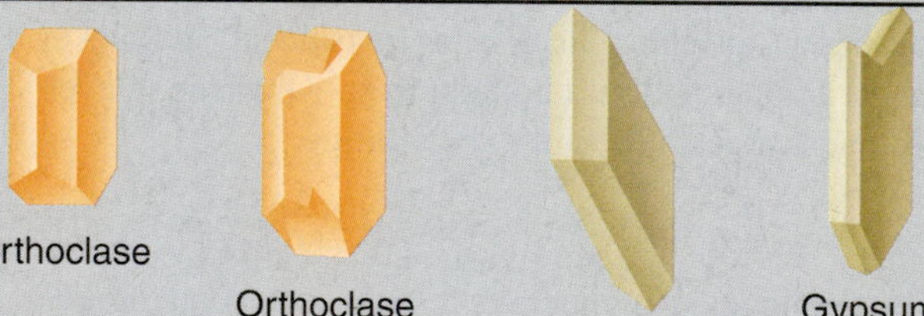 Orthoclase; Orthoclase (carlsbad twin); Gypsum; Gypsum (twinned)
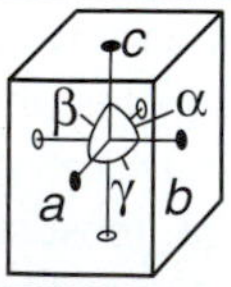 TRICLINIC	Three axes at oblique angles (α, β, and γ), all of unequal length. No rotational symmetry.	 Plagioclase

Figure 1.5

Characteristics of the six crystal systems and some examples.

*Colors have been added to the original and are not accurate. They are shown for illustrative purposes only. *Most laboratory collections of minerals for individual student use do not include crystals of these minerals. The collection may, however, contain incomplete single crystals, fragments of single crystals, or aggregates of crystals of one or more minerals. The best examples of these and other crystals may be seen on display in most mineralogical museums. **Staurolite is actually monoclinic but is also classified as pseudo-orthorhombic. Pseudo-orthorhombic means that staurolite appears to be orthorhombic because the angle β in the monoclinic system (see left-hand column under monoclinic) is so close to 90 degrees that in hand specimens it is not possible to discern that the angle β for staurolite is actually 89 degrees, 57 minutes.*

results is manifested in a **twinned crystal.** Some twins appear to have grown side by side (plagioclase), some are reversed or are mirror images (calcite), and others appear to have penetrated one another (fluorite, orthoclase, staurolite). Recognition of twinned crystals may be useful in mineral identification.

Special Properties

Magnetism

The test for **magnetism,** the permanent magnetic effect of naturally magnetic rocks, requires the use of a common magnet or magnetized knife blade. Usually, magnetite is the only mineral in your collection that will be attracted by a magnet.

Double Refraction (Birefringence)

If an object appears to be double when viewed through a transparent mineral, the mineral is said to have **double refraction.** Calcite is the best common example (see fig. 1.14).

Taste

The distinctive saline **taste** of halite is an easy means of identifying the mineral. Few minerals are soluble enough to possess this property. (*For obvious sanitary reasons, do not use the taste test on your laboratory hand specimens.*)

Odor

Some common minerals have a characteristic **odor** (smell) associated with them. Exhaling your breath on a kaolinite specimen will dampen the surface, causing the mineral to exude a musty or damp earthy odor. The streak of sphalerite will give off a "rotten egg" odor.

Feel

The **feel** of a mineral is the impression gained by handling or rubbing it. Terms used to describe feel are common descriptive adjectives such as **soapy, greasy, smooth, rough,** and so forth.

Chemical Reaction

Calcite will effervesce (bubble) when treated with cold dilute (1N) hydrochloric acid ($CaCO_3 + 2HCl \leftrightarrow CaCl_2 + CO_2 \uparrow + H_2O$). This **chemical reaction** is one example of many that occur in nature. (NOTE): Your laboratory instructor will provide the proper dilute acid if you are to use this test. *Take care to keep acid off of your skin and clothing. In the event of a spill, wash in water quickly and notify your lab instructor.*

EXERCISE 1

Name

Section Date

Identification of Common Minerals

Your instructor will provide you with a variety of minerals to be identified. Take time to examine the minerals and review the various physical properties described in the previous pages. Select several samples and examine them for luster, color, hardness, and streak, and compare their specific gravity (G) using the heft test.

Study table 1.2, which is divided into three groups:

Group I, Nonmetallic luster, light-colored;
Group II, Nonmetallic luster, dark-colored; and
Group III, Metallic luster.

Some minerals have physical properties that make their identification relatively easy. For example, graphite is soft, feels greasy, and marks both your hands and paper. Galena is "heavy," shiny, and has perfect cubic cleavage. Calcite has perfect rhombohedral cleavage, is easily scratched by a knife, reacts with cold dilute HCl, and in transparent specimens shows double refraction.

When you feel that you have an understanding of the various physical properties and the tests that you must apply to determine these properties, select a specimen at random from the group of minerals provided to you in the laboratory. Refer to figures 1.6 through 1.21 as an aid to identification. Be aware that your laboratory collection may contain some minerals that are not shown in the figures. Due to the normal variations within a single mineral species, some of the specimens may appear different from the same minerals shown in the figures in this manual.

Identification of the minerals listed in table 1.2 *follows an indentification scheme based on the sequential identification of luster, color, hardness, and cleavage.*

Using the worksheets for minerals provided, record your observations following the steps outlined below.

1. Carefully examine a single mineral specimen selected at random from the group of minerals provided to you in the laboratory. Assign it a number and record in the Worksheet for Minerals.
2. Determine whether the sample has a metallic or a nonmetallic luster.
3. Then determine whether it is light- or dark-colored. (The terms *dark* and *light* are subjective. A mineral that is "dark" to one observer may be "light" to another. This possibility is anticipated in table 1.2, where mineral specimens that could fall into either the "light" or "dark" categories are listed in both groups. The same is true for minerals that may exhibit either metallic or nonmetallic luster.)

4a. If the mineral falls into either Group I or II, proceed to test it first for hardness and then for cleavage. This will place the mineral with a small group of other minerals in table 1.2. Identification can be completed by noting other diagnostic general or specific physical properties.

4b. If the mineral falls into Group III (table 1.2), test it for hardness, cleavage, and streak and note general and special properties such as color until the mineral fits the description of one of those given in table 1.2 under Group III.

5. To assist you in confirming your identification, refer to the expanded mineral descriptions in table 1.3.
6. Your laboratory instructor will advise you as to the procedure to use to verify your identification.
7. Refer to table 1.3 to learn about occurrence, economic value, and uses of each mineral. The chemical groupings and composition of some of the common minerals are presented in table 1.4.

Name ______________________ Section ______________________ Date ______________________

Worksheet for Minerals

Sample #	Luster	Color Light/Dark	Hardness	Cleavage Angles	Streak	Special Properties	Mineral Name	Chemical Composition
4								

Name Section Date

Worksheet for Minerals

Sample #	Luster	Color Light/Dark	Hardness	Cleavage Angles	Streak	Special Properties	Mineral Name	Chemical Composition

Table 1.2 Simplified Mineral Identification Key

Group	Hardness	Cleavage	Properties	Mineral
Group I Nonmetallic luster, light-colored	Hard (scratches glass)	Shows cleavage	Vitreous luster. Color varies from white or cream to pink. Hardness 6.0. Cleavage two planes at nearly 90 degrees. Cleavage planes usually show muscle fiber appearance. Streak white. G = 2.6. Crystals common. Grains have glossy appearance.	K-FELDSPAR (Orthocase, Sanidine, Microcline)
			Vitreous luster. Color varies from white to gray or reddish to reddish brown. Hardness 6.0–6.5. Cleavage two planes at nearly 90 degrees. Cleavage planes show striations. Streak white. G = 2.6–2.8. Striations diagnostic. Some samples may show a play of colors.	PLAGIOCLASE
			Vitreous luster. Color white, gray, or pale brown. Hardness 6.0–7.0. Cleavage good in one direction. G = 3.2. Characterized by slim elongated crystals in parallel or subparallel alignment.	SILLIMANITE
			Vitreous luster. Color pale to dark green to yellow green. Hardness 6.0–7.0. Cleavage poor. Streak white. G = 3.3–3.5. Commonly in granular masses of striated prisms.	EPIDOTE
		No cleavage	Vitreous luster. Colorless or white, but almost any color can occur. Hardness 7.0. Cleavage none. Conchoidal fracture. Streak white. G = 2.65. Hexagonal crystals with striations perpendicular to the long dimension of the crystal common. Also massive.	QUARTZ
			Waxy or dull luster. Color varies from white to pale yellow, brown, or gray. Hardness 7.0. Cleavage none. Streak white. G = 2.6. Characterized by conchoidal fracture with sharp edges.	CHALCEDONY (Flint/Chert)
			Waxy luster. Variegated banded colors. Hardness 7.0. Cleavage none. Streak white G = 2.6.	CHALCEDONY (Agate)
			Vitreous luster. Color commonly olive-green, sometimes yellowish. Hardness 6.5–7.0. Cleavage indistinct. Streak white to gray. G = 3.2–4.4. Commonly in granular masses.	OLIVINE
			Vitreous luster. Color green, yellow, pink, blue. Hardness 7.0–7.5. Cleavage none. Streak white. G = 3.0–3.2. Characterized by elongated transparent striated prisms with triangular cross sections.	TOURMALINE
	Soft (does not scratch glass)	Shows cleavage	Vitreous luster. Colorless, also white, gray, yellow, or red. Hardness 2.5. Perfect cubic cleavage. Streak white. G = 2.2. Table salt taste.	HALITE
			Vitreous luster. Colorless and transparent, white, variety of colors possible. Hardness 3.0. Perfect rhombohedral cleavage. Streak white to gray. G = 3.0. Effervesces in cold dilute HCl; double refraction in transparent varieties.	CALCITE
			Vitreous to pearly luster. Colorless, white, pink, gray, greenish, or yellow-brown. Hardness 3.5–4.0. Rhombohedral cleavage. Streak white. G = 2.9–3.0. Crystals common, twinning common. Reaction with cold dilute HCl only when powdered.	DOLOMITE
			Vitreous to pearly luster. Colorless to white, gray, yellowish orange, or light brown. Hardness 2.0. Cleavage good in one direction producing thin sheets. Fracture may be fibrous. Streak white. G = 2.3. Crystals common, twinning common.	GYPSUM (Selenite)
			Pearly to greasy to dull luster. Color usually pale green, also shades of white or gray. Hardness 1.0. Perfect basal cleavage. Streak white. G = 2.6–2.8. Soapy feel.	TALC
			Vitreous to silky or pearly luster. Colorless to shades of green, gray, or brown. Hardness 2.5–3.5. Perfect basal cleavage yielding thin flexible and elastic sheets. Streak white. G = 2.8–2.9.	MUSCOVITE
			Greasy waxlike to silky luster. Color varies, shades of green most common. Hardness 2.0–3.0. Cleavage imperfect. Fibrous parting. Streak white. G = 2.5–2.6.	ASBESTOS (Chrysotile)
			Vitreous luster. Colorless but wide range of colors possible. Hardness 4.0. Perfect octahedral cleavage (4 planes). Streak white. G = 3.2.	FLUORITE
		No cleavage	Dull to earthy luster. Color white, often stained. Hardness 2.0–2.5. No cleavage apparent in common massive varieties. Streak white. G = 2.6–2.7. Earthy smell when damp.	KAOLINITE
			Pearly to greasy or dull luster. Color pale green or shades of gray. Hardness 1.0. No apparent cleavage in massive varieties. Streak white. G = 2.6–2.8. Soapy feel.	TALC
			Earthy luster. White or various colors. Hardness 3.0, may be less. No apparent cleavage in massive varieties. Streak white, G = 3.0. Effervesces in cold dilute HCl.	CALCITE
			Earthy luster. White or various colors. Hardness 3.5–4.0, but apparent may be less. No apparent cleavage. Streak white. G = 2.9–3.0. Reacts with cold dilute HCl only when powdered.	DOLOMITE
			Earthy luster. White color. Hardness 2.0, but apparent may be less. No apparent cleavage in massive varieties. Streak white. G = 2.3. Massive fine-grained variety called *alabaster*, fibrous variety called *satin spar*.	GYPSUM (Alabaster) (Satin Spar)
			Vitreous luster. Color yellow. Hardness 1.5–2.5. Cleavage poor; brittle fracture. Streak pale yellow. G = 1.5–2.5.	SULFUR

Table 1.2 Simplified Mineral Identification Key *(Continued)*

Group	Hardness	Cleavage	Properties	Mineral
Group II Nonmetallic luster, dark-colored	Hard (scratches glass)	Shows cleavage	Vitreous luster. Color dark green to black. Hardness 5.5–6.0. Cleavage two planes at nearly 90 degrees. Streak white to gray. G = 3.2–3.6. May exhibit parting.	AUGITE
			Vitreous luster. Color dark green to black. Hardness 5.0–6.0. Cleavage two planes with intersections at 56 and 124 degrees. Streak white to gray. G = 3.0–3.5. Six-sided crystals common.	HORNBLENDE
			Vitreous luster. Color varies from white, gray, to reddish or reddish brown. Hardness 6.0–6.5. Cleavage two planes at nearly 90 degrees. Striations on cleavage planes. Streak white. G = 2.6–2.8. Some forms exhibit play of colors on cleavage surfaces.	PLAGIOCLASE
			Vitreous to dull luster. Color brown to gray-brown. Hardness 7.0. Cleavage fair. Streak gray. G = 3.7–3.8. Characterized by prismatic crystals sometimes forming twins in the shape of crosses.	STAUROLITE
		No cleavage	Vitreous luster. Color varies but commonly brown. Hardness 9.0. Cleavage none. G = 4.0. Barrel-shaped hexagonal crystals with striations on basal faces common.	CORUNDUM
			Vitreous to resinous luster. Color varies but dark red to reddish brown common. Hardness 7.0–7.5. Cleavage none. Streak white or shade of the mineral color. G = 3.4–4.2. Fracture may resemble a poor cleavage. Brittle.	GARNET
			Vitreous luster. Color commonly olive-green, sometimes yellowish. Hardness 6.5–7.0. Cleavage indistinct. Streak white or gray. G = 3.2–4.4. Commonly in granular masses.	OLIVINE
			Vitreous luster. Color gray to gray-black. Hardness 7.0. Cleavage none. Streak white. G = 2.65. Conchoidal fracture. Crystals common; also a variety of massive forms.	QUARTZ
			Waxy to dull luster. Color red to red-brown or brown. Hardness 7.0. Cleavage none. Streak white to gray. G = 2.6.	CHALCEDONY (Jasper)
			Waxy or dull luster. Color dark gray to black. Hardness 7.0. Cleavage none. Streak white. G = 2.6. Characterized by conchoidal fracture with sharp edges.	CHALCEDONY (Flint/Chert)
			Vitreous luster. Color green to black. Hardness 7.0–7.5. Cleavage none. Streak white. G = 3.0–3.2. Characterized by elongated transparent striated prisms with triangular cross sections.	TOURMALINE
	Soft (does not scratch glass)	Shows cleavage	Vitreous to pearly luster. Color dark green, brown, to black. Hardness 2.5–3.0. Perfect basal cleavage forming thin elastic sheets. Streak white to gray. G = 2.7–3.4.	BIOTITE
			Resinous luster. Color yellow-brown to dark brown. Hardness 3.5–4.0. Cleavage perfect in six directions (dodecahedral). Streak brown to light yellow or white. G = 3.9–4.1. Cleavage faces common; twinning common.	SPHALERITE
			Vitreous to earthy luster. Color green to greenish black. Hardness 2.0–3.0. Perfect basal cleavage forming flexible nonelastic sheets. Streak white to pale green. G = 2.3–3.3. May have slippery feel.	CHLORITE
			Vitreous luster. Colorless but wide range of colors possible. Hardness 4.0. Perfect octahedral cleavage (four planes). Streak white. G = 3.2.	FLUORITE
		No cleavage	Submetallic to earthy luster. Color red to red-brown. Hardness 5.0–6.0 but apparent may be as low as 1. Cleavage none. Streak red. G = 5.0–5.3. Earthy appearance.	HEMATITE "Soft Iron Ore"
			Vitreous to subresinous luster. Color varies; green, blue, brown, purple. Hardness 5.0. Cleavage poor basal. Streak white. G = 3.1–3.2. Crystals common.	APATITE
			Earthy luster. Color varies; yellow, yellow-brown to brownish black. Apparent hardness 1.0. No cleavage apparent to earthy masses. Streak brownish yellow to orange-yellow. G = 3.3–4.3. Earthy masses.	GOETHITE (Limonite)

Table 1.2		Simplified Mineral Identification Key *(Continued)*	
Group III Metallic luster	Black, green-black or dark green streak	Metallic luster. Color black. Hardness 5.5–6.0. Cleavage none. Streak black. G = 5.2. Strongly magnetic.	MAGNETITE
		Metallic luster. Color dark gray to black. Hardness 1.0. Perfect basal cleavage. Streak black. G = 2.1–2.2. Greasy feel, smudges fingers when handled.	GRAPHITE
		Metallic luster. Color brass-yellow. Hardness 6.0–6.5. Cleavage none. Streak greenish or brownish black. G = 4.8–5.0. Cubic crystals with striated faces common. *"Fool's gold."*	PYRITE
		Metallic luster. Color brass-yellow, often tarnished to bronze or purple. Hardness 3.5–4.0. Cleavage none. Streak greenish black. G = 4.1–4.3. Usually massive.	CHALCOPYRITE
		Bright metallic luster. Color shiny lead-gray. Hardness 2.5. Perfect cubic cleavage. Streak lead-gray. G = 7.6. Cleavage, high G, and softness diagnostic.	GALENA
	Red streak	Metallic luster. Color steel-gray. Hardness 5.0–6.0. Cleavage none. Streak red to red-brown. G = 5.3. Often micaceous or foliated. Brittle.	HEMATITE (Specularite)
		Metallic luster. Color copper. Hardness 3.0–3.5. Cleavage none. Streak copper. G = 8.8–8.9. Malleable.	COPPER
	Yellow, brown, or white streak	Metallic to dull luster. Color yellow-brown to dark brown, may be almost black. Hardness 5.0–5.5. Cleavage perfect parallel to side pinacoid. Streak brownish yellow to orange-yellow. G = 4.3. Brittle.	GOETHITE (Limonite)
		Submetallic to resinous luster. Color yellow to yellow-brown to dark brown. Hardness 3.5–4.0. Perfect cleavage in six directions (dodecahedral). Streak brown to light yellow to white. G = 3.9–4.1. Cleavage faces common; twinning common.	SPHALERITE

NOTE: Values for hardness and specific gravity have in most cases been rounded to the nearest tenth and have been revised to reflect the data in *Dana's New Mineralogy*, 8th ed., by Richard V. Gaines, H. Catherine W. Skinner, Eugene E. Foord, Bryan Mason, and Abraham Rosenzweig (New York, NY: John Wiley and Sons, Inc., 1997).

Figure 1.6 Quartz (Rock crystal).

Figure 1.7 Rose quartz.

Figure 1.8 Smoky quartz (crystal).

Figure 1.9 Chalcedony (Chert).

Figure 1.10 K-feldspar.

Figure 1.11 Plagioclase.

Figure 1.12 Gypsum (Selenite crystal).

Figure 1.13 Talc.

Figure 1.14 Calcite. (Note double refraction.)

Figure 1.15 Fluorite.

Figure 1.16 Biotite.

Figure 1.17 Olivine.

Figure 1.18 Hematite (Specularite).

Figure 1.19 Hematite.

Figure 1.20 Goethite (Limonite).

Figure 1.21 Pyrite (crystals).

Table 1.3	Mineral Catalog
APATITE Ca, F Phosphate Hexagonal	**Luster:** nonmetallic; vitreous to subresinous. **Color:** varies, greenish yellow, blue, green, brown, purple, white. **Hardness:** 5.0. **Cleavage:** poor basal; fracture conchoidal. **Streak:** white. **G** = 3.1–3.2. **Habit:** crystals common. Also, massive or granular forms. **Uses:** important source of phosphorus for fertilizers, phosphoric acid, detergents, and munitions.
ASBESTOS (Fiberous Serpentine) Mg, Al Silicate Monoclinic	**Luster:** nonmetallic; greasy or waxlike in massive varieties, silky when fibrous. **Color:** varies, shades of green most common. **Hardness:** 2.0–3.0. **Cleavage:** none. **Streak:** white. **G** = 2.5–2.6. **Habit:** occurs in massive, platy, and fibrous forms. **Uses:** wide industrial uses, especially in roofing and fire-resistant materials.
AUGITE Ca, Mg, Fe, Al Silicate Monoclinic	**Luster:** nonmetallic; vitreous. **Color:** dark green to black. **Hardness:** 5.5–6.0. **Cleavage:** good, two planes at nearly 90 degrees; may exhibit well-developed parting. **Streak:** white to gray. **G** = 3.2–3.6. **Habit:** short, stubby eight-sided prismatic crystals. Often in granular crystalline masses. Most important ferromagnesium mineral in dark igneous rocks. **Uses:** no commercial value. *NOTE:* The *pyroxene* group of minerals contains over 15 members. Augite is an example of the monoclinic members of this group.
BIOTITE K, Mg, Fe, Al Silicate Monoclinic	**Luster:** nonmetallic; vitreous to pearly. **Color:** dark green, brown, or black. **Hardness:** 2.5–3.0. **Cleavage:** perfect basal forming elastic sheets. **Streak:** white to gray. **G** = 2.7–3.4. **Habit:** crystals common as pseudohexagonal prisms, more commonly in sheets or granular crystalline masses. Common accessory mineral in igneous rocks, also important in some metamorphic rocks. **Uses:** commercial value as insulator and in electrical devices.
CALCITE $CaCO_3$ Hexagonal	**Luster:** nonmetallic; vitreous (may appear earthy in fine-grained massive forms). **Color:** colorless and transparent or white when pure; wide range of colors possible. **Hardness:** 3.0. **Cleavage:** perfect rhombohedral. **Streak:** white to gray. **G** = 2.7. **Habit:** crystals common. Also massive, granular, oolitic, or in a variety of other habits. Effervesces in cold dilute HCl. Strong double refraction in transparent varieties. Common and widely distributed rock-forming mineral in sedimentary and metamorphic rocks. **Uses:** chief raw material for portland cement; wide variety of other uses.
CHALCEDONY (Cryptocrystalline Quartz) SiO_2 Hexagonal	**Luster:** nonmetallic; waxy or dull. **Color:** varies. Chalcedony is a group name for a variety of extremely fine-grained, very diverse forms of quartz including *agate,* banded forms; *carnelian* or *sard,* red to brown; *jasper,* opaque and generally red or brown; *chert* and *flint,* massive, opaque, and ranging in color from white, pale yellow, brown, gray, to black and characterized by conchoidal fracture with sharp edges; *silicified wood,* reddish or brown showing wood structure. **Hardness:** 7.0. **Cleavage:** none. **Streak:** white. **G** = 2.6. **Habit:** occurs in a wide variety of sedimentary rocks and veins, cavities, or as dripstone. **Uses:** various forms used as semiprecious stones.
CHALCOPYRITE $CuFeS_2$ Hexagonal	**Luster:** metallic, opaque. **Color:** brass-yellow, often tarnished to bronze or purple. **Hardness:** 3.5–4.0. **Cleavage:** none, fracture uneven. **Streak:** greenish black. **G** = 4.1–4.3. **Habit:** may occur as small crystals but usually massive. **Uses:** most important as common copper ore mineral.
CHERT	See Chalcedony
CHLORITE Mg, Fe, Al Silicate Monoclinic	**Luster:** nonmetallic; vitreous to earthy. **Color:** green to green-blue. **Hardness:** 2.5–3.0. **Cleavage:** perfect basal forming flexible nonelastic sheets. **Streak:** white to pale green. **G** = 2.3–3.3. **Habit:** occurs as foliated masses or small flakes. Common in low-grade schists and as alteration product of other ferromagnesian minerals. **Uses:** no commercial value.
COPPER Cu Cubic (Isometric)	**Luster:** metallic. **Color:** copper. **Hardness:** 3.0–3.5. **Cleavage:** none; fracture hackly; malleable. **Streak:** copper. **G** = 8.8–8.9. **Habit:** distorted cubes and octahedrons; dendritic masses. **Uses:** ore of copper for pipes, electrical wire, coins, ammunition, brass.

Table 1.3	Mineral Catalog *(Continued)*
CORUNDUM Al_2O_3 Hexagonal	**Luster:** nonmetallic, adamantine to vitreous. **Color:** varies; yellow, brown, green, purple; gem varieties blue (sapphire) and red (ruby). **Hardness:** 9.0. **Cleavage:** none, basal parting common with striations on parting planes. **Streak:** white, **G** = 4.0. **Habit:** barrel-shaped crystals common, frequently with deep horizontal striations. **Uses:** wide uses as an abrasive.
DOLOMITE $CaMg(CO_3)_2$ Hexagonal	**Luster:** nonmetallic, vitreous to pearly. **Color:** colorless, white, gray, greenish, yellow-brown; other colors possible. **Hardness:** 3.5–4.0. **Cleavage:** rhombohedral. **Streak:** white. **G** = 2.9–3.0. **Habit:** crystals common. Twinning common. Fine-grained; massive and granular forms common also. Distinguished from calcite by fact that it effervesces in cold dilute HCl only when powdered. Widespread occurrence in sedimentary rocks. **Uses:** variety of industrial uses as flux, as a source of magnesia for refractory bricks, and as a source of magnesium or calcium metal.
EPIDOTE (Group) Complex Silicate Monoclinic	**Luster:** nonmetallic; vitreous. **Color:** pale to dark green to yellow-green. **Hardness:** 6.0–7.0. **Cleavage:** poor; fracture brittle. **Streak:** white. **G** = 3.3–3.5. **Habit:** massive or striated prisms. **Uses:** gemstone.
FLINT/JASPER	See Chalcedony
FLUORITE CaF_2 Cubic (Isometric)	**Luster:** nonmetallic, vitreous transparent to translucent. **Color:** colorless when pure; occurs in a wide variety of colors: yellow, green, blue, purple, brown, and shades in between. **Hardness:** 4.0. **Cleavage:** perfect octahedral (four directions parallel to faces of an octahedron). **Streak:** white. **G** = 3.2. **Habit:** twins fairly common. Common as a vein mineral. **Uses:** industrial use as a flux in steel and aluminum metal smelting; source of fluorine for hydrofluoric acid.
GALENA PbS Cubic (Isometric)	**Luster:** bright metallic. **Color:** lead-gray. **Hardness:** 2.5. **Cleavage:** perfect cubic. **Streak:** lead-gray. **G** = 7.6. **Habit:** crystals common, easily identified by cleavage, high specific gravity, and softness. **Uses:** chief lead ore.
GARNET Fe, Mg, Ca, Al Silicate Cubic (Isometric)	**Luster:** nonmetallic; vitreous to resinous. **Color:** varies but dark red and reddish brown most common; white, pink, yellow, green, black depending on composition. **Hardness:** 7.0–7.5. **Cleavage:** none. **Streak:** white or shade of mineral color. **G** = 3.4–4.2. **Habit:** crystals common. Also in granular masses. **Uses:** some value as an abrasive. Gemstone varieties are pyrope (red) and andradite (green).
GOETHITE (Limonite) FeO (OH) Orthorhombic	**Luster:** varies, crystals adamantine; metallic to dull in masses, fibrous varieties may be silky. **Color:** dark brown, yellow-brown, reddish brown, brownish black, yellow. **Hardness:** 5.0–5.5, but apparent may be as low as 1.0. **Cleavage:** perfect parallel to side pinacoid, fracture uneven. **Streak:** brownish yellow to orange-yellow. **G** = 4.3. **Habit:** crystals uncommon. Usually massive or earthy as residual from chemical weathering, or stalactic by direct precipitation. Often in radiating fibrous forms. This species includes the common brown and yellow-brown ferric oxides collectively called *limonite*. **Uses:** commercial source of iron ore.
GRAPHITE C Hexagonal	**Luster:** metallic to dull. **Color:** dark gray to black. **Hardness:** 1.0. **Cleavage:** perfect basal. **Streak:** black. **G** = 2.1–2.2. **Habit:** characteristic greasy feel, marks easily on paper. Crystals uncommon, usually as foliated masses. Common metamorphic mineral. **Uses:** wide industrial uses due to high melting temperature (3,000°C) and insolubility in acid. Used also as lead in pencils.
GYPSUM $CaSO_4{\cdot}2H_2O$ Monoclinic	**Luster:** nonmetallic; vitreous to pearly; some varieties silky. **Color:** colorless to white, gray, yellowish orange, light brown. **Hardness:** 2.0. **Cleavage:** good in one direction producing thin sheets; fracture conchoidal in one direction, fibrous in another. **Streak:** white. **G** = 2.3. **Habit:** crystals common and simple in habit; twinning common. Varieties include *selenite,* coarsely crystalline, colorless and transparent; *satinspar,* parallel fibrous structure; and *alabaster,* massive fine-grained gypsum. Occurs widely as sedimentary deposits and in many other ways. **Uses:** mined for use in wallboard, plaster, and filler for paper products.

Table 1.3	Mineral Catalog *(Continued)*
HALITE NaCl Cubic (Isometric)	**Luster:** nonmetallic, vitreous. Transparent to translucent. **Color:** colorless, also white, gray, yellow, red. **Hardness:** 2.5. **Cleavage:** perfect cubic. **Streak:** white. **G** = 2.2. **Habit:** crystals common, also massive or coarsely granular. Characteristic taste of table salt. **Uses:** widely used as source of both sodium and chlorine and as salt for table, pottery glaze, and industrial purposes.
HEMATITE Fe_2O_3 Hexagonal	**Luster:** metallic in form known as specularite and in crystals; submetallic to dull in other varieties. **Color:** steel-gray in specularite, dull to bright red in other varieties. **Hardness:** 5.0–6.0, but apparent may be as low as 1.0. **Cleavage:** none; basal parting fracture uneven. **Streak:** red-brown. **G** = 5.3. **Habit:** crystals uncommon. May occur in crystalline, botryoidal, or earthy masses. Specularite commonly micaceous or foliated. **Streak:** characteristic. **Uses:** common iron ore.
HORNBLENDE (Amphibole) Ca, Na, Mg, Fe, Al Silicate Monoclinic	**Luster:** nonmetallic, vitreous. **Color:** dark green, dark brown, black. **Hardness:** 5.0–6.0. **Cleavage:** perfect on two planes meeting at 56° and 124°. **Streak:** gray or pale green. **G** = 3.0–3.5. **Habit:** long, six-sided crystals common. **Color:** usually darker than other minerals in amphibole group. *Tremolite-Actinolite,* also a member of the amphibole group, is lighter in color and commonly is fibrous or asbestiform, may range in color from white to green, and has white streak. **Uses:** no commercial use. *Nephrite,* the amphibole form of jade, is used for jewelry.
ILMENITE $FeTiO_3$ Hexagonal	**Luster:** metallic to submetallic. **Color:** Black. **Hardness:** 5.0–6.0. **Cleavage:** none, conchoidal fracture. **Streak:** Black. **G:** 4.72. **Habit:** compact, massive, crystals usually thick tabular. **Uses:** the major source of titanium, used principally in the manufacture of titanium dioxide for paint pigments, and metallic titanium is used in aircraft and space vehicle construction.
KAOLINITE Hydrous Aluminosilicate Triclinic	**Luster:** nonmetallic, dull to earthy. **Color:** white, often stained by impurities to red, brown, or gray. **Hardness:** 2.0–2.5. **Cleavage:** perfect basal but rarely seen because of small grain size. **Streak:** white. **G** = 2.6–2.7. **Habit:** found in earthy masses. Earthy odor when damp. *NOTE:* Kaolinite is used here as an example of the clay minerals. It normally is not possible to distinguish the various clay minerals on the basis of their physical properties. Other clay minerals include *montmorillonite* (smectite), *illite,* and *vermiculite.* **Uses:** kaolinite is used as a paper filler and ceramic; montmorillonite for drilling muds; illite has no industrial use; and vermiculite is mined and processed for use as a lightweight aggregate, potting soils, and as insulation.
K-FELDSPAR (Orthoclase, Sanidine, Microcline) $K(AlSi_3O_8)$ Monoclinic (Orthoclase, Sanidine) Triclinic (Microcline)	**Luster:** nonmetallic, vitreous. **Color:** varies, white, cream, or pink; *sanidine,* the high-temperature variety, is usually colorless. **Hardness:** 6.0. **Cleavage:** two planes at nearly right angles. **Streak:** white. **G** = 2.6. **Habit:** crystals not common. Has glossy appearance. Distinguished from other feldspars by absence of twinning striations. *NOTE: Microcline* variety is triclinic. When light green the color is diagnostic, more commonly white, green, pink. Occurrence helpful; most K-feldspar in pegmatites is microcline. **Uses:** commonly used in ceramics, glassmaking, and scouring and cleansing products.
LIMONITE	See Goethite
MAGNETITE $FeFe_2O_4$ Cubic (Isometric)	**Luster:** metallic. **Color:** black. **Hardness:** 5.5–6.0. **Cleavage:** none, some octahedral parting. **Streak:** black. **G** = 5.2. **Habit:** usually in granular masses. Strongly magnetic, some specimens show polarity (lodestones). Widespread occurrence in a variety of rocks. **Uses:** used commercially as iron ore.
MUSCOVITE K, Al Silicate Monoclinic	**Luster:** nonmetallic, vitreous to silky or pearly. **Color:** colorless to shades of green, gray, or brown. **Hardness:** 2.5–3.5. **Cleavage:** perfect basal yielding thin sheets that are flexible and elastic; may show some parting. **Streak:** white. **G** = 2.8–2.9. **Habit:** usually in small flakes or lamellar masses. Commercial deposits are generally found in granite pegmatites. **Uses:** variety of industrial uses.
OLIVINE $(Mg, Fe)_2SiO_4$ Orthorhombic	**Luster:** nonmetallic, vitreous. **Color:** olive-green to yellowish; nearly pure Mg-rich varieties may be white (forsterite) and nearly pure Fe-rich varieties brown to black (fayalite). **Hardness:** 6.5–7.0. **Cleavage:** indistinct. **Streak:** white or gray. **G** = 3.2–4.4. **Habit:** usually in granular masses. Crystals uncommon. A mineral of basic and ultrabasic rocks. **Uses:** forsterite variety used for refractory bricks.

Table 1.3	Mineral Catalog *(Continued)*
PLAGIOCLASE Ranges in composition from Albite, $NaAlSi_3O_8$, to Anorthite, $CaAl_2Si_2O_8$ Triclinic	**Luster:** nonmetallic, vitreous. **Color:** white or gray, reddish, or reddish brown. **Hardness:** 6.0–6.5. **Cleavage:** two planes at close to right angles, twinning striations common on basal cleavage surfaces. **Streak:** white. **G** = 2.6–2.8. **Habit:** crystals common for Na-rich varieties, uncommon for intermediate varieties, rare for anorthite. Twinning common. Twinning striations on basal cleavage useful to distinguish from orthoclase. Some varieties show play of colors. **Uses:** sodium-rich varieties mined for use in ceramics.
PYRITE FeS_2 Cubic (Isometric)	**Luster:** metallic. **Color:** brass-yellow, may be iridescent if tarnished. **Hardness:** 6.0–6.5. **Cleavage:** none, conchoidal fracture. **Streak:** greenish or brownish black. **G** = 4.8–5.0. **Habit:** crystals common. Usually cubic with striated faces. Crystals may be deformed. Massive granular forms also. Most widespread sulfide mineral. Known as "fool's gold." **Uses:** source of sulfur for sulfuric acid. *NOTE: Marcasite* (FeS_2) is orthorhombic, usually paler in color, and is commonly altered.
QUARTZ SiO_2 Hexagonal	**Luster:** nonmetallic, vitreous. **Color:** typically colorless or white, but almost any color may occur. **Hardness:** 7.0. **Cleavage:** none, conchoidal fracture. **Streak:** white, but difficult to obtain on streak plate. **G** = 2.65. **Habit:** prismatic crystals common with striations perpendicular to the long dimension; also a variety of massive forms. Color variations lead to varieties called smoky quartz, rose quartz, milky quartz, and amethyst. Common mineral in all categories of rocks. **Uses:** wide variety of commercial uses include glassmaking, electronics, and construction products.
SILLIMANITE Al_2SiO_5 Orthorhombic	**Luster:** nonmetallic; vitreous. **Color:** white, gray, or pale brown. **Hardness:** 6.0–7.0. **Cleavage:** good, one direction. **G** = 3.2. **Habit:** slender prisms. **Uses:** high-temperature ceramics.
SPHALERITE ZnS Cubic (Isometric)	**Luster:** usually nonmetallic, some varieties submetallic, most commonly resinous. **Color:** yellow, yellow-brown to dark brown. **Hardness:** 3.5–4.0. **Cleavage:** perfect dodecahedral (six directions at 120 degrees). **Streak:** brown to light yellow or white. **G** = 3.9–4.1. **Habit:** crystals common as distorted tetrahedra or dodecahedra. Twinning common. Also massive or granular. **Uses:** important zinc ore.
STAUROLITE Complex Silicate Monoclinic	**Luster:** nonmetallic. **Color:** brown to gray-brown. **Hardness:** 7.0. **Cleavage:** good, one direction. **G** = 3.7–3.8. **Habit:** prismatic; forms twins in shape of crosses. **Uses:** gemstone crosses called "fairy crosses."
SULFUR S Orthorhombic	**Luster:** nonmetallic; vitreous to earthy. **Color:** yellow. **Hardness:** 1.5–2.5. **Cleavage:** poor; fracture brittle. **G** = 2.1. **Habit:** transparent to translucent crystals; earthy masses. **Uses:** wide industrial uses, especially drugs, sulfuric acid, and insecticides.
TALC Mg Silicate Monoclinic	**Luster:** nonmetallic, pearly to greasy or dull. **Color:** usually pale green, also white to silver-white or gray. **Hardness:** 1.0. **Cleavage:** perfect basal, massive forms show no visible cleavage. **Streak:** white. **G** = 2.6–2.8. Usually foliated masses or dense fine-grained dark gray to green aggregates (soapstone). **Habit:** crystals extremely rare. Soapy feel is diagnostic. **Uses:** commercial uses in paints, ceramics, roofing, paper, and toilet articles.
TOURMALINE Complex Borosilicate Hexagonal	**Luster:** nonmetallic; vitreous. **Color:** black, green, yellow, pink, blue. **Cleavage:** none; fracture brittle. **Hardness:** 7.0–7.5. **G** = 3.0–3.2. **Habit:** elongated opaque to transparent striated prisms with triangular cross sections. **Uses:** gemstone; crystals used in radio transmitters.
Zircon $ZrSiO_4$ Tetragonal	**Luster:** brilliant. **Color:** usually shade of brown, colorless. **Hardness:** 7.5. **Cleavage:** prismatic, poor. **Streak:** colorless. **G:** 4.68. **Habit:** crystals common. **Uses:** source of metallic zirconium used in the construction of nuclear reactors, source of zirconium oxide, one of the most refractory substances known, used as a gemstone when transparent, and mineral grains in rocks are commonly used for radioactive age determinations.

Table 1.4 Chemical Grouping and Composition of Some Common Minerals

Chemical Group		Example: Mineral Name	Example: Chemical Composition*
ELEMENTS		Copper Graphite Diamond Sulfur	Cu C C S
OXIDES		Quartz Hematite Magnetite Goethite (Limonite) Corundum	SiO_2 Fe_2O_3 Fe_3O_4 FeO(OH) Al_2O_3
SULFIDES		Pyrite Chalcopyrite Galena Sphalerite	FeS_2 $CuFeS_2$ PbS ZnS
SULFATES		Anhydrite Gypsum	$CaSO_4$ $CaSO_4 \cdot 2H_2O$
CARBONATES		Calcite Dolomite	$CaCO_3$ $CaMg(CO_3)_2$
PHOSPHATES		Apatite	$Ca_5(PO_4)_3F$
HALIDES		Halite Fluorite	NaCl CaF_2
SILICATES	OLIVINE GROUP	Olivine	$(Mg,Fe)_2SiO_4$
SILICATES	AMPHIBOLE GROUP	Hornblende Asbestos (Fibrous Serpentine)	Ca,Na,Mg,Fe,Al Silicate Mg,Al Silicate
SILICATES	PYROXENE GROUP	Augite	Ca,Mg,Fe,Al Silicate
SILICATES	MICA GROUP	Muscovite Biotite Chlorite Talc Kaolinite	K,Al Silicate K,Mg,Fe,Al Silicate Mg,Fe,Al Silicate Mg Silicate Al Silicate
SILICATES	FELDSPAR GROUP	K-feldspar Plagioclase (Ab, An) Albite (Ab) Anorthite (An)	$K(AlSi_3O_8)$ Mixture of Ab and An $NaAlSi_3O_8$ $CaAl_2Si_2O_8$
SILICATES	GARNET GROUP	Garnet	Fe,Mg,Ca,Al Silicate
SILICATES	ANDALUSITE GROUP	Sillimanite Staurolite	Al_2SiO_5 Hydrated Fe,Mg,Al Silicate
SILICATES	EPIDOTE GROUP	Epidote	Hydrated Ca,Fe,Al Silicate
SILICATES	SILICA GROUP	Quartz	SiO_2
SILICATES	TOURMALINE GROUP	Tourmaline	Hydrated Na,Li,Ca,Fe,Mn,Cr,V

*Some common elements and their symbols:
Al-Aluminum, C-Carbon, Ca-Calcium, Cl-Chlorine, Cu-Copper, F-Fluorine, Fe-Iron, H-Hydrogen, K-Potassium, Mg-Magnesium, Na-Sodium, O-Oxygen, P-Phosphorus, Pb-Lead, S-Sulfur, Si-Silicon, Zn-Zinc

Rocks

Background

As a part of the study of rocks in the laboratory or in the field, some understanding of the origin and occurrence of the rocks being studied is essential. A **rock** is a naturally occurring aggregate of minerals (granite, shale) or a body of undifferientiated mineral matter (obsidian) or lithified organic matter (coal). This section deals with the identification and classification of rocks based on the study of hand specimens with reference also to their origin and occurrence.

Three major categories of rocks are recognized. They are **igneous,** rocks formed by the cooling and crystallization of molten material within or on the surface of the earth; **sedimentary,** rocks formed from sediments derived from preexisting rocks, by precipitation from solution, or by the accumulation of organic materials; and **metamorphic,** rocks resulting from the change of preexisting rocks into new rocks with different textures and mineralogy as a result of the effects of heat, pressure, chemical action, or combinations of these. All types of these three categories of rocks can be observed at the earth's surface today. We can also observe some of the processes that lead to their formation. For example, the eruption of volcanoes produces certain types of igneous rocks. We can observe weathering, erosion, transportation, and deposition of sediments that upon **lithification**—the combination of processes (including compaction and cementation) that convert a sediment into an indurated rock—produce certain types of sedimentary rocks. Other observable processes of rock formation on the surface of the earth include the deposition of dripstone in caves and the growth of coral reefs in the oceans.

The earth's crust is dynamic and is subject to a variety of processes that act upon all types of rocks. Given time and the effect of these processes, any one of these rocks can be changed into another type. This relationship is the basis of the **rock cycle,** represented in figure 1.22. The heavy arrows indicate the normal complete cycle, while the open arrows indicate how this cycle may be interrupted. Keep in mind that the schematic presented in figure 1.22 is greatly simplified and that a given rock observed today represents only the last phase of the cycle from which the rock has been derived.

Igneous Rocks

Igneous rocks are aggregates of minerals that crystallize from molten material that is generated mostly within the earth's mantle. The heat required to generate this molten material comes from within the earth. The temperature of the earth increases from the surface inward at an average

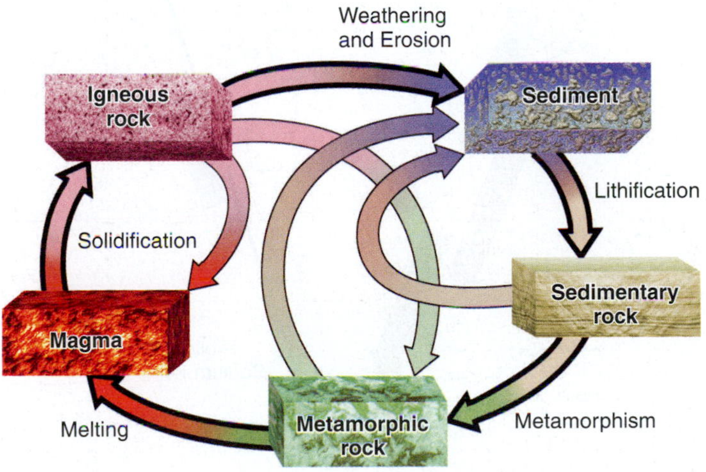

Figure 1.22
The rock cycle (shown schematically).

rate of 20°C to 30°C per kilometer (km) as determined by temperature measurements in deep mines or from drill holes. This increase in temperature with depth, known as the **geothermal gradient,** varies from place to place. In areas of tectonic or volcanic activity, the geothermal gradient may rise much faster than under stable continental interiors. The temperature required to melt rock depends on the composition of the material, the pressure, and the presence or absence of water. Partial melting of rock materials may occur in some instances.

While the geothermal gradient varies from place to place, at a depth of 35 km to 40 km, the temperature may be sufficient to melt rock. The molten material, **magma,** is a complex solution of silicates plus water vapor plus other gases. This magma may solidify before it reaches the surface, but other magma may reach the surface, where it is extruded as **lava.** The rocks formed by solidification of magma within the crust or mantle are known as **intrusive** igneous rocks, and those that form at or on the surface are called **extrusive** igneous rocks.

The composition of a given magma depends on the composition of the rocks that were melted to form the magma. The magma formed tends to rise toward the surface of the earth. As it rises, cooling and crystallization begin. Ultimately, the molten mass solidifies into solid rock. The type of igneous rock formed depends on a number of factors, including the original composition of the magma, the rate of cooling, and the reactions that occurred within the magma as cooling took place.

Intrusive rock masses have been studied in detail, and it is recognized that there is an orderly sequence in which the various minerals crystallize. This sequence of crystallization (commonly called **Bowen's reaction series** after the geologist who first proposed it) has been verified in principle.

The reaction series presented in figure 1.23 gives some suggestion as to the formation of the various types of igneous rocks, and it helps to explain why the association of some minerals (such as olivine and quartz) is rare in nature. The first minerals to form tend to be low in silica. The plagioclase side of the diagram is labeled as a **continuous reaction series.** This means that the first crystals to form, calcium-rich plagioclase (anorthite), continue to react with the remaining liquid as cooling continues. This process involves the substitution of sodium and silicon for calcium and aluminum in the plagioclase without a change in the crystal structure. As this process continues, the composition of the plagioclase becomes increasingly sodium- and silica-rich, and the last plagioclase formed is albite.

The **discontinuous reaction series** consists of common ferromagnesian minerals (olivine, pyroxene, amphibole, biotite) found in igneous rocks. The first mineral to crystallize in this series is olivine. As the cooling continues, the olivine crystals react with the remaining liquid and form pyroxenes at the expense of olivine. As this process of interaction between the crystals and the silica-rich liquid continues, the pyroxene reacts to form amphibole, and the final member of this discontinuous series is biotite, which forms at the expense of amphibole.

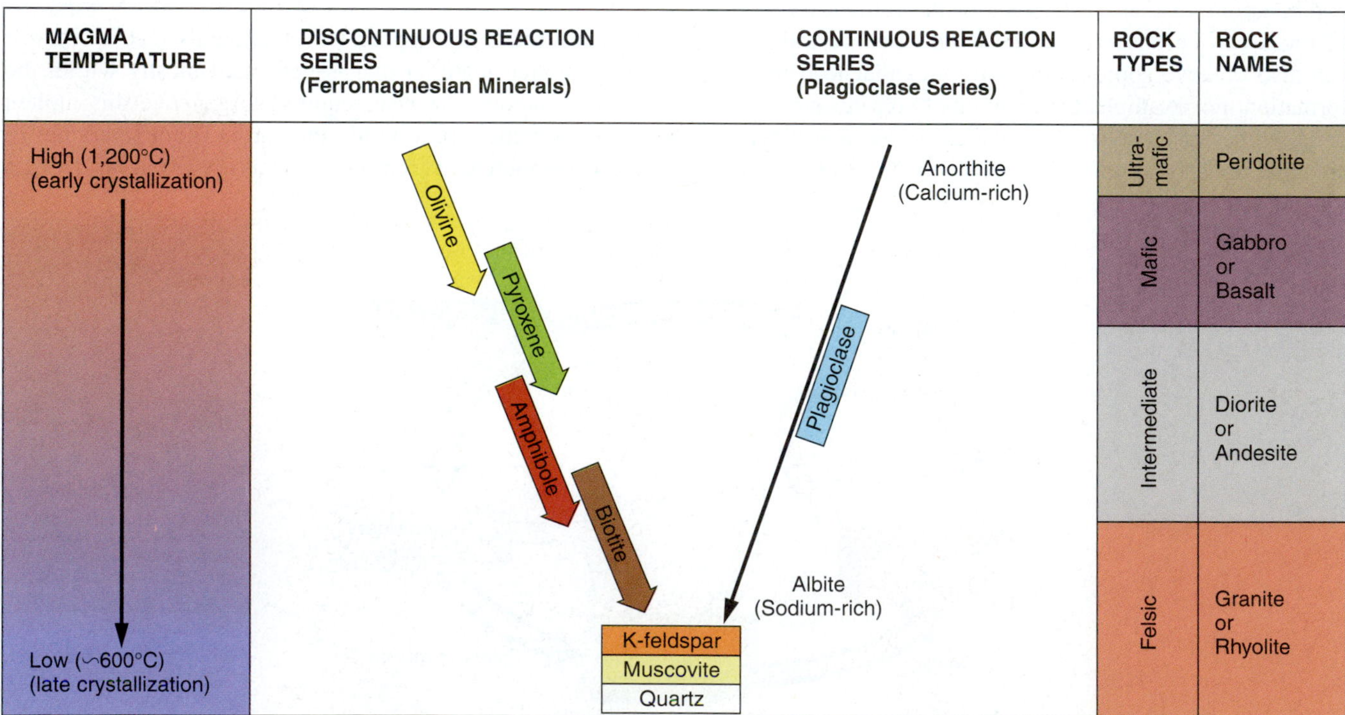

Figure 1.23

Reaction series for igneous rock formation from a magma.

If the original magma is low in silica and high in iron and magnesium, the magma may solidify before the complete series of reactions has occurred. The resulting **mafic** rocks are high in magnesium and iron, low in silica, and are usually dark-colored (see fig. 1.23 and table 1.7). Olivine, pyroxene, and calcium-rich plagioclase are the common mineral associations of mafic rocks. Conversely, magma originally high in silica and low in ferromagnesian elements may reach the final stages of the reaction series, and the rocks formed are composed of potassium feldspar, quartz, and muscovite. These **felsic** rocks are usually light-colored (see fig. 1.23 and table 1.7). The rocks that form between felsic and mafic are intermediate in composition and color (see fig. 1.23 and table 1.7).

Keep in mind that this reaction series is idealized. Changes in the original composition of the magma occur in nature by **fractional crystallization** (the removal of crystals from the liquid magma by settling, floating or filtering), by **assimilation** of part of the country rock through which the magma is rising, and/or by the **mixing** of two magmas of differing composition.

Occurrence

As magma works its way toward the surface of the earth, it encounters preexisting rock. The rock mass intruded by the magma is referred to as **country rock,** and it is not uncommon for pieces of the country rock to be engulfed by the magma. Fragments of country rock included within igneous rocks are called **inclusions** or **xenoliths.**

Intrusive and extrusive igneous rocks may assume any number of geometric forms. Figure 1.24 illustrates several of the more common shapes of igneous rock masses found in nature. Table 1.5 shows the relationship of igneous rock types to their modes of occurrence.

Table 1.5 Relationship of Igneous Rock Types to Their Modes of Occurrence in the Earth's Crust

	Rock Types	Some Modes of Occurrence
Extrusive	Pumice Scoria	Pyroclastics Crusts on lava flows, pyroclastics
Extrusive	Obsidian	Lava flows
Extrusive	Rhyolite Andesite Basalt	Lava flows (also shallow intrusives such as dikes and sills)
Intrusive	Rhyolite porphyry Andesite porphyry Basalt porphyry	Dikes, sills, laccoliths, intruded at medium to shallow depths
Intrusive	Granite Diorite Gabbro Peridotite	Batholiths and stocks of deep-seated intrusive origin

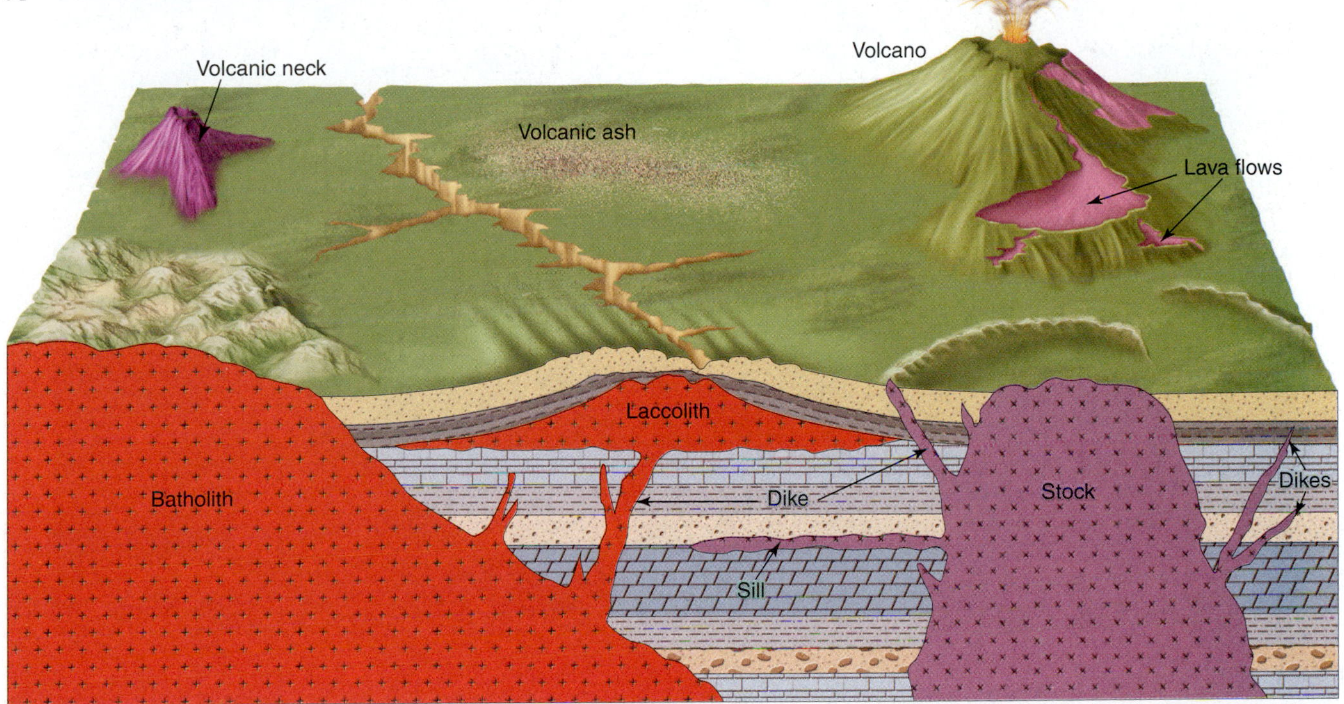

Figure 1.24
Block diagram showing the occurrence of igneous rocks.

The largest intrusive igneous rock mass **(pluton)** is a **batholith.** Batholiths are usually coarse-grained in texture and granitic in composition. By definition, a batholith crops out over an area of more than 100 square kilometers (about 36 square miles). A **stock,** by definition, crops out over an area less than 100 square kilometers (less than 36 square miles). A **dike** is a tabular igneous intrusion whose contacts cut across the trend of the country rock (figs. 1.24 and 1.25). A **sill** is also tabular in shape, but its contacts lie parallel to the trend of the country rock. A **laccolith** is similar to a sill but is generally much thicker, especially near its center where it has caused the country rock to bulge upward. These five major types of intrusives—batholiths, stocks, dikes, sills, and laccoliths—can be grouped into two main categories based on the relationship of their contacts to the trend of the enclosing country rock. The contacts of batholiths, stocks, and dikes all cut across the trend of the country rock and hence are called **discordant igneous plutons.** Sills and laccoliths, on the other hand, have contacts that are parallel to the trends of the country rock and are called **concordant igneous plutons.** Table 1.6 summarizes these relationships.

Figure 1.24 shows several types of extrusive volcanic rocks including a volcano, pyroclastics, and lava flows. The volcanic neck is an example of an eroded volcano.

Table 1.6 Relationship Between Concordant or Discordant Igneous Bodies and Their Modes of Occurrence

Relationship of Igneous Rock Contact to Country Rock	Modes of Occurrence
Concordant	Laccolith, sill
Discordant	Batholith, stock, dike

Textures of Igneous Rocks

The **texture** of a rock is its appearance that results from the size, shape, and arrangement of the mineral crystals or grains in the rock. The texture of igneous rocks is a reflection of the mineralogy and the cooling history of the magma or lava from which they were formed. Igneous rocks with a coarse-grained texture **(phaneritic)** were formed where the cooling rate was slow and larger crystals or grains could form. Igneous rocks with exceptionally large crystals or

Figure 1.25

Intersecting dark Cambrian basalt dikes intruding pink Precambrian granite at Mill Creek, Oklahoma. The height of the vertical exposure is about 100 feet.

Photo by Robert Rutford.

grains exhibit a **pegmatitic** texture. Igneous rocks with a fine-grained texture **(aphanitic)** were formed where cooling was rapid. A **porphyritic** texture reflects a two-stage cooling history. The larger crystals formed first under conditions of slow cooling, but before the magma or lava turned to rock, it migrated to a zone of faster cooling where the remainder of the melt solidified. **Vesicular** textures are indicative of lava flows in which trapped gases produced the **vesicles** while the lava was still molten. A **glassy** texture reflects an extremely rapid rate of cooling. The surfaces of lava flows commonly possess a **glassy texture.** A **pyroclastic** texture results from an explosive volcanic eruption. The texture of igneous rocks can be described in terms of one of the following:

Phaneritic (Coarse-Grained)

An igneous rock in which the constituent minerals are macroscopic in size has a phaneritic (coarse-grained) texture. The dimensions of the individual crystals or grains range from about 1.0 mm to more than 5.0 mm (figs. 1.26–1.30 and 1.41).

Pegmatitic (Very Coarse-Grained)

Igneous rocks with exceptionally coarse mineral constituents, usually over 3.0 cm in size, exhibit a pegmatitic (very coarse-grained) texture (fig. 1.30). Individual minerals may be over 10 meters (m) in length.

Aphanitic (Fine-Grained)

An aphanitic (fine-grained) igneous rock is composed of mineral crystals or grains that are microscopic in size; that is, they cannot be discerned with the naked or corrected eye (fig. 1.31).

Porphyritic

A porphyritic igenous rock is one in which the macroscopic mineral crystals or grains are embedded in a matrix of microscopic crystals or grains (figs. 1.33, 1.35, and 1.37), or macroscopic crystals or grains of one size range occur in a matrix of smaller macroscopic crystals or grains. The larger crystals or grains are called **phenocrysts,** and the collection of smaller grains in which they are embedded is called the **groundmass.**

Vesicular

A vesicular texture is characterized by the presence of vesicles—tubular, ovoid, or spherical cavities in the rock (figs. 1.32, 1.38, and 1.39). These "holes" in the rock are a result of gas bubbles being trapped in the rock. The size of the vesicles ranges from less than 1.0 mm to several centimeters in diameter. A rock may be so vesicular that its density is affected. For example, pumice will float on water. Vesicles that are filled with secondary mineral matter are called **amygdules,** and the texture then is called **amygdaloidal.**

Glassy

An igneous rock with a glassy texture is one that lacks crystalline structure resulting from the cooling of a magma or lava so rapidly that crystal growth did not occur (fig. 1.40). The material lacks regular internal atomic structure.

Pyroclastic

Pyroclastic describes the fragmental texture of an igneous rock consisting of fragments of rocks and glassy particles expelled aerially from a volcanic vent during an explosive eruption (fig. 1.36).

Mineralogic Composition of Igneous Rocks

The minerals of igneous rocks are grouped into two major categories, **primary** and **secondary.** Primary minerals are those that are crystallized from the cooling magma. Secondary minerals are those formed after the magma has solidified; they include minerals formed by chemical alteration of the primary minerals or by the deposition of new minerals in an igneous rock. Secondary minerals are not important in the classification of igneous rocks, but may be useful in naming rocks.

Primary minerals consist of two types—**essential minerals** and **accessory minerals.** The essential minerals are those that must be present in order for the igneous rock to be assigned a specific position in the classification scheme (i.e., given a specific name). Accessory minerals are those that may or may not be present in a rock of a given type, but the presence of an accessory mineral in a given rock may affect the name of the rock. For example, the essential minerals of a granite are quartz and K-feldspar (K-feldspar is a common notation for the group of potassium-rich feldspars of which orthoclase is the most common). If a particular granite contains an accessory mineral such as biotite or hornblende, the rock may be called a biotite granite or a hornblende granite, respectively. Another example of a common accessory mineral is zircon $Zr(SiO_4)$. Its presence does not affect the name of the rock but it may be useful in determining the radioactive date of the rock (see Exercise 7).

The essential minerals contained in the common igneous rocks are quartz, K-feldspar, plagioclase, pyroxene (commonly augite), amphibole (commonly hornblende), and olivine. The key to igneous rock identification is the ability of the observer to recognize the presence or absence of quartz and to distinguish between K-feldspar and plagioclase. Color is of little help in the matter because quartz, K-feldspar, and plagioclase can all occur in the same shade of gray. The distinction between quartz and the feldspars is made by the fact that quartz has no cleavage; macroscopic quartz crystals do not exhibit shiny cleavage faces as the feldspars do.

The distinction between K-feldspar and plagioclase is more difficult because both have cleavage surfaces that show up as shiny surfaces in phaneritic hand specimens. A pink-colored feldspar is usually K-feldspar, but a white or gray feldspar may be either K-feldspar or plagioclase. Plagioclase, however, has characteristic striations that may be visible on cleavage surfaces, especially if the crystals are several millimeters in size. Figure 1.11 shows a large fragment of a plagioclase crystal with striations on it. Striations on smaller plagioclase crystals may be difficult but not impossible to detect. K-feldspar usually has characteristic muscle fiber appearance on cleavage surfaces (see fig. 1.10).

Colors of Igneous Rocks

The mineral constituents of an igneous rock impart a characteristic color to it. Hence, rock color is used as a first-order approximation in establishing the general mineralogic composition of an igneous rock. As already pointed out in Exercise 1, color is a relative and subjective property when modified only by the adjective **light,** or **intermediate,** or **dark.** All observers would agree that a white rock is "light-colored," a black rock is "dark-colored," and a rock with half of its constituent minerals white and half of them black is a rock of "intermediate color." Mineral constituents are not all black or white, however. Some are pink, gray, and other colors, a fact that adds to the difficulty encountered in igneous rock classification for the beginning student. Nevertheless, the terms **light, intermediate,** and **dark** are useful terms, especially in the classification of hand specimens with an aphanitic texture. For example, the rocks in figures 1.26, 1.27, and 1.34 are light-colored. The rock in figure 1.30 is intermediate in color, and the rocks in figures 1.29 and 1.31 are dark-colored.

Igneous Rock Classification

Table 1.7 is a chart that shows the names of about 20 common igneous rocks and their corresponding textures and mineralogic compositions. This classification scheme is a simplified version of a more complex classification system in which the number of rock names is three to four times the number contained in table 1.7. Some simplification at the level of the beginning student is a pedagogical necessity, and if you should find a rock in your laboratory collection that does not fit easily into one of the categories shown in the table, it may be because the table has been so simplified. For example, **pegmatite** (see fig. 1.30) is an example of a very coarse-grained igneous rock commonly associated with the

Table 1.7 Simplified Classification and Identification Chart for Hand Specimens of Common Igneous Rocks

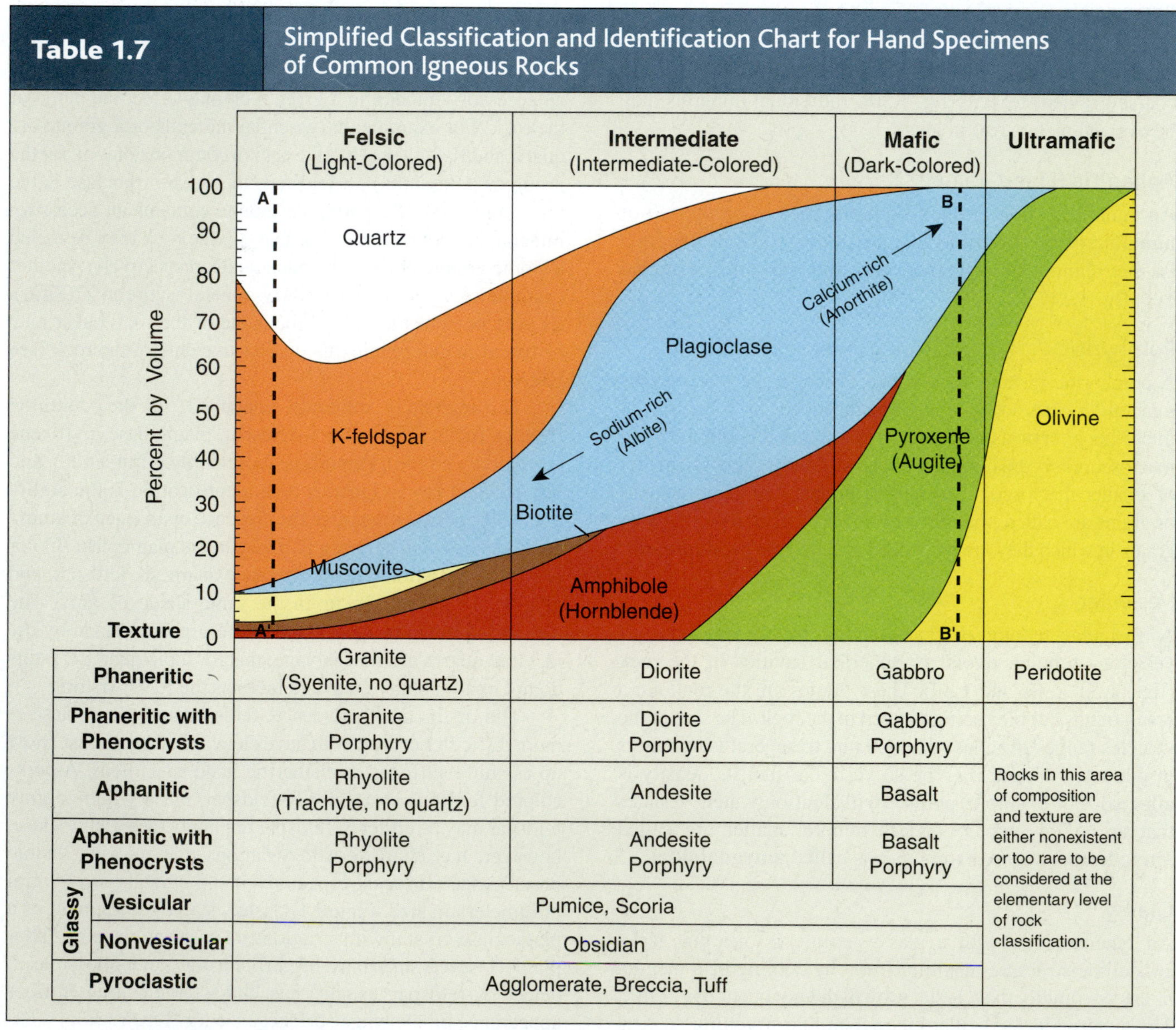

Texture	Felsic (Light-Colored)	Intermediate (Intermediate-Colored)	Mafic (Dark-Colored)	Ultramafic
Phaneritic	Granite (Syenite, no quartz)	Diorite	Gabbro	Peridotite
Phaneritic with Phenocrysts	Granite Porphyry	Diorite Porphyry	Gabbro Porphyry	Rocks in this area of composition and texture are either nonexistent or too rare to be considered at the elementary level of rock classification.
Aphanitic	Rhyolite (Trachyte, no quartz)	Andesite	Basalt	
Aphanitic with Phenocrysts	Rhyolite Porphyry	Andesite Porphyry	Basalt Porphyry	
Glassy: Vesicular	Pumice, Scoria			
Glassy: Nonvesicular	Obsidian			
Pyroclastic	Agglomerate, Breccia, Tuff			

margins of plutons. In general, the mineralogic composition of pegmatites is granitic, but they may contain extremely large crystals of uncommon minerals such as spodumene, beryl, tourmaline, or topaz. This type of rock texture does not easily fit into a simplified rock classification table and thus has not been included in table 1.7.

In table 1.7, the lower left-hand side of the table presents the textural categories, while the compositional and color categories of the rocks are displayed across the top of the chart. The upper (colored) part of the chart shows the mineral constituents of the various rock types. By determining the approximate percentages of the minerals in a hand specimen of a rock, the position of those percentages on the chart allows the student to name the unknown rock. For example, a phaneritic rock consisting of 32% quartz, 57% K-feldspar, 2% plagioclase, and 9% accessory minerals (5% muscovite, 2% biotite, and 2% amphibole) would be represented by the dashed vertical line (A–A′) toward the left side of the chart, and thus, the rock is granite (fig. 1.26). Given the percentages of the accessory minerals, a more complete name might well be muscovite granite.

The vesicular rocks are those with a texture that results from the expansion of gases during the fluid stage of the lava. Common varieties are vesicular basalt (fig. 1.32); scoria, often coarsely vesicular (fig. 1.39); and pumice, usually a light-colored vesicular glassy rock of rhyolitic composition (fig. 1.38).

A group of igneous rocks that does not fit easily into the general classification scheme are the **pyroclastic rocks,** those that are accumulations of the material ejected from explosive-type volcanoes. The lavas of these volcanoes are characterized by **high viscosity** (they do not flow easily) and high silica content. They are rhyolitic or andesitic in composition. Their mineral constituents may be difficult to determine.

The volcanic ash generated from an eruption is known as **tuff** when it becomes consolidated into a rock. Light-colored tuff is called **rhyolite tuff** (fig. 1.36), and tuff of intermediate color is called **andesite tuff.** In some tuffs, small, beadlike fragments of volcanic glass occur. These features are called **lapilli,** and the rock containing them is called **lapilli tuff.**

A rock composed of the angular fragments from a volcanic eruption is a **volcanic breccia.** A volcanic rock composed of volcanic bombs and other rounded fragments is known as an **agglomerate.**

Name ____________________

EXERCISE 2

Section ____________________ Date ____________________

Identification of Common Igneous Rocks

The identification of igneous rocks is based on their texture and mineralogy. Color is also useful, but it usually is a reflection of the mineral composition of the rock. A foliated texture, the parallel arangement of mineral grains in a rock, is most usually a texture of metamorphic rocks (see fig. 1.61; table 1.8), and is rare in igneous rocks.

Your instructor will provide a collection of rock specimens for you to identify. Take time to examine the specimens and review the physical properties as described in the previous pages.

Using table 1.8, separate the igneous rocks from the collection of specimens. Have your laboratory instructor verify your identification before you continue.

Figures 1.26 through 1.41 show some examples of similar-size specimens of common igneous rocks. Although these photographs may be helpful in identification, remember that the color of igneous rocks can vary considerably. For example, compare figures 1.26 and 1.27. Recognition of the texture and mineralogy is the key to the identification of igneous rocks. (NOTE: The scale to the right of each specimen will help you to estimate grain size and texture.)

Using the worksheets provided on the following pages to record your observations, follow these steps:

1. Group the specimens into the textural categories described for you in the text. Note the variation in grain size for the phaneritic (coarse-grained) specimens. Separate out those specimens with diagnostic features such as phenocrysts or vesicles.
2. For the phaneritic (coarse-grained) specimens, identify the minerals present in each specimen.
3. Before trying to identify and name each of the rock specimens in your collection, review tables 1.7 and 1.9.
   (a) Find line B–B′ at the right-hand side of table 1.7. Determine the percentage of each of the minerals in the rock this line represents.

   Quartz ________%
   K-feldspar ________%
   Plagioclase ________%
   Pyroxene ________%
   Olivine ________%
   Amphibole ________%
   Biotite ________%
   Muscovite ________%

   If a rock of this mineralogy has an aphanitic texture, it is a ____________________.

   (b) For another rock, you have determined that the minerals present are as follows:

Mineral	%
Quartz	10%
K-feldspar	16%
Plagioclase	49%
Amphibole	25%
Pyroxene	0%
Olivine	0%
Biotite	0%
Muscovite	0%

   If a rock of this mineralogy has a phaneritic texture, it is a ____________________.

   If it has phenocrysts, it would be called a

   ____________________.
4. Using tables 1.7 and 1.9, proceed to identify and name each specimen in your collection. Note that you may have more than one specimen of a given rock type. If you identify several specimens as granite, try to identify accessory minerals in each so that the name you assign is more definitive than just "granite." For example, a granite containing biotite should be called a *biotite granite.*
5. Your laboratory instructor will advise you as to the procedure to be used to verify your identification.

Table 1.8 Simplified Chart for Classification for Hand Specimens of Common Rocks

Texture			Diagnostic Features*				Composition*	Rock Category+
Foliated			Minerals Aligned				Silicates; Carbonates; Oxides	Metamorphic
Nonfoliated	Crystalline (Nonfragmental)		Randomly Oriented Crystals				Quartz; Feldspars; Micas; Pyroxenes; Olivine	Igneous
							Garnet; Graphite; Epidote; Serpentine; Staurolite	Metamorphic
							Carbonates; Gypsum; Halite	Sedimentary
			Interlocking Crystals				Silicates; Oxides	Igneous
							Carbonates; Oxides; Quartz; Serpentine	Metamorphic
	Noncrystalline	Clastic (Fragmental)	Fossils	Present	Undeformed		Silicates; Carbonates; Oxides	Sedimentary
					Deformed		Silicates; Carbonates; Oxides	Metamorphic
				Absent	Undeformed		Silicates; Carbonates; Oxides; Rock Fragments	Sedimentary
					Deformed	Glassy	Glass; Silicates; Oxides; Rock Fragments	Igneous
						Nonglassy	Silicates; Carbonates; Oxides	Metamorphic
					Glassy/Aphanitic		Silicates; Oxides; Rock Fragments	Igneous
		Amorphous (Very Fine-Grained)	Glassy				Glass	Igneous
			Nonglassy				Quartz; Carbonates	Sedimentary

*Multiple properties must be used for accurate classification.
+See tables 1. 7, 1.9, 1.12, and 1.14 for rock names.

Table 1.9 Simplified Classification and Identification Chart for Hand Specimens of Common Igneous Rocks

Texture		Composition and/or Diagnostic Features		Rock Name	Uses
Phaneritic		Feldspar > Mafic Minerals	K-feldspar > Plagioclase	Granite*# (Syenite, no quartz)	Decorative stone; road material
			K-feldspar < Plagioclase	Diorite*#	Decorative stone; road material
		Feldspar < Mafic Minerals	Olivine < 50%	Gabbro*#	Decorative stone
			Olivine > 50%	Peridotite*#	Source of Pt, Pd, Ni, Co
Aphanitic		Felsic		Rhyolite*+ (Trachyte, no quartz)	Decorative stone
		Intermediate		Andesite*+	Decorative stone; road material
		Mafic		Basalt*+	Decorative stone; road material
Glassy	Fragmental (Pyroclastic)	< 2 mm Grains		Volcanic Tuff	Decorative stone; garden containers
		> 2 mm Grains	Rounded	Agglomerate	Decorative stone
			Angular	Volcanic Breccia	Decorative stone
	(Nonfrag-mental)	Vesicular	> 50% Vesicles	Pumice	Lightweight concrete; soap
			< 50% Vesicles	Scoria	Gardening; road material
		Nonvesicular		Obsidian	Artifacts; surgical knives

*If phenocrysts are present, add porphyritic in front of rock name.
#If crystals are very large, add pegmatite after rock name.
+If vesicles are present, add vesicular in front of rock name.

Figure 1.26 Pink granite (phaneritic texture).

Figure 1.27 White granite (phaneritic texture).

Figure 1.28 Diorite (phaneritic texture).

Figure 1.29 Gabbro (phaneritic texture).

Figure 1.30 Pegmatite (pegmatitic texture).

Figure 1.31 Basalt (aphanitic texture).

Figure 1.32 Vesicular basalt (vesicular aphanitic texture).

Figure 1.33 Basalt porphyry (porphyritic aphanitic texture).

Name Section Date

Worksheet for Igneous Rocks

Sample #	Color	Texture	Essential Minerals	Accessory Minerals	Rock Name

Sedimentary Rocks

Origin

Sedimentary rocks are derived from preexisting materials through the work of mechanical (physical) or chemical processes, collectively known as **weathering,** under conditions normal at the surface of the earth, or they may be composed of accumulations of organic debris. Mechanical (physical) weathering, often called **disintegration,** produces rock and mineral detritus (fragments) that are transported by gravity, water, wind, or ice and deposited elsewhere on the earth's surface as **clastic sediments.** Chemical weathering, **decomposition,** may dissolve rock material and make it available in solution in streams, rivers, and groundwater. Decomposition may also result in the formation of new minerals by processes such as hydration, oxidation, and carbonation. An example of such processes is the weathering of feldspars to produce clay minerals.

After sediment has been deposited, it may be compacted and cemented into a sedimentary rock. Common cements are silica, calcium carbonate, and iron oxide. This process of changing the soft sediment into rock is known as **lithification,** and it may involve a number of changes resulting from heat, pressure, biological activity, and reaction with circulating groundwaters carrying other materials in solution. The sum of these changes is called **diagenesis.** Sedimentary rocks display various degrees of lithification. In this manual, we will be concerned only with those sedimentary rocks that are sufficiently lithified to permit their being displayed and handled as coherent hand specimens.

Sedimentary rocks provide a number of clues as to their history of transportation and deposition. Some of these features may be present in the hand specimens that will be available in the laboratory. **Sedimentary structures** include such features as bedding, cross-bedding, graded bedding, ripple marks, and mud cracks, among others. In addition, some rock types are characteristic of specific environments of deposition.

Sedimentary rock classification is based on texture and mineralogic composition. Both features are related to the origin and lithification of the original sediment, but the origin cannot be inferred from a single hand specimen. Therefore, the classification scheme used will emphasize the physical features and mineralogy of the rock rather than its exact mode of origin.

Occurrence of Sedimentary Rocks

Sedimentary rocks are formed in a wide variety of sedimentary environments. Any places that sediments accumulate are sites of future sedimentary rocks. **Sedimentary environments** are grouped into three major categories: **continental, mixed (continental and marine),** and **marine.** Within each of these broad categories, several subcategories exist. An abbreviated classification of sedimentary environments is given in table 1.10. This table is not intended to show all of the possible environments of deposition but rather to illustrate the environments in which most of the sedimentary rock types you will study here might have originated.

Table 1.10 Simplified and Abbreviated Classification of Sedimentary Environments and Some of the Rock Types Produced in Each

Sedimentary Environment	Sedimentary Rock Type
Continental	
Desert	Sandstone
Glacial	Tillite
River beds	Sandstone, Conglomerate
River floodplains	Siltstone
Alluvial fans	Arkose, Conglomerate, Sandstone
Lakes	Shale, Siltstone, Marl
Swamps	Lignite, Coal
Caves, Hot Springs	Travertine
Mixed (Continental and Marine)	
Littoral (between high and low tides)	Sandstone, Coquina, Conglomerate, Shale
Deltaic	Sandstone, Siltstone, Shale
Marine	
Neritic (low tide to edge of shelf)	Sandstone, Shale, Arkose, Reef limestone, Calcarenite
Bathyal (400 to 4,000 m depth)	Chalk, Rock salt, Rock gypsum, Shale, Limestone, Wacke
Abyssal (depths more than 4,000 m)	Diatomite, Shale, Chert

Because the earth's surface is dynamic and everchanging, sedimentary environments in a given geographic location do not remain constant throughout geologic time. Areas that are now above sea level may have been covered by the sea at various times during the geologic past. Glaciers that existed in the geologic past have since disappeared. Lakes have formed or have dried up. As the depositional environment in a given locality changes through time, the kind of sediment deposited there changes in response to the new environmental conditions.

Sediments that accumulate in a particular sedimentary environment are generally deposited in layers that are horizontal or almost horizontal. The lateral continuity of these **strata** or **beds** reflects the areal extent and uniformity of the environment in which they were deposited. The thickness of a particular bed is a function of the length of **time** and the **rate** at which the sediment accumulated. Separating the beds are **bedding planes,** which represent a slight or major change in the depositional history. These horizontal surfaces tend to be surfaces along which the sedimentary rocks break or separate. Rates of deposition vary widely, and a layer of shale 10 feet thick may represent a longer period of geologic time than a sandstone layer 100 feet thick.

When the environment of deposition changes in a particular region, the nature of the sediments accumulating there also changes. A bed of sandstone on top of a bed of shale reflects a change in sedimentary conditions from one in which clay was deposited to one in which sand was deposited. Both the underlying shale and the overlying sandstone represent the passage of geologic time.

Textures of Sedimentary Rocks

Texture is the appearance of a rock that results from the size, shape, and arrangement of the mineral grains in the rock. The texture of sedimentary rocks can be described in terms of one of the following: clastic, crystalline, amorphous, oolitic, or bioclastic.

Clastic Texture

Rocks with a **clastic** texture are derived from *detritus* (rock, mineral, and fossil fragments) that have been transported, deposited, and lithified. (NOTE: In some cases, the term *detritus* is used synonymously with *clastic* as a textual term. We have chosen to use *detritus* as the source and *clastic* as the resulting texture.)

The size of the individual particles is one of the chief means of distinguishing sedimentary rock types, and clastic particles are named according to their dimensions (i.e., their average grain diameters). Table 1.11 shows the names and sizes of sedimentary particles that will be useful in the classification of sedimentary rocks. This is a simplified version of a much more detailed classification system.

Particles larger than about 0.25 mm can be distinguished with the naked or corrected eye (fig. 1.42). (Grains of ordinary table

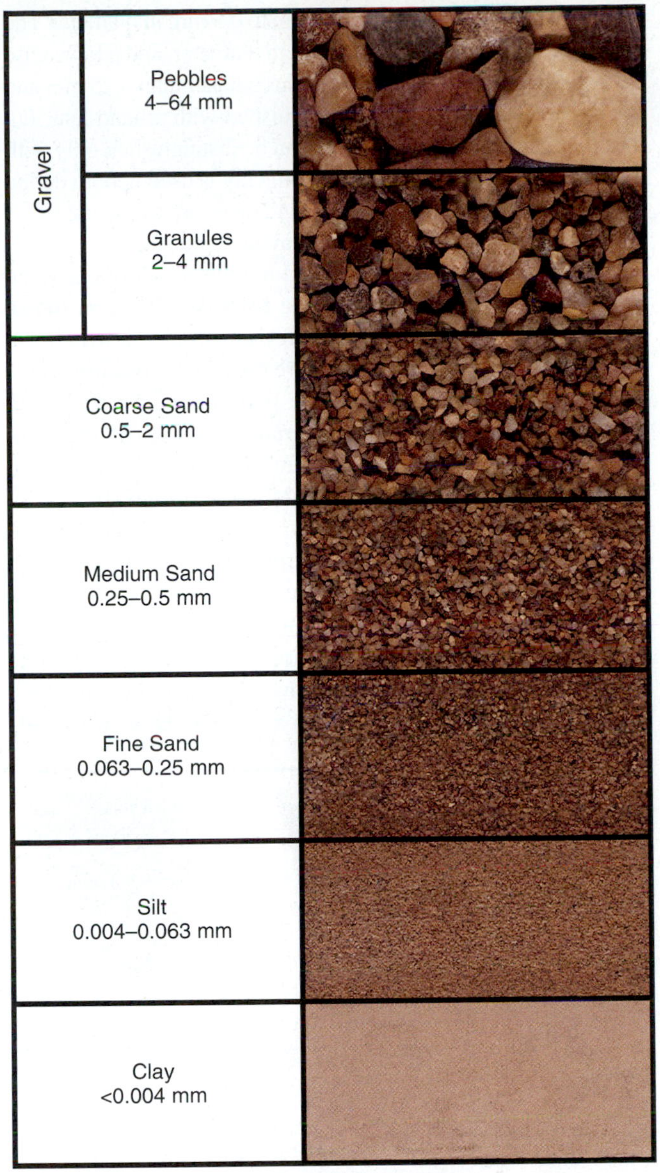

Figure 1.42
Photograph of clastic particles, ranging in size from clay to pebbles.

Table 1.11 Simplified Particle Size Range, Clastic Sediment Name, and Associated Sedimentary Rock Type

Particle Size Range	Sediment	Rock
Over 256 mm (10 in)	Boulder	Conglomerate (rounded frag[illegible]ts) or breccia (angular fragments)
2 to 256 mm (0.08 to 10 in)	Gravel	Conglomerate or bre[illegible]
0.063 to 2 mm (0.025 to 0.08 in)	Sand	Sandstone
0.004 to 0.063 mm (0.00015 to 0.025 in)	Silt	Siltstone*
Less than 0.004 mm (less than 0.00015 in)	Clay	Shale (claystone*)

*Both siltstone and claystone are also known as mudstone, commonly called *shale* if the rock shows a tendency to split on parallel planes.

salt range in size from about 0.25 mm to 0.5 mm in diameter. The dot in the letter "i" is about 0.5 mm in diameter, and a lowercase "l" is about 2.5 mm high.) Sand grains smaller than 0.25 mm and the larger silt grains can be distinguished with a hand lens, but the smaller grains of silt and clay can be distinguished only with a microscope. Be aware that the term **clay** is used here to define grain size rather than mineralogy, and take care to use the terms **"clay mineral"** and **"clay size"** to avoid confusion.

Sedimentary rocks composed mainly of silt or clay-size particles are said to have a **dense texture.** The term **dense** is also a general textural term for any rock in which the individual mineral components are crowded close together.

The shapes and sizes of the particles in the rock also influence texture. The mineral grains or rock fragments in a specimen with a clastic texture should be examined to determine whether they are **angular** or **rounded** (fig. 1.43). Particles freshly broken tend to be angular but become smoother and rounder as they are transported by wind, water, and ice. The rule of thumb is that the greater the distance transported, the smoother and rounder the particles.

Clastic texture is also influenced by the range in particle size in the rock. Rocks with a narrow range in grain size (many quartz sandstones or beach sands) are said to be **well-sorted;** those with a wide range in grain size (glacial) are said to be **poorly sorted.** Sorting is a function of mode of transportation and distances from the source (fig. 1.44).

Crystalline Texture

Crystalline texture is characteristic of a sedimentary rock composed of interlocking crystals. If the individual crystals are less than 0.25 mm in diameter, the rock has a **dense texture** (i.e., micrite) as far as macroscopic examination of a hand specimen is concerned.

Amorphous Texture

Amorphous texture is a very compact texture found in rocks composed of finely divided **noncrystalline** material deposited by chemical precipitation.

Oolitic Texture

Oolitic texture is formed of spheroidal particles less than 2 mm in diameter, called **ooliths.** The ooliths are usually composed of calcium carbonate or silica. They form by deposition of the material out of solution onto a nucleus as concentric layers (much like the formation of a hailstone), and they are cemented together into a coherent rock.

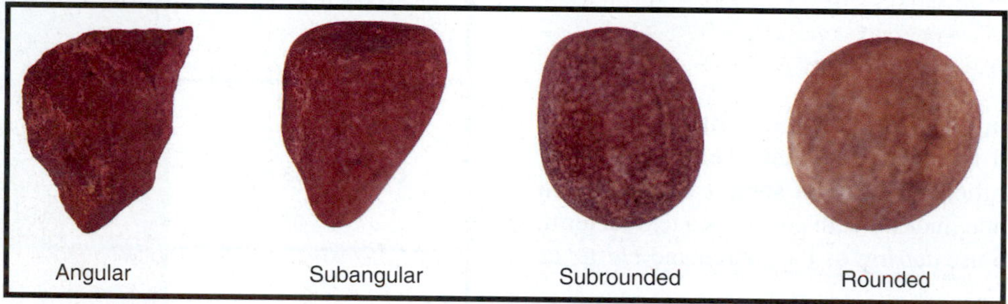

Figure 1.43

Examples of particle shapes.

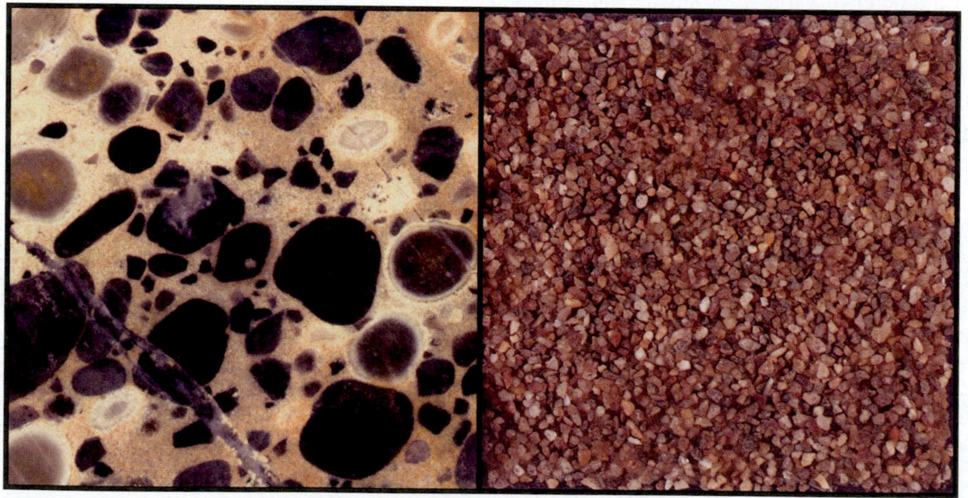

A. Poorly-Sorted **B. Well-Sorted**

Figure 1.44

(A) Examples of poorly-sorted sediment (B) well-sorted sediment.

Photos by Judy Taylor (A), James L. Carter (B).

Bioclastic Texture

Bioclastic texture is produced by the aggregation of fragments of organic remains, the most common of which are shell fragments or plant fragments. Rocks that contain fragments, molds, or casts of organic remains or **fossils** (e.g., shells, bones, teeth, leaves, seeds) or other recognizable evidence of past life (e.g., footprints, leaf prints, worm burrows, etc.) are said to be **fossiliferous.** Fossils are commonly embedded in a matrix of sandstone, shale, limestone, or dolomite.

Sedimentary Structures

Sedimentary rocks may contain features that are characteristic of their environment of deposition. These include **mud cracks, graded bedding, ripple marks, cross-bedding,** and **sole marks,** among others. Even in hand specimens, many of these features may be evident and provide some indication of the sedimentary environment in which the rock was formed. For example, the **mud cracks** found in rocks formed in exactly the same way as the mud cracks we see formed on dry mud today (fig. 1.45A).**Graded bedding** (fig. 1.45B) is the gradual vertical shift from coarse to fine clastic material in the same bed.

Oscillation (symmetric) ripple marks (fig. 1.46A) are characteristic of sediments deposited where there was a forward and backward movement of water, such as one might find in a standing body of water affected by wave action. **Cu[illegible]rent (asymmetric) ripple marks** (fig. 1.46B) indicate that [illegible] sediment was deposited by running water or by wind. **Cro[illegible]bedding** is characteristically laid down at an angle to the h[illegible]ontal, such as on the lee side of a sand dune, and while the [illegible]or bedding is horizontal, there is a subset that is at an angle [illegible] 1.46C). **Sole marks** occur when there has been some distu[illegible]nce of the top of a soft sediment that is then preserved whe[illegible]dditional material is deposited. These markings are sometime[illegible]cks and trails of bottom-dwelling organisms.

A number of these sedimentary structures [illegible]useful in determining if a sequence of rocks is right-side-up ([illegible]ging in an upward direction). Graded bedding, oscillation ripple marks, many ichnofossils, mudcracks, etc., can be helpful in establishing the correct sequence of rocks from "old" to "young."

Composition of Sedimentary Rocks

The major mineral components of sedimentary rocks are generally less varied than those in igneous rocks. The weathering of preexisting rocks involves chemical reactions that act upon the minerals within the rocks. Calcite, for example, may be dissolved and go into solution. In other minerals, these reactions may result in a complete alteration of the mineral and new minerals may be formed.

The stability of minerals at the surface of the earth varies greatly. Recall the reaction series discussed in the section on igneous rocks (see fig. 1.23). During the weathering process, those minerals that formed at high temperatures are the least

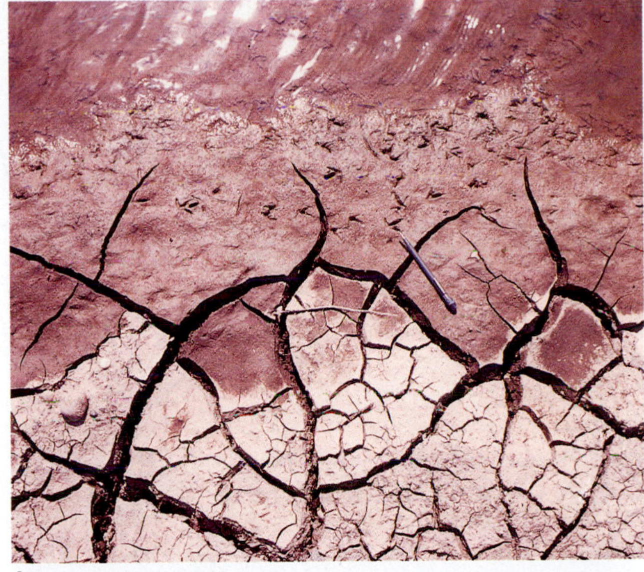

A

B

C

Figure 1.45

(A) Mud cracks formed in a modern clay-rich sediment (plan view). Note also the bird tracks and pencil for scale. (B) Graded bedding, Klondike Mountain Formation, Republic, Washington. (C) A slab of Ferron Sandstone, Utah, exhibiting current ripple marks.

Photos by: (A) R. Rutford. (B) D. R. Gaylord. (C) Janok Bhattacharya.

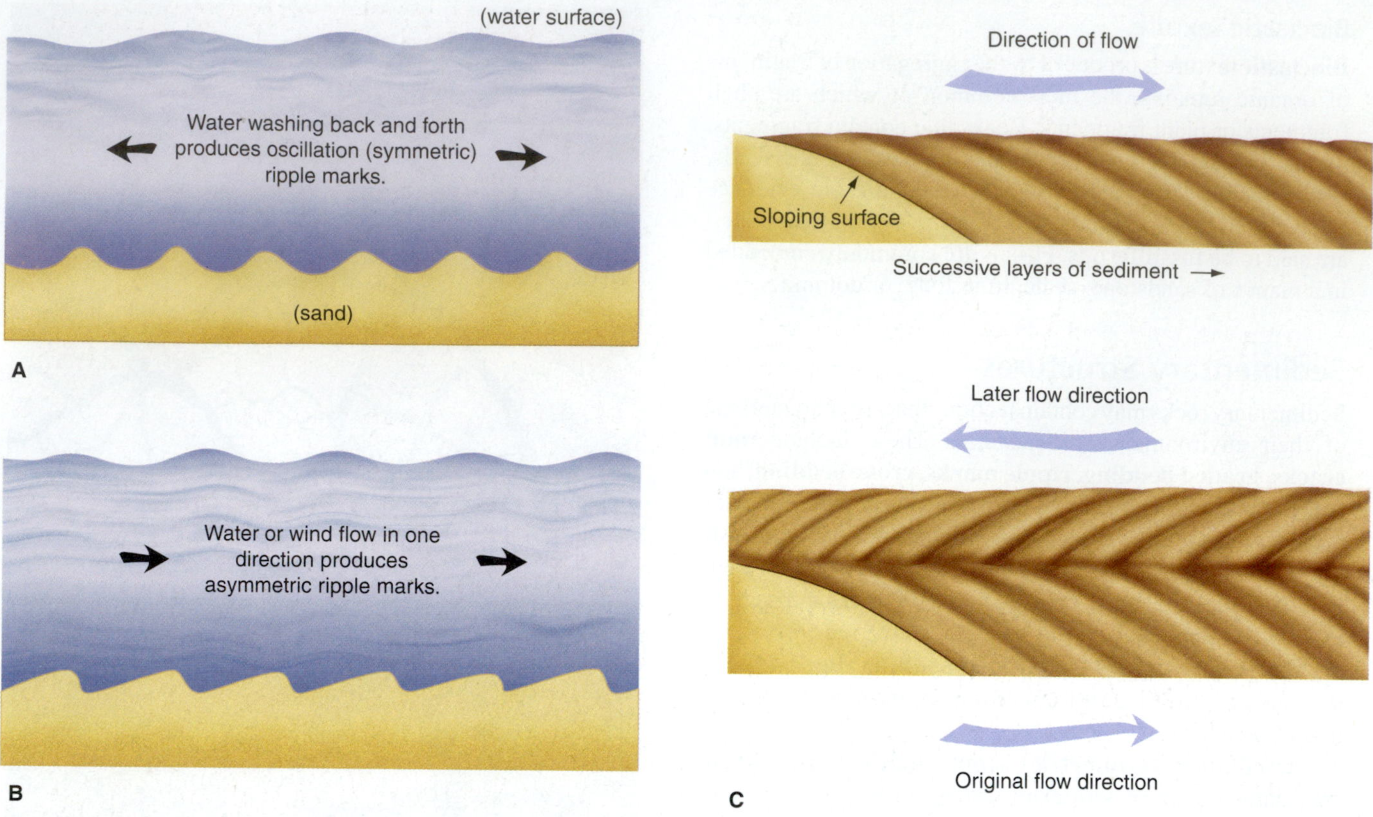

Figure 1.46

(A) Schematic of the formation of oscillation ripple marks. (B) Schematic of the formation of current ripple marks. (C) Schematic example of cross-bedding formation.

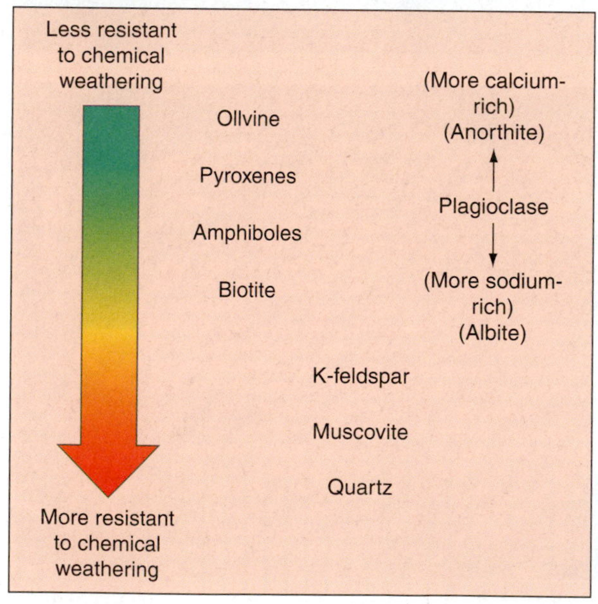

Figure 1.47

Susceptibility of minerals to chemical weathering is inversely related to Bowen's reaction series.

stable at the surface of the earth and are most susceptible to chemical weathering. Figure 1.47 relates the Bowen's reaction series to mineral stability. Thus, it is not surprising that olivine grains are relatively rare in sedimentary rocks and that quartz is common. The end products of the chemical weathering of most silicate minerals are quartz and clay minerals. A great many chemical reactions occur during the weathering, transportation, deposition, and lithification of sediments, all of which have some influence on the final composition of the rock.

Because some of the minerals that occur are often microscopic in size, they cannot always be identified easily. Tests for hardness and chemical composition are employed where visual inspection alone fails to identify the primary substance of which a sedimentary rock is composed.

A discussion of materials that are common in sedimentary rocks follows. Some are shown in the suite of sedimentary rocks in figures 1.48 through 1.55.

Quartz

Quartz, (silica, SiO_2) most commonly occurs as grains in sandstone. Quartz is the most ubiquitous of all minerals in sedimentary rocks because of its hardness and chemical stability in the weathering environment. Other forms of silica include **chert, flint** and **jasper,** dense, cryptocrystalline

Figure 1.48 Ferruginous sandstone (clastic texture).

Figure 1.49 Arkose (clastic texture).

Figure 1.50 Conglomerate (clastic texture).

Figure 1.51 Shale (clastic texture).

Figure 1.52 Breccia (clastic texture).

Figure 1.53 Fossiliferous limestone (bioclastic texture).

Figure 1.54 Chalk (bioclastic texture).

Figure 1.55 Coquina (bioclastic texture).

forms of quartz, and **diatomite,** a porous accumulation of the remains of siliceous plants of microscopic size.

Carbonate Minerals

Calcite, $CaCO_3$, and dolomite, $CaMg(CO_3)_2$, are two common minerals that occur as the major constituents of **limestone** (calcite) and **dolostone** (dolomite) or as the cementing material in a wide variety of clastic sediments. A sedimentary rock containing calcite in any form will effervesce (fizz) strongly when a drop of cold dilute HCl is placed on it. Powdered dolostone (dolomite) will react weakly with the same cold dilute HCl. It is important to distinguish whether the grains or crystals are reacting with the acid or whether it is the cement that is a carbonate.

The names of sedimentary rocks in which a carbonate mineral is present but is not a major constituent are prefixed by the term **calcareous,** which designates the presence of significant amounts of calcite or dolomite. For example, a rock made up of sand-sized quartz grains cemented with calcite would be called a **calcareous quartz sandstone.**

Calcite occurs in the crystalline form in **crystalline limestone.** Many shell fragments are composed of calcite, and the resulting rocks may range from **fossiliferous limestone** (fig. 1.53) to rocks called **coquina** composed of almost 100% shell fragments (fig. 1.55). Most ooliths are composed of calcite. Thus, there are numerous types of rocks called limestone ranging from fine-grained crystalline micrite to coarse-grained coquina.

Clay Minerals

The term **clay minerals** refers to a group of **silicate** minerals that have layered atomic structures. Clay minerals are not to be confused with **clay-size particles,** which are smaller than 0.004 mm in diameter (table 1.11). All clay minerals occur as clay-size particles, but not all clay-size particles are clay minerals.

Although one clay mineral, kaolinite, is white, most clay minerals are green to gray and impart a dark color to the sedimentary rock in which they occur. Clay minerals are most common in dense, fine-grained rocks such as **mudstone, shale** (distinguished by its **fissility,** the tendency to part in thin layers), and **wacke,** a "dirty sandstone" that contains more than 10% clay minerals. Many limestones contain appreciable amounts of clay. The adjective used to describe a rock containing some clay is **argillaceous,** as, for example, an **argillaceous limestone.**

Feldspar Minerals

Compared to quartz, calcite, and clay minerals, feldspars do not occur in great abundance in sedimentary rocks. Under certain circumstances, however, some sedimentary rocks may contain significant amounts of feldspar. A feldspar-rich sandstone is called **arkose** (fig. 1.49). The corresponding adjective for rocks with some feldspar is **arkosic.** Feldspar also occurs in variable amounts in wacke.

Heavy Minerals

Heavy minerals generally have a specific gravity greater than 2.9. Minerals such as garnet, hornblende, ilmenite, magnetite, tourmaline, and zircon, among others, may be found in sedimentary rocks, but they rarely exceed 1% of the total volume. These minerals may be useful in determining the source of the sediments that make up the rock.

Evaporite Minerals

Evaporite minerals are formed by chemical precipitation as a result of evaporation from an aqueous solution. Common evaporate minerals are halite (NaC1) and gypsum ($CaSO_4 \cdot 2H_2O$). Both are formed by chemical precipitation from an aqueous solution. Gypsum may occur as selenite crystals in sedimentary rocks, but in this form it usually formed after the sediment was deposited and lithified.

Rock Fragments

Some sedimentary rocks contain very coarse detrital constituents such as pebbles, cobbles, or even boulders. These coarse materials are usually **rock fragments** rather than single minerals. **Conglomerates** are clastic sedimentary rocks containing rounded pebbles or cobbles (fig. 1.50). If the coarse rock fragments are angular, the rock name is sedimentary **breccia** (fig. 1.52).

Biogenic Constituents

Biogenic constituents are those produced directly by the physiological activities of organisms and may make up significant amounts of a given sedimentary rock. These materials may include shell fragments, teeth, bones, plant remains, or other organic debris.

Classification of Sedimentary Rocks

The great variety of sedimentary environments in nature and the gradations between them are responsible for the large diversity of sedimentary rock types. Because of this, a classification scheme encompassing all possible varieties would be unduly complex for the beginning student.

The threefold classification of sedimentary rocks presented in table 1.12 is highly simplified and is based on the major constituents of sedimentary rocks: **detrital materials, chemical precipitates** and **biogenic materials.** This arrangement is a logical and useful guide for an understanding of the similarities and differences in sedimentary rock types encountered by the beginning student.

The identification scheme for use with table 1.12 is presented as a part of Exercise 3.

Table 1.12 Simplified Classification and Identification Chart for Hand Specimens of Common Sedimentary Rocks

Origin	Textural Features and Particle Size	Composition and/or Diagnostic Features	Rock Name	
Detrital Materials	Clastic (Pebbles and granules embedded in matrix of cemented sand grains)	Angular rock or mineral fragments.	SEDIMENTARY BRECCIA	
		Rounded rock or mineral fragments.	CONGLOMERATE	
	Clastic (Coarse sand and granules)	Angular fragments of feldspar mixed with quartz and other mineral grains. Pink feldspar common.	SANDSTONE	ARKOSE
	Clastic (sand size particles)	Rounded to subrounded quartz grains. Color: white, buff, pink, brown, tan.	SANDSTONE	QUARTZ SANDSTONE
		Calcite and/or dolomite grains. Light-colored.	SANDSTONE	CALCARENITE
	Clastic (pooly sorted sand mixed with clay-size particles)	Quartz and other mineral grains mixed with clay. Color: dark gray to gray-green.	SANDSTONE	WACKE ("Dirty Sandstone")
	Clastic (Fine-grained. Silt and clay-size particles)	Mineral constituents (commonly quartz) rarely identifiable with a hand lens. Usually well-stratified. Color: varies.	MUDSTONE	SILTSTONE
		Mineral constituents not identifiable. May be soft enough to be scratched with fingernail. Usually well-stratified. Fissile (tendency to separate in thin layers). Color: varies.	MUDSTONE	SHALE
		Mineral constituents not identifiable. May be soft enough to be scratched with fingernail. Massive (earthy). Color: varies.	MUDSTONE	CLAYSTONE
Chemical Precipitates	Dense (Crystalline or Oolitic)	$CaCO_3$; effervesces with cold dilute HCl. May contain fossils. Varieties include crystalline (micrite), oolitic, and fossiliferous. Color: white to black.	LIMESTONE	
	Dense or Crystalline	$CaMg(CO_3)_2$; powder effervesces weakly with cold dilute HCl. May contain fossils (*fossiliferous*). Color: varies, but commonly similar to limestones. Stratification generally absent in hand specimens.	DOLOSTONE	
	Dense (Porous)	$CaCO_3$; effervesces freely with cold dilute HCl. Color: varies. Contains irregular dark bands.	TRAVERTINE	
	Dense (Amorphous)	Scratches glass, conchoidal fracture. Color: varies.	CHERT	
	Crystalline	$CaSO_4 \cdot 2H_2O$; commonly can be scratched with fingernail. Color: varies; commonly pink, buff, white.	ROCK GYPSUM	
		NaCl. Salty taste. Color: white to gray. Crystalline. May contain fine-grained impurities in bands or thin layers.	ROCK SALT	
Biogenic Materials	Earthy	$CaCO_3$; effervesces freely with cold dilute HCl; easily scratched with fingernail. Microscopic organisms. Color: white.	CHALK	
		Soft. Resembles chalk but does not react with HCl. Commonly stratified. Color: gray to white. Microscopic siliceous plant remains.	DIATOMITE	
	Bioclastic	$CaCO_3$; calcareous shell fragments in a massive or crystalline matrix. Effervesces freely with cold dilute HCl.	FOSSILIFEROUS LIMESTONE	
		$CaCO_3$; calcareous shell fragments cemented together. Little or no matrix material.	COQUINA	
	Fibrous	Brown plant fibers. Soft, porous, low specific gravity.	PEAT	
	Dense	Brownish to brown-black. Harder than peat.	LIGNITE	
		Black, dull luster. Smudges fingers when handled.	BITUMINOUS COAL	

EXERCISE 3

Name
Section   Date

Identification of Common Sedimentary Rocks

The identification of sedimentary rocks is based on their texture, particle size, mineralogy, and composition.

Your instructor will provide a collection of rock specimens for you to identify. Take time to examine the specimens so that you have some familiarity with the various rock types.

Use table 1.8 to assist you in identifying the sedimentary rocks in the collection of specimens provided.

Have your laboratory instructor verify your identification before you continue.

Figures 1.48 through 1.55 show some examples of similar-size specimens of common sedimentary rocks. Remember there is a wide diversity of sedimentary rocks and these examples may not exactly represent the rock types provide to you. (NOTE: The scale bar to the right will help you to estimate grain size and texture.)

As you complete the following parts of the exercise, use the worksheets provided on the following pages to record your observations. Follow these steps:

1. For each sample specimen determine if:
   (a) It is composed of mineral or rock fragments (detrital materials) or,
   (b) It is generally light-colored and crystalline, oolitic, or dense (chemical precipitates) or,
   (c) It is composed of fossil materials from organisms or plants (biogenic materials).
2. For each specimen in 1a, estimate the grain size and shape. Then determine the major constituent (i.e., quartz grains, rock fragments, etc.). Test with a drop of cold dilute HCl if you suspect the presence of carbonate. Using the appropriate part of table 1.12, determine the rock name for each specimen in this group.
3. For those specimens in 1b, test with a drop of cold dilute HCl (remembering that dolomite will react with cold dilute HCl only when the dolomite is in a powdered form).
   (a) If there is a reaction, examine the specimen for other diagnostic features such as ooliths, possibly some fossil remains, etc. Determine the rock name from table 1.12 for each specimen in this category.
   (b) If there is no reaction, test for hardness. (A taste test may be useful under appropriate circumstances, but as noted on page 7, for obvious sanitary reasons, do not use the taste test on laboratory hand specimens.)
4. For those specimens in 1c, sort out the light-colored, fine-grained rocks.
   (a) Apply a drop of cold dilute HCl to the light-colored, fine-grained specimens. Test for hardness. Examine them for other features. Determine the rock name from table 1.12 for each specimen in this category.
   (b) For the remaining specimens in this category, the composition and other diagnostic features should allow you to determine the rock name from table 1.12.
5. Where possible, apply the appropriate adjective to the sample; for example, *fossiliferous* limestone, *calcareous* quartz sandstone, or *ferruginous* sandstone. Be as specific as possible. This will require special attention to *all* the features of the specimens.
6. Your laboratory instructor will advise you as to the procedure to be used to verify your identification.

Metamorphic Rocks

All rocks are subject to processes of change that occur at the surface of the earth or within the earth. As we have seen, the processes with which we are familiar are those that take place at the surface of the earth and combine to form sedimentary rocks. **Metamorphic** rocks are formed at varying depths within the crust when preexisting rocks (**protoliths**) are changed physically or chemically under conditions of high temperature, high pressure, or both. The process of **metamorphism,** literally a "change in form," usually takes place deep beneath the earth's surface and acts on all rocks—igneous, sedimentary, and metamorphic (refer to the rock cycle diagram, fig. 1.22).

As noted in the discussion of igneous rocks, the geothermal gradient is one source of heat for metamorphism. The other source is the heat from igneous magmas as they rise upward into country rock. Metamorphism always occurs below the melting point of the rocks involved. In some cases, very high temperatures may occur, resulting in partial melting of some materials. The result is a **migmatite,** a rock of mixed igneous and metamorphic origin.

Pressure derives from two sources. The first is the **confining pressure** on the deeply buried rocks resulting from the weight of the overlying rocks. Confining pressure is equal in all directions and is related to the depth of burial. The second is **directed pressure** (differential stress), which as the name suggests is greater in one particular direction. An example of directed pressure is that associated with the process of **orogeny** or **mountain building.**

The process of metamorphism is aided by the presence of water-rich fluids in the rocks. While it is possible for chemical elements to migrate through the country rock without fluids being present, their movement is greatly facilitated by fluids.

The mineralogy and texture of metamorphic rocks provide some insight into the temperature and pressure environment in which they were formed. **Metamorphic facies** are defined by the typical assemblage of minerals found in the rocks formed at specific combinations of temperature and pressure. **Metamorphic grade** refers to the intensity of metamorphism in a given rock.

Metamorphic Environments

The metamorphism associated with the intrusion of a magma into the country rock (sometimes referred to as the host rocks) results in **contact metamorphism.** Both the heat and the chemical constituents that emanate from the magma produce mineralogical changes in the host rocks. Contact metamorphism is most intense at or near the zone of contact between the magma and the host rock, and the effects of contact metamorphism progressively diminish in a direction away from the contact zone. This **aureole** or halo of metamorphism surrounding the magma is characterized by the formation of high-temperature minerals close to the contact zone and progressively lower-temperature minerals as the distance from the contact zone increases. Normally, there is no evidence of the reorientation of the minerals. The effect of migration of materials from the magma into the host rock, **metasomatism,** may or may not occur. The size of the aureole depends on the temperature and size of the pluton, mineralogy of the host rock, and the presence or absence of fluids.

By far the largest volume of metamorphic rocks is produced by those processes associated with **regional metamorphism.** Regional metamorphism is a result of large-scale orogenic or mountain-building events involving movement and deformation of the earth's crust on a regional scale. During periods of mountain building, large segments of crustal rocks are deformed. These deformed areas occur in zones or belts hundreds of miles wide and thousands of miles long (the Andean Mountain belt is an example). Rocks in these belts are subjected to stretching (tensional) and squeezing (compressional) stresses that cause physical and mineralogical changes in the rock. They are subjected to extreme stresses, and rocks at great depth may deform as a plastic rather than as a brittle solid. This accounts for the fact that many rocks deformed under conditions of regional metamorphism have a texture, called **foliation,** that is characterized by a parallel arrangement of platy minerals such as the micas or the common orientation of the long axis of such minerals as hornblende (see figure 1.61).

A third but less important metamorphic environment is that of **dynamic** or **deformational metamorphism** associated with fault zones such as the San Andreas Fault in California (discussed in Exercise 25). In this environment, high pressure and low temperatures are present, and the materials commonly are composed of broken and distorted fragments of rocks on either side of the fault zone and minerals that form only at low temperatures and high pressure. If coarse-grained, these rocks are known as **fault breccia;** if fine-grained and exhibiting flow structure, **mylonite.**

Metamorphic Facies and Grade

Contact and regional metamorphism produce differing degrees of change in preexisting rocks. As briefly discussed, the aureole associated with contact metamorphism is defined by a decrease in degree of metamorphism as the distance from the pluton increases. The same is true of regional metamorphism; rocks on the outer margins of a mountain belt that has been subjected to a single period of deformation may be only slightly metamorphosed. The metamorphic rocks in the center of the belt may be so deformed that the texture and mineralogy of the original or parent rock have been obliterated.

The study of metamorphic rocks has led to the identification of **metamorphic facies:** the recognition of an assemblage of minerals in the rocks that formed during metamorphism under specific environmental conditions of temperature and pressure. The assemblages formed depend also on the mineralogy of the parent rock, but the same parent rock will yield the same assemblages at the same combination of temperature, pressure, and fluids. The commonly recognized metamorphic facies are named for characteristic minerals or characteristic rock types associated with them (fig. 1.56).

If we consider the effects of contact metamorphism on a single rock type, the resulting facies are relatively simple

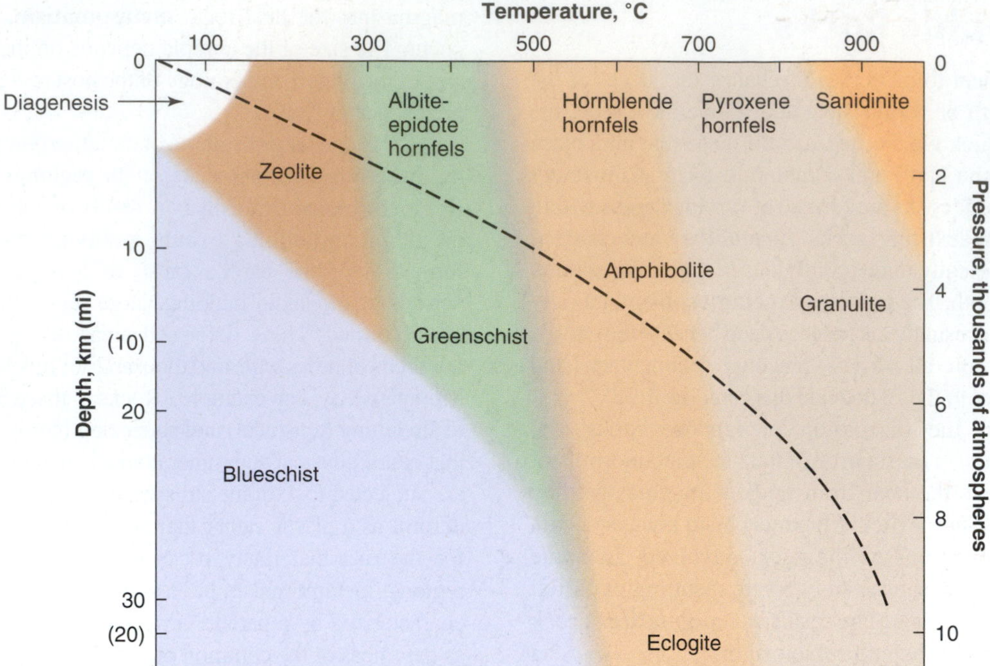

Figure 1.56

Metamorphic facies as functions of temperature and pressure (depth). Dashed curve is average continental geothermal gradient.

because temperature is the single important variable. The facies recognized are named for specific minerals: **sanidinite,** the high-temperature facies, is characterized by the presence of the mineral sanidine, a variety of K-feldspar, grading outward from the heat source through the **hornfels** facies to the **zeolite** or low-temperature facies. The high pressure and low-to-moderate temperature facies is known as "blueschist." Rocks formed in this facies tend to be schistose in texture and contain minerals that give the rocks a bluish color, thus the name.

The facies relationships associated with regional metamorphism are more complex due to the fact that not only are there two variables, both temperature and pressure, but the size of the area affected is such that there is a wide variety of rock types involved. Not all facies may be present in a given region. Figure 1.56 is a schematic diagram of the temperature and pressure relationships of the commonly recognized metamorphic facies.

Metamorphic grade is the measure of the intensity of metamorphism to which a given rock has been subjected. **Low-grade** metamorphism occurs in the marginal area; **high-grade** metamorphism occurs where the effects of temperature and pressure have been most intense; and **intermediate-grade** metamorphism lies in between. Table 1.13 is a simplified schematic showing the relationship between metamorphic grade, rock type, and the **index minerals** associated with each. Index minerals are those minerals that are stable over a specific range of temperature and pressure conditions and that are useful in determining metamorphic grade (fig. 1.57).

The metamorphism of shale, a common sedimentary rock, provides a simple way of illustrating metamorphic grade. The sequence of increasing metamorphic grade shows the gradual change from shale to slate to phyllite to schist to gneiss (pronounced "nice") to granulite (see table 1.13). In this simplified presentation, slate represents low-grade metamorphism; phyllite, low to intermediate; schist, intermediate to high; and gneiss and granulite, high to very high grade metamorphism. As metamorphic grade increases, there are corresponding mineralogic and textural changes that occur within the rocks.

Table 1.13 Simplified Table Relating Rock Type, Metamorphic Grade, and Index Minerals

Rock Type	Metamorphic Grade	Index Mineral
Slate	Low	
Phyllite	Low to intermediate	Chlorite
Schist	Intermediate to high	Garnet
Gneiss	High	Sillimanite
Granulite	Very high	

Texture and Composition of Metamorphic Rocks

Metamorphic textures consist of two main types, **foliated** and **nonfoliated.** The mineral constituents of foliated metamorphic rocks are oriented in a parallel or subparallel

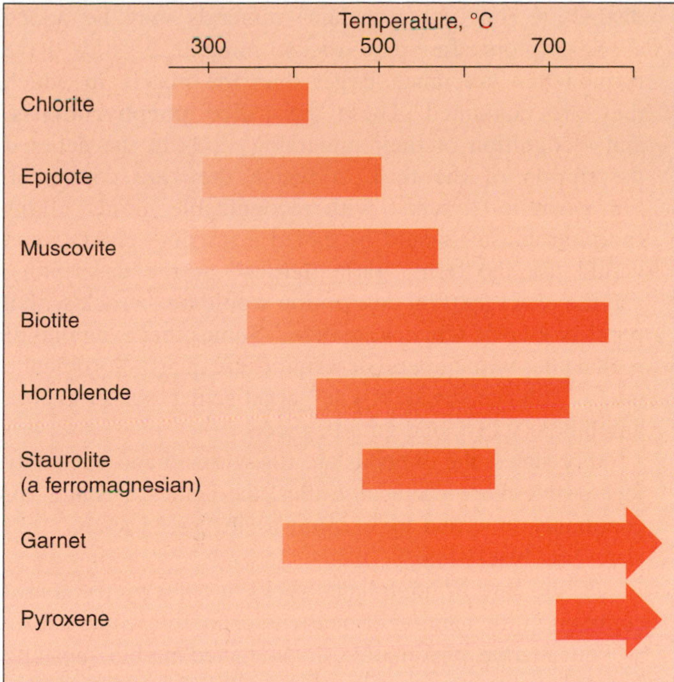

Figure 1.57
Approximate temperature ranges over which some representative index minerals are stable.

arrangement. Foliated metamorphic rocks are generally associated with regional metamorphism. The nonfoliated rocks exhibit no preferred orientation of mineral grains and are commonly composed of a single mineral.

Foliated Textures

Five kinds of foliated textures are recognized. In order of increasing metamorphic grade, these are **slaty, phyllitic, schistose, gneissic** (pronounced "nice-ick"), and **granulitic.**

Slaty Texture

This texture is caused by the parallel orientation of platy microscopic grains. The name for the rock with this texture is **slate** (fig. 1.58), and the rock is characterized by a tendency to separate along parallel planes. This feature is a property known as **slaty cleavage.** (Slaty cleavage or rock cleavage is not to be confused with cleavage in a mineral, which is related to the internal atomic structure of the mineral.)

Phyllitic Texture

This texture is formed by the parallel arrangement of platy minerals, usually micas, that are barely macroscopic (visible to the naked or corrected eye). The parallelism is often wavy, or crenulated. The predominance of micaceous minerals imparts a sheen to the hand specimens. A rock with a phyllitic texture is called a **phyllite** (fig. 1.59).

Figure 1.58
Slate, showing characteristic slaty cleavage.

Schistose Texture

This is a foliated texture resulting from the subparallel to parallel orientation of platy minerals such as chlorite or micas. Other common minerals present are quartz and amphiboles. A schistose texture lies between the parallel platy appearance of phyllite and the distinct banding of gneissic texture. The average grain size of the minerals is generally smaller than in a gneiss (see discussion that follows). A rock with schistose texture is called a **schist** (fig. 1.60).

Gneissic Texture

This is a coarsely foliated texture in which the minerals have been segregated into discontinuous bands, each of which is dominated by one or two minerals. These bands range in thickness from 1 millimeter to several centimeters. The individual mineral grains are macroscopic and impart a striped appearance to a hand specimen (fig. 1.61). Light-colored bands commonly contain quartz and feldspar, and the dark bands are commonly composed of hornblende and biotite. Accessory minerals are common and are useful in applying specific names to these rocks. A rock with a gneissic texture is called a **gneiss.**

Granulitic Texture

This texture is a medium to coarse, even-grained (mosaic) texture that is almost or completely lacking in water-bearing minerals (anhydrous). Quartz, two feldspars, almandine garnet, and, in one group, pyroxenes are the dominant minerals. Quartzo-feldspathic granulites usually show a distinct foliation of thin lenticular aggregates of quartz grains. In some rocks, the foliation is accentuated by discontinuous thin films of biotite. Foliation is virtually lacking in pyroxene granulites. A rock with a granulitic texture is called a **granulite** (fig. 1.62).

Nonfoliated Texture

Metamorphic rocks with no visible preferred orientation of mineral grains have a nonfoliated texture. Nonfoliated rocks commonly contain equidimensional grains of a single mineral such as quartz, calcite, or dolomite. Examples of such rocks are **quartzite** (fig. 1.63), formed from a quartz sandstone, and **marble** (fig. 1.64), formed from a limestone or dolomite (dolostone). Conglomerate that has been metamorphosed may retain the original textural characteristics of the parent rock, including the outlines and colors of the larger grain sizes such as granules and pebbles. However, because metamorphism has caused recrystallization of the matrix, the metamorphosed conglomerate is called **metaconglomerate** (fig. 1.65). In some cases, the metamorphism has deformed the shape of the granules or pebbles; in this case, the rock is called a **stretch-pebble conglomerate.**

Quartzite and metamorphosed conglomerate can be distinguished from their sedimentary equivalents by the fact that they break **across** the quartz grains, not around them. Marble has a crystalline appearance and generally has larger mineral grains than its sedimentary equivalents.

Hornfels is a fine-grained (dense-textured), nonfoliated rock usually of contact metamorphic origin. Hornfels has a nondescript appearance because it is usually some medium to dark shade of gray, is lacking in any structural characteristics, and contains few if any recognizable minerals in hand specimen.

Anthracite coal (fig. 1.66) is the metamorphic equivalent of bituminous coal. A sequence similar to that presented for the metamorphism of shale would show the change from peat to lignite to bituminous coal to anthracite coal and, under extreme conditions, to the mineral graphite.

The Naming of Metamorphic Rocks

Foliated metamorphic rocks are named according to their texture (e.g., slate, phyllite, schist, or gneiss). In addition to the root name, the name or names of the dominant or distinctive (but not abundant) minerals may be added as descriptors. In some rocks, mineral crystals have formed that are much larger than the matrix in which they are contained. These are called **porphyroblasts,** and recognition of their mineralogy aids in the detailed description of the rock in which they are contained. For example, a schist with recognizable garnet grains (porphyroblasts) would be called a **garnet schist** (some would use the term "garnetiferous" as the descriptor), and the exact textural description would be "a rock with a porphyroblastic schistose texture." Some other examples of rock names with mineral descriptors are quartz-hornblende gneiss, chlorite schist, biotite-garnet schist, garnet gneiss, amphibole schist, and granite gneiss.

For slates, the minerals are fine-grained and not easily identifiable macroscopically; thus, the names of slates are commonly modified by a color, as in "green slate," "red slate," or "black slate."

In the case of nonfoliated rocks, a color prefix is also common in the naming of quartzite or marble. Such names as white marble, pink marble, or variegated marble (meaning a marble containing streaks of several different colors) are commonplace.

Classification of Metamorphic Rocks

The classification of metamorphic rocks is based on texture and composition, the basic distinction being between foliated and nonfoliated textures (table 1.14). **The rock names given are only the root names,** because the chart would become too large and cumbersome if all possible varieties of gneiss, schist, marble, and quartzite were listed. You are encouraged to utilize as many descriptors as applicable in naming metamorphic rocks.

Figure 1.59 Phyllite (foliated texture).

Figure 1.60 Garnet schist (foliated texture).

Figure 1.61 Gneiss (foliated texture).

Figure 1.62 Granulite (granulitic texture).

Figure 1.63 Quartzite (nonfoliated texture).

Figure 1.64 Marble (nonfoliated texture).

Figure 1.65 Metaconglomerate (nonfoliated texture).

Figure 1.66 Anthracite coal (nonfoliated texture).

Table 1.14 Simplified Classification and Identification Table for Hand Specimens of Common Metamorphic Rocks

Texture		Composition and/or Diagnostic Features	Rock Name	Protolith
Foliated*	Increasing Temperature and Pressure (↓)	Slaty texture (slaty cleavage apparent). Dense, microscopic grains. Color variable; black and dark gray common. Also occurs in green, dark red, and dark purple colors.	SLATE	Shale/Mudstone
		Phyllitic texture. Fine-grained to dense. Micaceous minerals are dominant. Has a "shiny" appearance.	PHYLLITE	Shale/Mudstone
		Schistose texture. Medium to fine-grained. Common minerals are chlorite, biotite, muscovite, garnet, and dark elongate silicate minerals. Feldspars commonly absent. Recognizable minerals used as part of rock name. Porphyroblasts common. Has a "sparkling" appearance.	SCHIST	Shale/Mudstone
		Gneissic texture. Coarse-grained. Foliation present as macroscopic grains arranged in alternating light and dark bands. Abundant quartz and feldspar in light-colored bands. Dark bands may contain hornblende, augite, garnet, or biotite.	GNEISS	Shale/Mudstone/ Igneous rock
		Granulitic texture, medium to coarse, even-grained. Essentially anhydrous. Foliation present in light-colored quartzo-feldspathic rocks.	GRANULITE	Any rock type
Foliated or Nonfoliated		Medium to coarse-grained. Mostly crystals of amphibole, sometimes feldspar, mica, and talc.	AMPHIBOLITE	Mafic or ultramafic igneous rocks
Nonfoliated		Crystalline. Hard (scratches glass). Breaks across grains as easily as around them. Color variable; white, pink, buff, brown, red, purple.	QUARTZITE	Sandstone
		Dense, dark-colored; various shades of gray, gray-green, to nearly black.	HORNFELS	Any rock type
		Texture of conglomerate but breaks across coarse grains as easily as around them. Granules or pebbles are commonly granitic or jasper, chert, quartz, or quartzite. If pebbles are deformed, called *stretched-pebble conglomerate.*	METACONGLOMERATE	Conglomerate
		Crystalline. Composed of calcite or dolomite. Color variable; white, pink, gray, among others. Fossils in some varieties.	MARBLE	Limestone/Dolostone
		Microcrystalline texture, usually with smooth wavy surfaces. Composed of serpentine, sometimes with asbestos (crysotile). Color variable; shades of green, brown, or red most common.	SERPENTINITE	Mafic or ultramafic igneous rocks/ Dolostone
		Granulitic texture, medium to coarse, even-grained. Essentially anhydrous. Foliation virtually lacking in pyroxeneplagioclase-bearing rocks.	GRANULITE	Any rock type
		Black, shiny luster. Conchoidal fracture.	ANTHRACITE COAL	Bituminous coal

*Refer to Table 1.13 for metamorphic grade of each rock type.

EXERCISE 4

Name

Section Date

Identification of Common Metamorphic Rocks

The identification of metamorphic rocks is based on their texture and mineralogy. Take time to review the types of textures described on the previous pages.

Your instructor will provide a collection of rock specimens for you to identify. Use table 1.8 to separate the metamorphic rocks from the collection of specimens. Divide specimens into the two major textural categories—foliated and nonfoliated. Have your laboratory instructor verify your identification before you continue.

Figures 1.59 through 1.66 are examples of several types of metamorphic rocks and may assist you in identification. Remember that there are a wide variety of types, textures, and colors of metamorphic rocks, and your collection may include specimens that do no appear in these figures. (NOTE: The scale bar to the right of each specimen will help you to estimate grain size and texture.)

Using the worksheets provided on the following pages to record your observations. follow these steps:

1. Arrange the foliated specimens in sequence from coarse-grained to fine-grained or dense.
   (a) Work first with the coarse-grained rocks to apply a rock name. Examine each specimen carefully to identify specific minerals that are present either in abundance or as porphyroblasts. Use these when applying a specific rock name to the specimens of gneissic or schistose texture. Some examples that may be in your collection are

   GNEISS:
   - Biotite gneiss
   - Biotite hornblende gneiss
   - Quartz feldspar gneiss
   - (Others possible)

   SCHIST:
   - Garnet schist
   - Muscovite schist
   - Biotite schist
   - Tourmaline mica schist
   - Staurolite schist
   - Hornblende schist
   - Garnet mica schist
   - (Others possible)

   (b) The fine-grained to dense rocks with foliation should be examined on the basis of texture and then color. Descriptors are usually not possible with phyllite, and the slates are generally described by color. In some cases, they may contain enough calcium carbonate to react with cold dilute HCl and would be called "calcareous slate."
   (c) Identify and apply rock names to each of the foliated specimens. For each, give an indication of the metamorphic grade represented.
2. Work next with the nonfoliated specimens. Remember the following:
   (a) Marble is composed mainly of calcite or dolomite; hence, it is usually softer than glass and reacts with cold dilute HCl in the same way that limestone (calcite) or dolostone (dolomite) does.
   (b) Quartzite and metaconglomerate are rich in silica (quartz) and scratch glass.
   (c) Graphite as a rock has the same characteristics as the mineral graphite.
   (d) Identify and apply rock names to all specimens. Where possible, use descriptors such as color, fossil content, shape of pebbles, and so forth.
3. On the worksheet, you are asked to identify the protolith of each of the rocks you identify. The protolith of a metamorphic rock is the rock from which it was formed by metamorphism. For example, the protolith for quartzite would be quartz sandstone. For many of the common metamorphic rocks, it should be possible for you to determine the protolith.
4. Your laboratory instructor will advise you as to the procedure to be used to verify your identifications.

Name Section Date

Worksheet for Metamorphic Rocks

Sample #	Color	Texture	Grain Size	Diagnostic Minerals	Rock Name	Protolith

PART II

The Geologic Column and Geologic Time

Background

The goal of Part II is to acquaint you with the modes of occurrence of rocks and their age relationships. You may participate in organized field trips to observe firsthand how rocks occur in nature. To prepare you for this field experience, some basic geologic laws (principles) will be introduced. Simple geological diagrams will be used to show examples of some of these relationships that may be seen in the field.

The concept of geologic time will be introduced, and the relative and numerical age relationships of rock masses will be examined utilizing some of the basic laws, principles, assumptions, and data that are available to the geologist. An understanding of these relationships and the geologic diagrams presented here will enhance your appreciation of the rock strata and other geologic phenomena you will encounter in your travels.

Introduction

The age of rocks and the age of the earth have long been of interest to the scientific community. By the middle of the nineteenth century the precursors of the geologic time scale had begun to emerge. The application of the "principles" or "laws" that were the result of keen observations of sedimentary rock outcrops led to a generally accepted sequence of rock formations, from oldest to youngest with names applied to various units. Thus, the relative time scale that is now the basis for the numerical time scale was developed, and the names of the eras and periods that were applied early on are still in use today.

Geologists were not satisfied with the relative time scale and sought ways to get numeric ages. Various scenarios were proposed as a way to calculate absolute ages. One such method was to determine the total thickness of the sediments and to divide that by the rate of accumulation of those sediments. This method gave widely varying age estimates depending on the assumptions used to determine total thickness and rate of accumulation. Others suggested that the length of time to accumulate salt in the oceans could be used to determine the age of the earth. None of these gave reasonable numerical estimates of the age of the earth or of the rocks present.

The recognition that there was a horizontal and vertical continuity of sedimentary rock sequences was the first step in the understanding of relative age relationships. Soon the importance of the fossils contained in these rocks became a part of the process. Fossils are restricted to sedimentary rocks and while there are Archean sediments, only sediments younger than 542 million years preserve a fossil record diverse enough to be useful in understanding the relative age of these rocks.

The Geologic Column and Geologic Time

Relative Age Determinations

An understanding of the relationships between rock units and their ages is essential in the unraveling of the geologic history of a given area. The **relative chronology (relative ages)** of sedimentary rocks is determined by the application of seven basic geologic laws (sometimes called principles):

1. The **law of original horizontality,** which states that sediments deposited by water or wind are laid down in strata that are horizontal or nearly horizontal.
2. The **law of superposition,** which states that in any undisturbed sequence of sedimentary rocks, the layer at the bottom is older than the layer at the top of the sequence.
3. The **law of original (lateral) continuity,** which states that sedimentary rock units extend laterally until they thin or pinch out at their margins.
4. The **law of faunal assemblages,** which states that similar assemblages of fossils indicate similar geologic ages for rocks that contain them.
5. The **law of faunal succession,** which states that each geologic formation has a different aspect of life (fossils) from that in the formation above it and below it.
6. The **law of crosscutting relationships,** which states that any rock unit cutting another must be younger than the rock unit it cuts.
7. The **law of inclusions (xenoliths),** which states that fragments of rocks in another rock unit are older than the host rock containing them.

Through the application of the laws (principles) used to establish the relative ages of sedimentary strata around the world, a geologic time scale for all of earth history has been pieced together. The geologic time scale as used in North America is shown in table 2.1. The table is arranged with the oldest geologic age at the bottom and the youngest at the top.

Not all areas of the earth's surface were sites of deposition throughout all of geologic time. Some areas, especially mountain regions, contained preexisting rocks from which sedimentary materials were produced by natural decay and physical breakdown. These sediments were eroded and transported to sites of deposition by various geologic agents such as running water, glaciers, and wind.

A depositional site may change over geologic time into a site where erosion or no deposition (an *hiatus*) is taking place. Hence, a continuous sequence of strata representing all of geologic time is unlikely to occur at a single site. Gaps in the sedimentary record may occur during the passage of geologic time. Therefore, to construct a more complete geologic history of a particular area, we must piece together fragments of that history recorded in rocks from different localities.

The most thoroughly documented part of the geologic time scale, sometimes referred to as the **geologic column,** comes from the geologic strata deposited during the **Phanerozoic** eon. This segment of geologic time has been divided into subunits called eras, periods, and epochs. The three major eras in order of decreasing age are the **Paleozoic, Mesozoic,** and **Cenozoic.** Paleozoic means "early life," Mesozoic means "middle life," and Cenozoic means "recent life." Each of these is divided into periods, and the periods are further divided into epochs. The names of the eras are based generally on the fossils contained in strata formed during those eras. **A fossil is any evidence of past life such as bones, shells, leaf imprints, and the like.** The names of the periods and epochs are based on strata originally studied in Europe during the eighteenth and nineteenth centuries; hence, the names are chiefly European in origin.

Precambrian rocks, representing about 87% of earth history as we know it today, are those that were formed prior to the beginning of the Phanerozoic eon. In table 2.1, we have divided the Precambrian into the **Proterozoic** eon and the **Archean** eon. While the majority of Precambrian rocks are igneous or metamorphic in type and the field relationships where they are exposed are often complex, undeformed sedimentary rocks of Precambrian age are known from a variety of localities.

Table 2.1 is modified from that presented in "The Concise Geologic Time Scale" by Ogg, Ogg, and Gradstein (2008). The time scale was initially published in 2004 by The International Commission on Stratigraphy (ICS) and is also published in "The Concise Geologic Time Scale" volume.

PART II

The Geologic Column and Geologic Time

Background

The goal of Part II is to acquaint you with the modes of occurrence of rocks and their age relationships. You may participate in organized field trips to observe firsthand how rocks occur in nature. To prepare you for this field experience, some basic geologic laws (principles) will be introduced. Simple geological diagrams will be used to show examples of some of these relationships that may be seen in the field.

The concept of geologic time will be introduced, and the relative and numerical age relationships of rock masses will be examined utilizing some of the basic laws, principles, assumptions, and data that are available to the geologist. An understanding of these relationships and the geologic diagrams presented here will enhance your appreciation of the rock strata and other geologic phenomena you will encounter in your travels.

Introduction

The age of rocks and the age of the earth have long been of interest to the scientific community. By the middle of the nineteenth century the precursors of the geologic time scale had begun to emerge. The application of the "principles" or "laws" that were the result of keen observations of sedimentary rock outcrops led to a generally accepted sequence of rock formations, from oldest to youngest with names applied to various units. Thus, the relative time scale that is now the basis for the numerical time scale was developed, and the names of the eras and periods that were applied early on are still in use today.

Geologists were not satisfied with the relative time scale and sought ways to get numeric ages. Various scenarios were proposed as a way to calculate absolute ages. One such method was to determine the total thickness of the sediments and to divide that by the rate of accumulation of those sediments. This method gave widely varying age estimates depending on the assumptions used to determine total thickness and rate of accumulation. Others suggested that the length of time to accumulate salt in the oceans could be used to determine the age of the earth. None of these gave reasonable numerical estimates of the age of the earth or of the rocks present.

The recognition that there was a horizontal and vertical continuity of sedimentary rock sequences was the first step in the understanding of relative age relationships. Soon the importance of the fossils contained in these rocks became a part of the process. Fossils are restricted to sedimentary rocks and while there are Archean sediments, only sediments younger than 542 million years preserve a fossil record diverse enough to be useful in understanding the relative age of these rocks.

The Geologic Column and Geologic Time

Relative Age Determinations

An understanding of the relationships between rock units and their ages is essential in the unraveling of the geologic history of a given area. The **relative chronology (relative ages)** of sedimentary rocks is determined by the application of seven basic geologic laws (sometimes called principles):

1. The **law of original horizontality,** which states that sediments deposited by water or wind are laid down in strata that are horizontal or nearly horizontal.
2. The **law of superposition,** which states that in any undisturbed sequence of sedimentary rocks, the layer at the bottom is older than the layer at the top of the sequence.
3. The **law of original (lateral) continuity,** which states that sedimentary rock units extend laterally until they thin or pinch out at their margins.
4. The **law of faunal assemblages,** which states that similar assemblages of fossils indicate similar geologic ages for rocks that contain them.
5. The **law of faunal succession,** which states that each geologic formation has a different aspect of life (fossils) from that in the formation above it and below it.
6. The **law of crosscutting relationships,** which states that any rock unit cutting another must be younger than the rock unit it cuts.
7. The **law of inclusions (xenoliths),** which states that fragments of rocks in another rock unit are older than the host rock containing them.

Through the application of the laws (principles) used to establish the relative ages of sedimentary strata around the world, a geologic time scale for all of earth history has been pieced together. The geologic time scale as used in North America is shown in table 2.1. The table is arranged with the oldest geologic age at the bottom and the youngest at the top.

Not all areas of the earth's surface were sites of deposition throughout all of geologic time. Some areas, especially mountain regions, contained preexisting rocks from which sedimentary materials were produced by natural decay and physical breakdown. These sediments were eroded and transported to sites of deposition by various geologic agents such as running water, glaciers, and wind.

A depositional site may change over geologic time into a site where erosion or no deposition (an *hiatus*) is taking place. Hence, a continuous sequence of strata representing all of geologic time is unlikely to occur at a single site. Gaps in the sedimentary record may occur during the passage of geologic time. Therefore, to construct a more complete geologic history of a particular area, we must piece together fragments of that history recorded in rocks from different localities.

The most thoroughly documented part of the geologic time scale, sometimes referred to as the **geologic column,** comes from the geologic strata deposited during the **Phanerozoic** eon. This segment of geologic time has been divided into subunits called eras, periods, and epochs. The three major eras in order of decreasing age are the **Paleozoic, Mesozoic,** and **Cenozoic.** Paleozoic means "early life," Mesozoic means "middle life," and Cenozoic means "recent life." Each of these is divided into periods, and the periods are further divided into epochs. The names of the eras are based generally on the fossils contained in strata formed during those eras. **A fossil is any evidence of past life such as bones, shells, leaf imprints, and the like.** The names of the periods and epochs are based on strata originally studied in Europe during the eighteenth and nineteenth centuries; hence, the names are chiefly European in origin.

Precambrian rocks, representing about 87% of earth history as we know it today, are those that were formed prior to the beginning of the Phanerozoic eon. In table 2.1, we have divided the Precambrian into the **Proterozoic** eon and the **Archean** eon. While the majority of Precambrian rocks are igneous or metamorphic in type and the field relationships where they are exposed are often complex, undeformed sedimentary rocks of Precambrian age are known from a variety of localities.

Table 2.1 is modified from that presented in "The Concise Geologic Time Scale" by Ogg, Ogg, and Gradstein (2008). The time scale was initially published in 2004 by The International Commission on Stratigraphy (ICS) and is also published in "The Concise Geologic Time Scale" volume.

Table 2.1	Geologic Time Scale as Used in North America (Modified from the ICS International Stratigraphic Chart, 2008)						
EON		ERA	PERIOD	EPOCH	Approximate Age in Millions of Years (mya) Before Present	MAP SYMBOL	COMMON MAP COLOR
	Phanerozoic	CENOZOIC	Quaternary	Holocene		Q	Various shades of gray and yellow
					0.0118		
				Pleistocene			
					1.8		
			Neogene	Pliocene		Tpl	Various shades of orange, yellow-orange, and yellow
					5.3		
				Miocene		Tm	
					23.0		
			Paleogene	Oligocene		To	
					33.9		
				Eocene		Te	
					55.8		
				Paleocene		Tp	
					65.5		
		MESOZOIC	Cretaceous			K	Various shades of green
					145.5		
			Jurassic			J	Various shades of blue-green
					199.6		
			Triassic			T̄R	Various shades of blue
					251.0		
		PALEOZOIC	Permian			P or Cpm	Commonly blue, green, purple, pink, lavender, purple-gray
					299.0		
			Pennsylvanian			ℙ or Cp	
					318.1		
			Mississippian			M or Cm	
					359.2		
			Devonian			D	Various shades of purple, pink, lavender, tan, brown, red-brown, red
					416.0		
			Silurian			S	
					443.7		
			Ordovician			O	
					488.3		
			Cambrian			Є	
					542.0		
Precambrian	Proterozoic					pЄ	No standard color
					2,500		
	Archean						
					4,000		
		Hadean (informal)					
					4,600		

The geologic time scale is introduced here primarily to provide a broader context in which the development of simple geologic columns representing very small segments of geologic time can be placed. Other information in table 2.1, such as the abbreviations of geologic time units and standard colors used on geologic maps, is for general information only and has no application here. However, in Part V, where geologic maps are considered in detail, the usefulness of these abbreviations and map colors will become apparent.

Numerical Age Determinations

The numerical ages given in table 2.1 have been determined by radiometric age determinations on igneous rocks found within these sedimentary sequences. The time units of the geologic column are not of equal duration, as can be seen from the ages presented in table 2.1. The geologic time scale as it existed in the early part of the twentieth century was also subdivided by numerical ages but these dates lacked precision because they were based on less precise techniques.

There was no way to accurately measure the age of rocks until physicists recognized that there were radioactive isotopes of elements (**parent isotopes**) that undergo decay at a constant rate to form other isotopes (**daughter isotopes** of the same element). Igneous and metamorphic rocks present an opportunity to obtain numerical rock ages by utilizing the decay of certain isotopes of elements found in these rocks. Initially, entire igneous samples were ground into powders and analyzed for a "whole-rock" radiometric age. These first radiometric ages had large errors but were a good start toward building a numerical time scale. Radiometric age dating now utilizes very precise microchemical measurements of radiometric decay of individual isotopes (U to Pb for example) in mineral separates (e.g. zircon) from igneous and metamorphic rocks to obtain numerical ages of the rocks from which the **separates** were obtained. Relative ages of sedimentary sequences can be determined by dating separates from sedimentary strata.

One of the minerals utilized for these microchemical analyses is zircon, an accessory mineral that is common in intermediate (diorite, andesite) and felsic (granite, rhyolite) igneous rocks. In recent years the use of zircons for radiometric dating has become an important tool. Zircons are remarkably stable, durable, and resistant to weathering and erosion. When zircon crystallizes within a melt, zirconium may be replaced by small concentrations of uranium within the lattice. The uranium (parent isotope) begins the radiometric decay process to lead (daughter isotope). Zircon crystals are often zoned, indicating that the crystals grew in layers in the melt over a period of time. The U-Pb ages of the various layers of a zoned zircon crystal reveal that the rim is younger than the core (fig. 2.1). The U-Pb crystallization age from (fig 2.1) zircons can be confirmed by obtaining a K-Ar age on potassium feldspar crystals from the same rock.

Many metamorphic rocks, especially gneiss and schist, contain mica minerals (muscovite and biotite). Using the K-Ar radiometric decay series for mica, a numerical age for the crystallization for a gneiss or schist can be obtained. These ages are interpreted as the age of metamorphism of the rock.

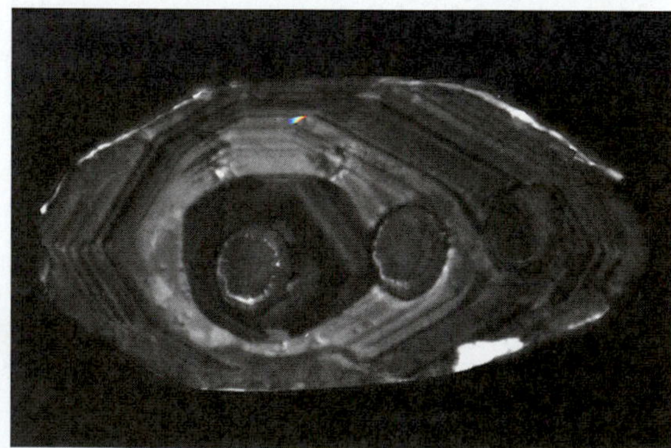

Figure 2.1
Zoned igneous zircon crystal showing various layers (zones). Circular spots are laser ablation pits (30 microns in diameter) used to date specific areas of the zircon.
© *James Carter.*

Sedimentary rocks may contain thin layers of volcanic ash that have settled into the sedimentary basin after a volcanic eruption. The ash may contain zircons which can be separated and dated using the U-Pb method to give a numerical age of the ash and the surrounding sediment. This can be useful as a means to correlate the ages of sediments across a basin or region. It also provides constraints on the ages of fossils in the sediments below and above the ash layer. Clastic sediments may contain small percentages of detrital heavy minerals. These detrital heavy minerals, specifically zircon, can be separated from clastic sediments and the U-Pb ages determined for 100 grains (see fig 2.16A and 2.16B).

The result is a spectrum of the ages of the zircons in the sediment that gives some indication of the rock ages in the source area of the sediments. The age of deposition of these sediments is also constrained by the zircon ages, as the age of deposition cannot be older than the numerical age of the youngest detrital zircon and a relative age of deposition can be determined.

The mathematical expression that relates radioactive decay to geologic time is

$t = (1/\lambda)\ln(1 + D/P)$, where

t is the age of the rock or mineral specimen
D is the number of atoms of a daughter product measured
P is the number of atoms of the parent isotope measured
ln is the natural logarithm (logarithm to the base e)
λ is the appropriate decay constant

The rate of radioactive decay can be expressed as half-lives, where the half-life of a radioactive isotope equals $\ln 2/\lambda$. The shape of the radioactive decay curve is the same for all radioactive isotopes (see fig. 2.15). The values for D and P are measured with a mass spectrometer. The age of a rock or mineral containing a radioactive isotope is calculated by multiplying the number of half-lives determined from the radioactive decay curve (see fig. 2.15) by the half-life of the decay pair (table 2.2). The parent–daughter and half-lives of some common decay pairs used in dating geologic samples are listed in table 2.2.

Table 2.2 Half-Life of Decay Pairs

Parent (P)	Daughter (D)	Half-Life ($t_{1/2}$)
^{14}C	^{14}N	5.7×10^3 years
^{40}K	^{40}Ar	1.3×10^9 years
^{235}U	^{207}Pb	0.7×10^9 years
^{238}U	^{206}Pb	4.5×10^9 years
^{87}Rb	^{87}Sr	48.8×10^9 years

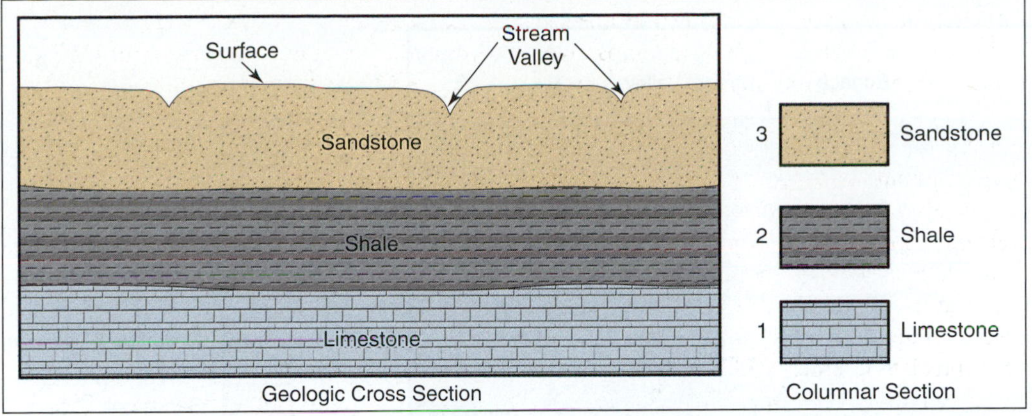

Figure 2.2
A simple geologic cross section and corresponding columnar section showing the relative ages of the three strata with the oldest (1) at the bottom and the youngest (3) at the top. Construction of the columnar section is based on the law of superposition.

Constructing a Columnar Section

We will now consider the means by which a geologic column for a given locality is constructed.

Figure 2.2 shows a series of rock strata in a **geologic cross section** with a corresponding **columnar section** to the right. The columnar section is constructed by applying the principle of superposition to the cross section (see page 56). It is apparent that the limestone in figure 2.2 is the oldest, the shale is younger than the limestone, and the sandstone is the youngest of the three formations. The process of sedimentation was continuous during the time it took for these three layers to be deposited; only the depositional environment changed. Figure 2.3 shows the symbols used to portray various rock types on geologic cross sections and columnar sections.

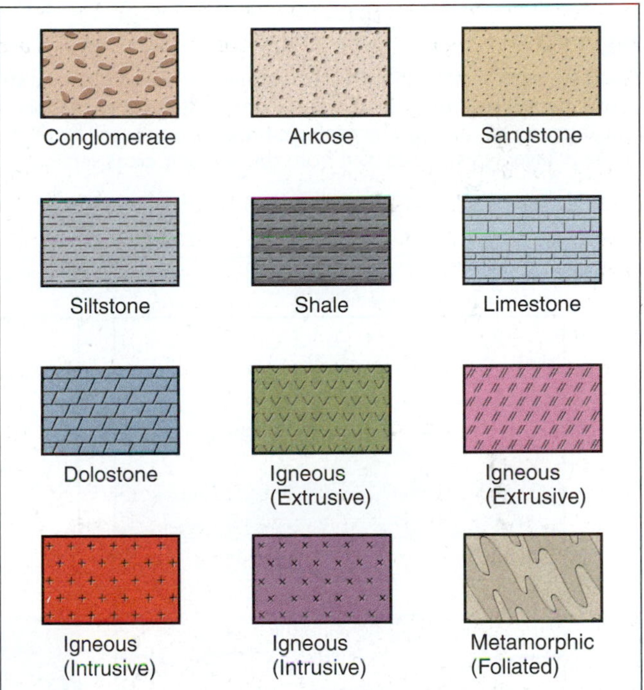

Figure 2.3
Symbols used in geologic cross sections and columnar sections. The colors are used to differentiate rock types only and bear no relationship to the colors used on geologic maps listed in table 2.1.

Unconformities

From the information available in figure 2.2, one can conclude that some time after the sandstone was formed, the marine environment in which it formed changed to an environment of erosion or nondeposition. One cannot deduce from figure 2.2 whether additional strata were laid on top of the sandstone. All that can be said is that a depositional environment gave way to an erosional one some time after the sandstone was formed. Now, if the sea that once covered the area underlain by the three strata in figure 2.2 invaded the area again and another sequence of strata was deposited, the situation would be depicted as in figure 2.4. The old erosional surface that is now buried beneath the younger strata is called an **unconformity,** a **surface of erosion** or **nondeposition.** It constitutes an unknown amount of geologic time that elapsed between the cessation of deposition of the older sequence of limestone-shale-sandstone and the deposition of the upper sequence of conglomerate-arkose-siltstone.

Three types of unconformities are recognized by geologists. The type shown in figures 2.4 and 2.5 is a **disconformity,** a surface that represents missing rock strata but the beds above

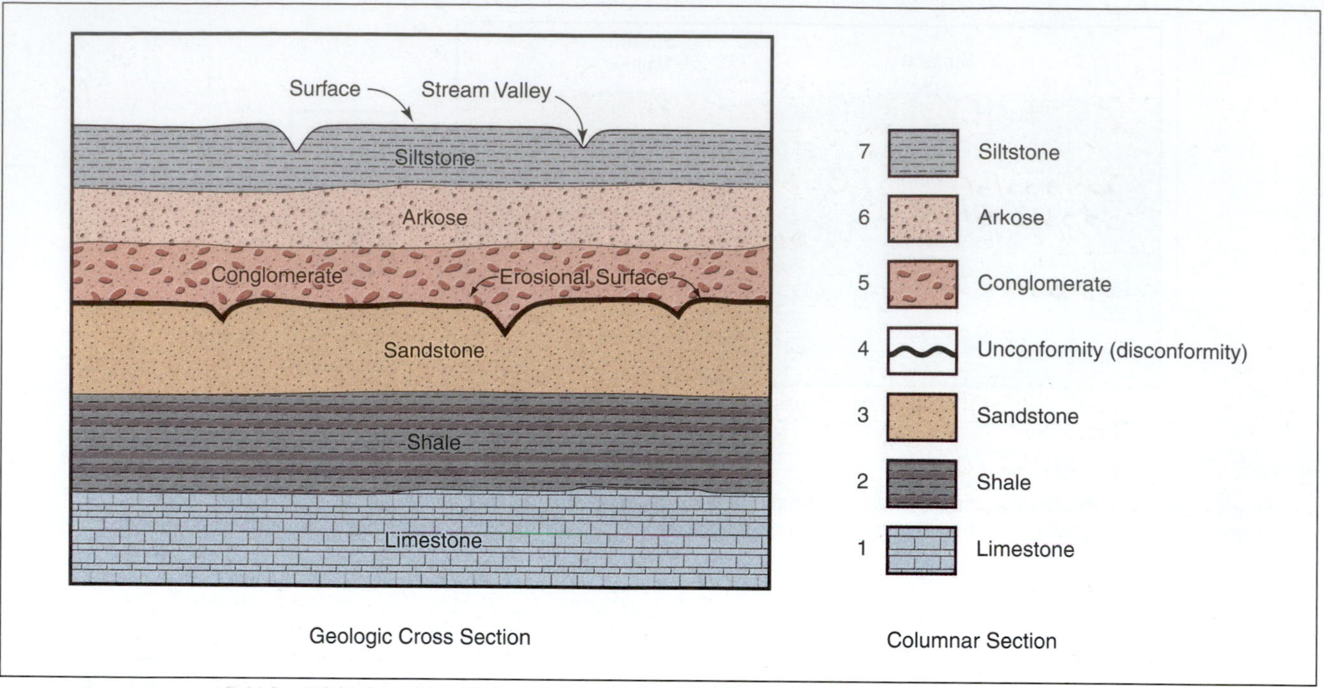

Figure 2.4

Geologic cross section showing two sequences of sedimentary strata separated by an unconformity. The columnar section at the right shows the sedimentary layers and unconformity of the cross section arranged in chronological order with the oldest at the bottom and the youngest at the top. The geologic time encompassed by the columnar section cannot be determined because only the relative ages of the rock layers can be deduced from the geologic cross section.

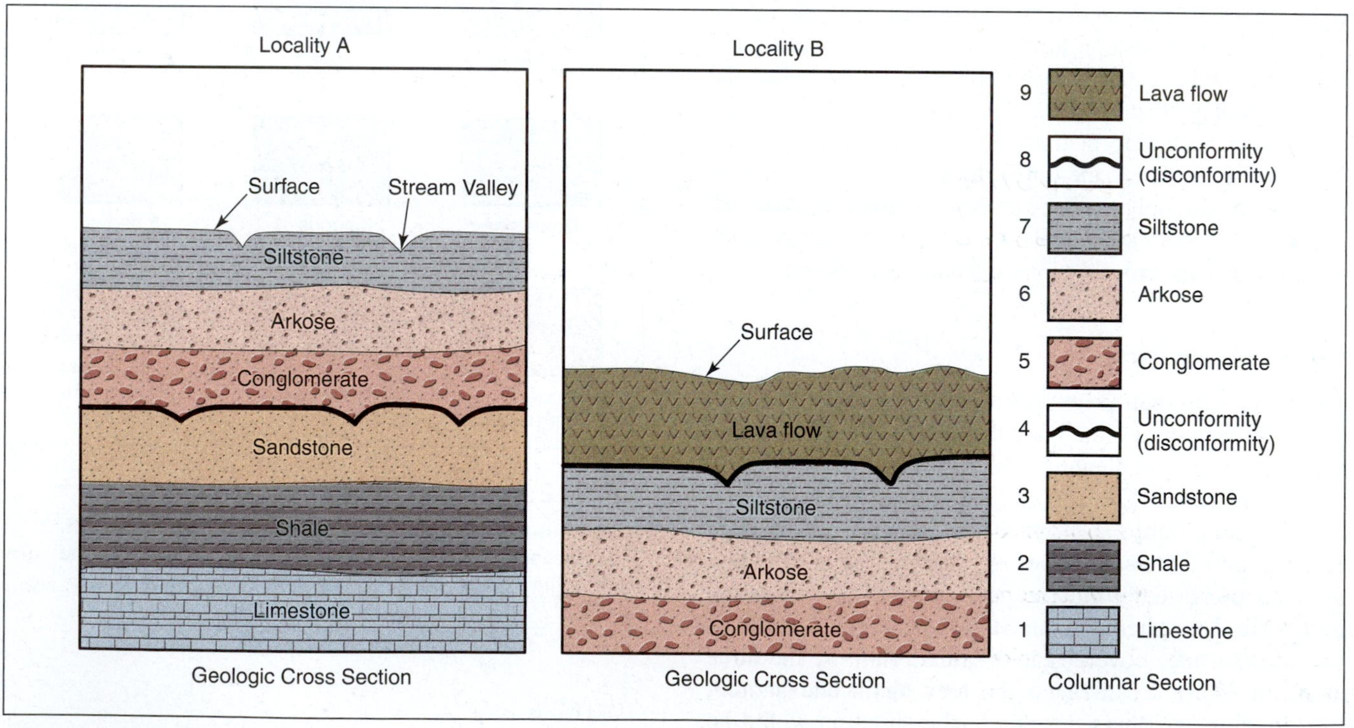

Figure 2.5

The geologic cross sections shown here are based on two localities, A and B, that are separated by a few miles in which no outcrops exist. By combining the stratigraphy of the two localities, a columnar section using information from both can be constructed. The columnar section is based on the assumption that the conglomerate-arkose-siltstone sequences in the two localities are of the same geologic age, in which case they are said to be correlated. Correlation in this case is based on a similar stratigraphic sequence whose constituent beds have lithologic similarities.

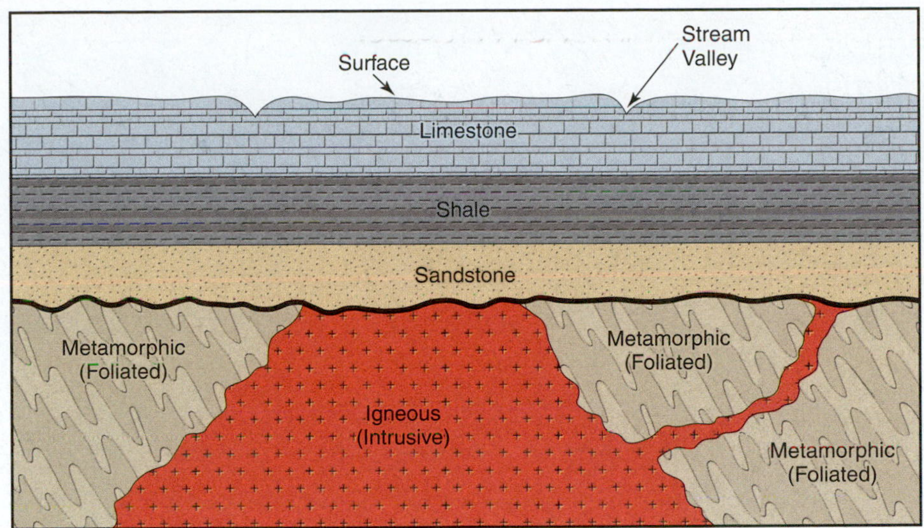

Figure 2.6
Nonconformity.

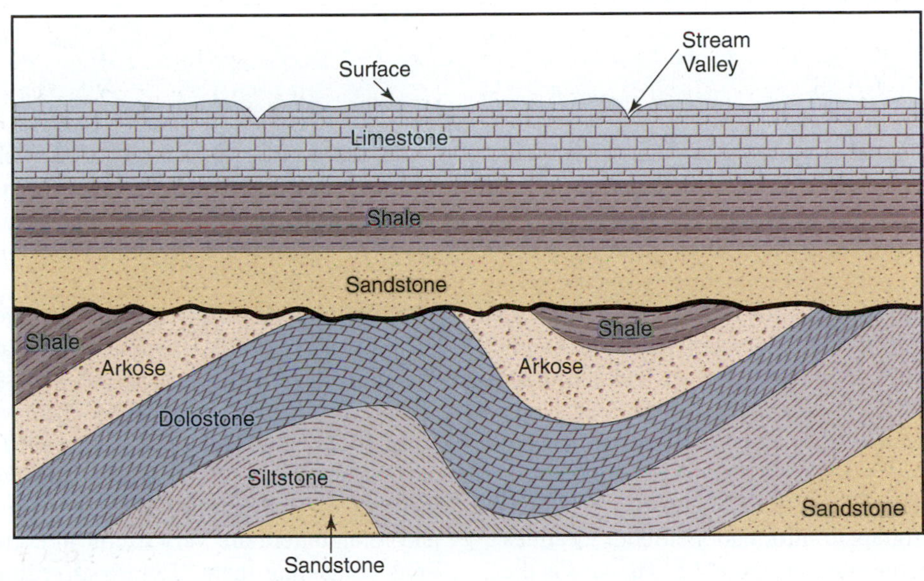

Figure 2.7
Angular unconformity.

and below are parallel to one another. A second type is a **nonconformity** (fig. 2.6), an unconformity in which an erosion surface on plutonic or metamorphic rocks has been covered by younger sedimentary or volcanic beds. A third type is an **angular unconformity** (figs. 2.7 and 2.8). This type of unconformity records a period of erosion of folded or tilted rocks followed by deposition of younger, flat-lying sedimentary or volcanic rocks.

In figure 2.4, the principle of superposition still applies in constructing the columnar section. The relative ages of all six formations and the disconformity are shown by the appropriate numbers next to the boxes in the columnar section. The disconformity represents a break of unknown geologic duration in the sedimentary record for this particular locality. It is assumed that this disconformity represents a period of erosion, but the actual length of time in years represented by the unconformity cannot be determined from the information in figure 2.4. All that can be done is to place it in its relative chronological position in the columnar section.

Correlation

In figure 2.4, the two periods of sedimentation and the disconformity between them all represent the passage of geologic time in the same geographic locality. A columnar section constructed therefrom is based only on the rocks that crop out in that particular area. However, by tracing certain formations

Figure 2.8

Angular unconformity (contact shown by red line) between the tilted Silurian Grit and the horizontal Carboniferous Limestones Series, the Midlands, UK. Width of outcrop in photo is approximately 7 feet.

Photo by Kent C. Nielsen.

from this locality to others nearby, it may be possible to extend the columnar section to include rocks older or younger.

As an example, consider figure 2.5 in which two geologic cross sections are shown from two locations separated by a few miles. Locality A contains a sedimentary sequence with six lithologically distinct formations. At locality B, only the upper sedimentary sequence crops out; the lower one is out of sight, presumably beneath the surface and hence not observable. At locality B, a lava flow lies on top of a siltstone. If, in fact, the siltstone at locality A and the siltstone at B are one and the same formation, they are said to be **correlated.** This being so, it is then possible to construct a columnar section from the stratigraphic information at both localities A and B, as shown to the right in figure 2.5. The lava is the youngest formation in the section, and two unconformities are present, one between the siltstone and the lava flow, and one between the conglomerate and the sandstone.

If fossils are present in any of the rock strata, they can be used as a basis for correlation as well. The application of the law of faunal assemblages and the law of faunal succession may assist greatly in correlating rock units from one area to another (fig. 2.9).

Fossils are most common in marine sediments of Phanerozoic age (542–0 million years), and the fossils most useful in correlation are called **index fossils** (fig 2.10). These organisms are characterized by their widespread (global) occurrence during a relatively narrow time span. Examples include invertebrates such as brachiopods, trilobites, ammonites, and graptolites. Microfossils such as conodonts, radiolarian, and foraminifera are very useful in correlation of the rock units containing them. Terrestrial sediments younger than Devonian may contain windblown plant remains such as pollen and spores.

PART III

Topographic Maps, Aerial Photographs, and Other Imagery from Remote Sensing

Background

The earth is the laboratory of geologists. They are interested not only in the materials of which the earth is made but also in the configuration of its surface. Two important tools are used by geologists: maps and remote sensing imagery.

A **map** is a representation of part of the earth's surface. Maps that show only the horizontal distribution of surface features of the earth and the human-made structures on that surface are of limited use to the geologist because geologic phenomena are three-dimensional. Therefore, a map that portrays the earth's surface in three dimensions is of particular value to geologists. Maps that meet this requirement are topographic maps. Standard practice is that north is at the top of the map. The first section in Part III of this manual deals with topographic maps.

Remote sensing is a process whereby the image of a feature is recorded by a camera or other device and reproduced in one form or another as a "picture" of the feature. One of the oldest tools of remote sensing is the camera, which produces images on a photosensitive film. When developed by chemical means, the so-called black-and-white photographs or true color photographs are the results.

The space age saw the introduction of many other kinds of remote sensing devices that produce new kinds of images such as "radar photographs," false color images, and a variety of other "pictures." These are useful not only to geologists but also to geographers, ecologists, foresters, soil scientists, meteorologists, and the like. The radar picture or image, for example, is made by recording the radiation from earth features in a way that can be resolved into a photolike picture. Radar images from earthorbiting satellites are used extensively to show cloud cover on televised weather reports and forecasts over most commercial television stations. Radar images made from aircraft are useful in the study of many geologic phenomena.

False color photos or images are made from records of electromagnetic radiation from the earth's surface. The images called "false color" show the earth's features in colors that are assigned by the person processing the data, and the colors are often different from the "true" colors that we see with our eyes. For example, vegetation is often assigned the color red on false color images. The use of false color is to enhance the differences in earth features resulting from variations in vegetation, soil, water, and rock types.

Photographs or other types of images made by cameras or other sensing devices installed in airplanes, manned spacecraft, or satellites are thus another kind of map that provide geologists with useful tools for analyzing and interpreting the components of a given landscape on earth as well as on the moon and on other planets in the solar system.

The second section of Part III introduces you to aerial photographs and false color images. (A radar map will be introduced in Part V.) The goal of Part III of this manual is to provide you with a rudimentary knowledge of topographic maps, aerial photographs, and false color images so that you will be able to apply this knowledge in your study of the landforms presented in Part IV.

Map Coordinates and Land Divisions

The earth's surface is arbitrarily divided into a system of reference coordinates called **longitude** and **latitude.** This coordinate system consists of imaginary lines on the earth's surface called **meridians** and **parallels** (fig. 3.1). Both of these sets of lines are best described by assuming the earth to be represented by a spherical globe with an axis of rotation passing through the North and South Poles. A meridian is one-half of a great circle drawn on this globe that passes through both poles. A **great circle** results when any plane passes through the center of the globe, but in this case, it must also pass through the poles. The intersection of the plane with the surface divides the globe into two equal halves—hemispheres—and the arc of the great circle is the shortest distance between two points on the surface of the spherical globe—thus the use of the term "great circle routes" when flying from one continent to another.

Meridians run in a true north–south direction and are spaced farther apart at the equator than at the poles. Meridians are labeled according to their positions in degrees from the zero, or prime, meridian, which by international agreement passes through Greenwich near London, England. The zero, or prime, meridian is commonly referred to as the **Greenwich meridian.** If meridional lines are drawn for each degree in an easterly direction from Greenwich (toward Asia) and in a westerly direction from Greenwich (toward North America), a family of great circles will be created. Each one of the meridians (one-half of a great circle from pole to pole) is labeled according to the number of degrees it lies east or west of the Greenwich, or zero, meridian. The 180° west meridian and the 180° east meridian are one and the same. The International Date Line generally follows this meridian, but because of national boundaries and other political factors, it does not follow it exactly.

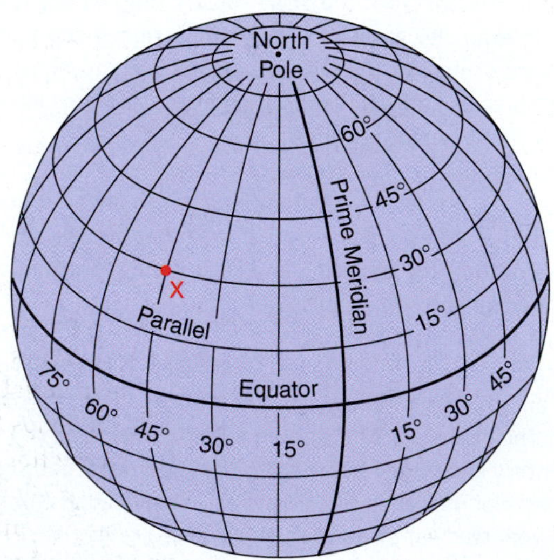

Figure 3.1

Generalized system of meridians and parallels. The location of point X is longitude 45° West, latitude 30° North.

Another great circle passing around the earth midway between the poles is the **equator.** It divides the earth into the Northern and Southern Hemispheres. A family of lines drawn parallel to the equator constitutes the second set of reference lines needed to accurately locate a point on the earth's surface. These lines form circles that are called the parallels of latitude. They are labeled according to their distance in degrees north or south of the equator. The parallel that lies halfway between the equator and the North Pole is at latitude 45° North, and the North Pole itself is at 90° North.

This system of meridians and parallels provides a means of accurately designating the location of any point on the globe. For example, Santa Monica, California, lies at about longitude 118°29′ West and latitude 34°01′ North. For increased accuracy in locating a point, degrees may be subdivided into 60 subdivisions known as **minutes,** indicated by the notation ′. Minutes may be further subdivided into 60 subdivisions known as **seconds,** indicated by the notation ″. A position description might read 64°32′32″ East, 44°16′18″ South.

Meridional lines converge toward the North or South Pole from the equator, and the length of a degree of longitude varies from 69.17 statute miles at the equator to zero at the poles. Latitudinal lines, on the other hand, are always parallel to each other. However, because the earth is not a perfect sphere but is slightly bulged at the equator, a degree of latitude varies from 68.7 statute miles at the equator to 69.4 statute miles at the poles. Thus, the area bounded by parallels and meridians is not a true rectangle. United States Geological Survey (U.S.G.S.) quadrangle maps are also bounded by meridians and parallels, but on the scale at which they are drawn, the convergence of the meridional lines is so slight that the maps appear to be true rectangles. The U.S.G.S. standard quadrangle maps embrace an area bounded by 7 1/2 minutes of longitude and 7 1/2 minutes of latitude. These quadrangle maps are called 7 1/2-minute quadrangles. Other maps published by the U.S.G.S. are 15-minute quadrangles, and a few of the older ones are 30-minute quadrangles.

Meridians always lie in a true north–south direction—that is, they always point to the North or South Pole. On U.S.G.S. topographic maps, a symbol in the margin shows the direction to **true north** and the relationship between true north, **magnetic north,** and **grid north** (figure 3.2). True north is the direction toward the North Pole, and on the symbol, it is marked by a line with a star at the outer (north) end. Magnetic north is the direction the north-seeking end of a magnetic compass needle will point, and it is marked with the notation MN. The magnetic poles are not coincident with the true north and south poles, and magnetic north is different from true north except on the meridian that passes through the north magnetic pole. The angle between true north and magnetic north is called the **magnetic declination,** and that angle is shown on the marginal symbol. The position of the magnetic pole varies with time, so there is a date given as to when the magnetic declination shown on the map was determined. Charts are available that provide the annual shift of the magnetic pole.

The third line on the marginal symbol represents grid north, and it is marked with the notation GN. The grid referred to here is the Universal Transverse Mercator System (UTM), an overlay system used by the National Geospatial Intelligence Agency (NGA) [formerly the National Imagery and Mapping Agency] to divide the entire earth into 60 sectors running from 84° North to 80° South, each 6° wide. The Universal Polar Stereographic Grid covers the polar regions. These systems allow the precise location of any point on the surface of the earth independent of longitude and latitude or any local system such as the Range and Township system in the United States. A full explanation of the Universal Transverse Mercator System and its use is beyond the scope of this manual, but you should be aware that it exists, that Global Positioning System (GPS) data can be read out in UTM grid notations, and that it is being adopted by many agencies for use in Geographic Information Systems (GIS) data collections.

In recent years, the use of satellites for positioning has become a way for geologists and others to locate themselves in the field. This system, known as the Global Positioning System (GPS), provides information on longitude, latitude, and elevation as well as UTM coordinates to the person on the ground equipped with a unit to read the satellite signals. Several satellites (four or more of the 27 available) are required to get an accurate position (from 1 cm to 10 m, depending on the type of GPS equipment). While this is an excellent way of locating a globally positioned point, some knowledge of maps is necessary to plot the location and record the information (geologic or otherwise) at that point.

Geographic Information Systems (GIS) are related systems. A GIS allows the geographic features in real-world locations to be digitally represented and stored in a database. This is a system of computer-based tools for end-to-end processing (capture, storage, retrieval, analysis, and display) of data using location on the earth's surface for interrelation in support of operation management, decision making, and science. GPS provides the location data and, as we shall see, remote sensing of a variety of types provides information about the earth's surface. A GIS model for a geologist might include data on rock type, topography, hydrology, soils, and vegetation that, superimposed on one another, allow the abstract representation of the data in the form of a map.

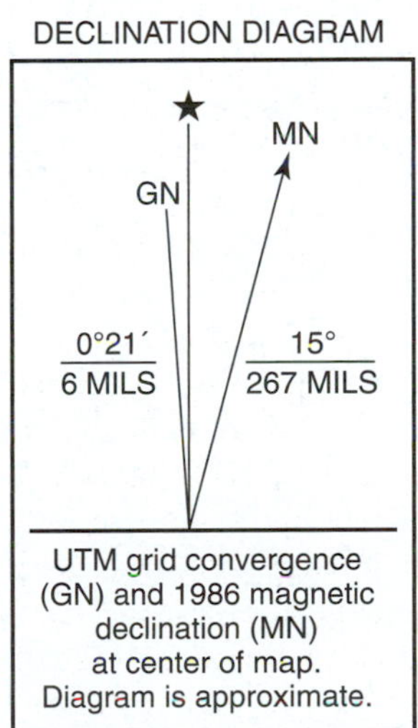

Figure 3.2

An example of the relationship between true north (★), magnetic north (MN), and grid north (GN).

Map Projection

While it is easy to draw the system of meridians and parallels on a globe, it is not possible to draw this system on a flat piece of paper (a map) without introducing some distortion. A wide variety of methods have been developed to reduce this distortion during the construction of a map. This process of constructing a map, the transferring of the meridians and parallels to a flat sheet of paper, is a geometric exercise called **projection,** and the resulting product is called a **map projection.** The map projection selected for use by the cartographer, one who draws maps, depends on the purpose of the map and the material to be presented (see box 3.1). Most of the U.S.G.S. maps found in this manual are drawn on a polyconic projection. This projection preserves neither shape nor area, but for the small area represented, the distortion of both is at a minimum. The 1:250,000 scale map of Greenwood (see fig. 4.14) uses a transverse Mercator projection, and figure 5.39 uses a Lambert conformal conic projection. The map used for figure 6.1 is a Mercator projection.

BOX 3.1

Map Projections

A map should show the spatial relationships of features at earth's surface as accurately as possible in terms of distance, area, and direction. These requirements can be met when the cartographer is dealing with a relatively small area, such as a county or a small state in the United States. Over this area, earth's curvature is minimal, and the surfaces can be mapped onto a piece of flat paper with little distortion. However, at a global scale, the problems are acute; the skin of a sphere will simply not lie flat without being distorted in some way.

No flat map can portray shape, distance, area, and direction over the spherical globe accurately. The different portraits of earth shown in box figure 1 are different map projections, each of which attempts to minimize overall distortion or to maximize the accuracy of one measure of space. There are no right or wrong map projections because they all contain distortion of one sort or another. In virtually all projections, linear scale is particularly not constant across the map, meaning that measurement of long distances will be inaccurate.

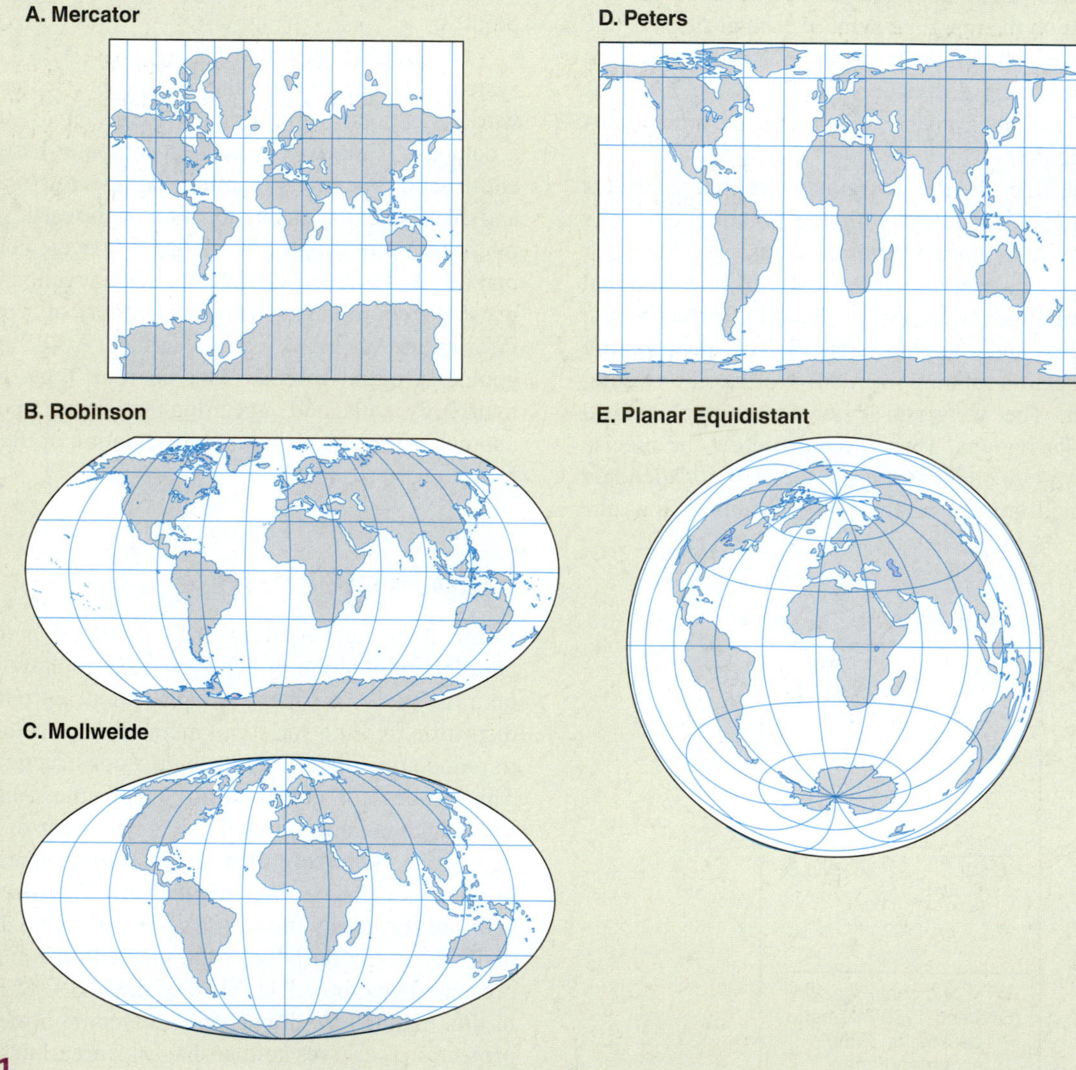

Figure 1
Examples of different map projections.

The simplest way to imagine constructing a map projection is to think of a transparent globe with a lightbulb inside it and a sheet of paper touching the globe in one of the three ways shown in box figure 2. The outline of the continents and the lines of latitude and longitude are silhouetted on the paper to form the map. In fact, most projections are not constructed as simply as this. They are calculated mathematically and may be a compromise among a number of methods and designed to preserve the best features of each. There are three main desirable properties in a map, and different projections attempt to preserve each one.

Equal-area projections, as the name suggests, maintain the areas of the continents in their correct proportions across the globe. They are used to depict distributions such as population, zones of climate, soils, or vegetation. The Robinson projection used for world maps is an equal area projection. The scale distortion in **conformal maps** is equal in the two main directions from the projection's origin but increases away from the origin. This means that conformal projections maintain the correct shape over small areas, but the outline of continents and oceans mapped over a larger area is distorted. **Azimuthal maps** are constructed around a point or *focus.* Lines of constant bearing or compass direction radiating from the focus are straight lines on an azimuthal map. Distortion of shape and area is symmetrical around the central point and increases away from it. The azimuthal projection is used most often in geography to portray the polar regions, which are distorted in projections designed for lower latitudes. Azimuthal maps can also be equal-area or conformal.

It is mathematically impossible to combine the properties of equal area and reasonably correct shape on one map. The commonly used Mercator projection is a conformal map and demonstrates graphically how poor this projection is for showing geographical distributions. The Mercator projection was developed in 1569 by a Flemish cartographer, Gerardus Mercator, to help explorers and navigators. It has the important property, for navigation, that a straight line drawn anywhere on the map, in any direction, gives the true compass bearing between these two points. However, the size of landmasses in the mid- and high-latitudes (toward the poles) is grossly distorted in this projection. Alaska, for example, appears to be the same size as Brazil, although Brazil is actually five times as large. Greenland appears similar in size to the whole of South America. This distortion is partly because the meridians, which actually come together toward the poles, are shown with uniform spacing throughout the Mercator map. Despite its unsuitability for most geographical purposes, the Mercator is still very widely used.

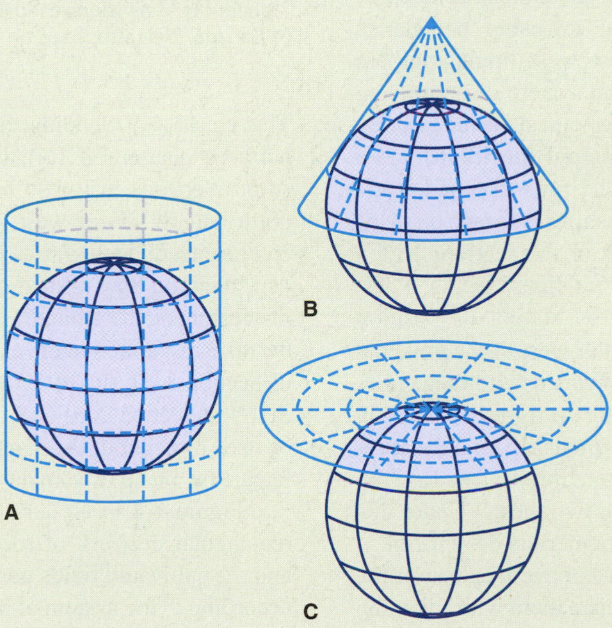

Figure 2

Methods of constructing map projections. (A) Cylindrical. (B) Conical. (C) Planar. The cylindrical and conical projections are "unwrapped" from the globe to give a flat map.

Range and Township

For purposes of locating property lines and land descriptions on legal documents, another system of coordinates is used in the United States and some parts of Canada (see inside back cover). In the United States, this grid system is called the U.S. Public Land Survey (U.S.P.L.S.). This system is tied into the latitude and longitude coordinate system but functions independently of it. The basic block of this system is the **section,** a rectangular block of land 1 mile long and 1 mile wide. An area containing 36 sections is called a **township.** Each township consists of 36 1-square-mile sections, six sections on a side, that are numbered according to the system shown in figure 3.3. One section of land contains 640 acres. While it was the intent of the original government land surveyors to make each section an exact square of land, many sections and townships are irregular in shape because of surveying errors and other discrepancies in laying out the network.

Examination of the maps on the inside back cover of this manual indicates that there are portions of the United States that do not use this grid system. The original 13 colonies had a system of land description, called metes and bounds, in place before the establishment of the U.S. Public Land Survey, and that system has continued. In a metes and bounds land description, the starting point and bearings along the sides are given as well as a statement of the markers used at the corners, which may be natural features such as a tree or a rock. A sketch map drawn to scale is often attached to the description. The other major exception to the U.S.P.L.S. is the state of Texas, where early Spanish land grants and later land grants made by the Mexican government and the Republic of Texas to settlers in the 1880s have resulted in a description system that is quite complex and difficult to describe.

The north–south lines marking township boundaries are called **range lines,** and the east–west boundaries are called **township lines.** The coordinate system of numbering townships has a reference or beginning point at the intersection of a meridian of longitude (the **principal meridian**) and a parallel of latitude (the baseline). The township and range lines marking the boundary of a township are shown on maps with heavy red lines; the boundaries of the sections within each township are shown with lighter red lines (see fig. 4.32). A particular township is identified by stating its position north or south of the baseline and east or west of the principal meridian. The system of numbering township and range lines is shown in figure 3.3. The letter *T* along the right-hand margin of the large map stands for the word **township,** and the letter *R* stands for the word **range.** The notation T. 1 S, R. 2 W. is read, "Township one south, Range two west." Under this system, each township has a unique numerical designation.

In the United States, while most states have a separate prime meridian and baseline, several states share them with adjoining states so that there are only 34 U.S.P.L.S. systems. Several principal meridians and baselines are used in the coterminous United States, so the township and range coordinate numbers are never very large **(see maps on inside of back cover).**

For purposes of locating either human-made or natural features in a given section, an additional convention is employed.

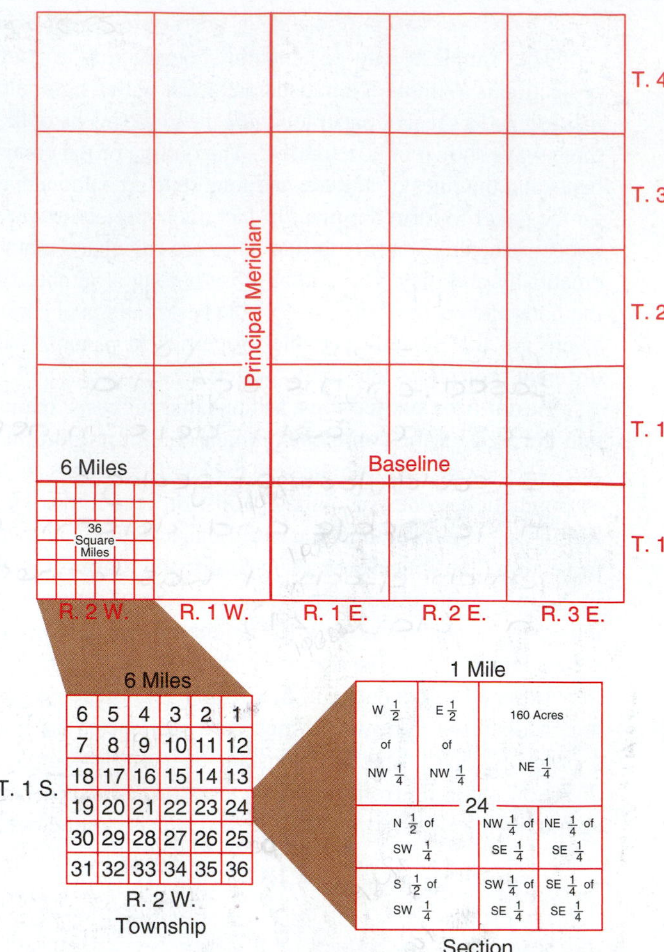

Figure 3.3

Standard land divisions used in the United States and some parts of Canada. (See also maps on inside back cover.)

This consists of dividing the section into quarters called the northeast quarter (NE1/4), the southwest quarter (SW1/4), and so on. Sections may also be divided into halves such as the north half (N1/2) or west half (W1/2). The quarter sections are further divided into four more quarters or two halves, depending on how refined one wants to make the description of a feature on the ground. For example, an exact description of the 40 acres of land in the extreme southeast corner of the map of Section 24 in figure 3.3 would be as follows: SE1/4 of the SE1/4 of Section 24, T. 1 S., R. 2 W. This style of notation will be used throughout the manual in referring you to a particular feature on a map used in an exercise.

Figure 3.4 is an aerial photograph of rural Iowa. The rectangular network of roads tends to follow section lines, and the cultivated fields generally conform in shape and size according to the system of land divisions shown in figure 3.3.

Because the maps used in this manual are only *selected parts* of standard U.S.G.S. quadrangle maps, the township and range lines numbering system, magnetic declination, and other data usually found on the lower margin of the maps may not be included. Where necessary, these data will be supplied in the text of the exercise.

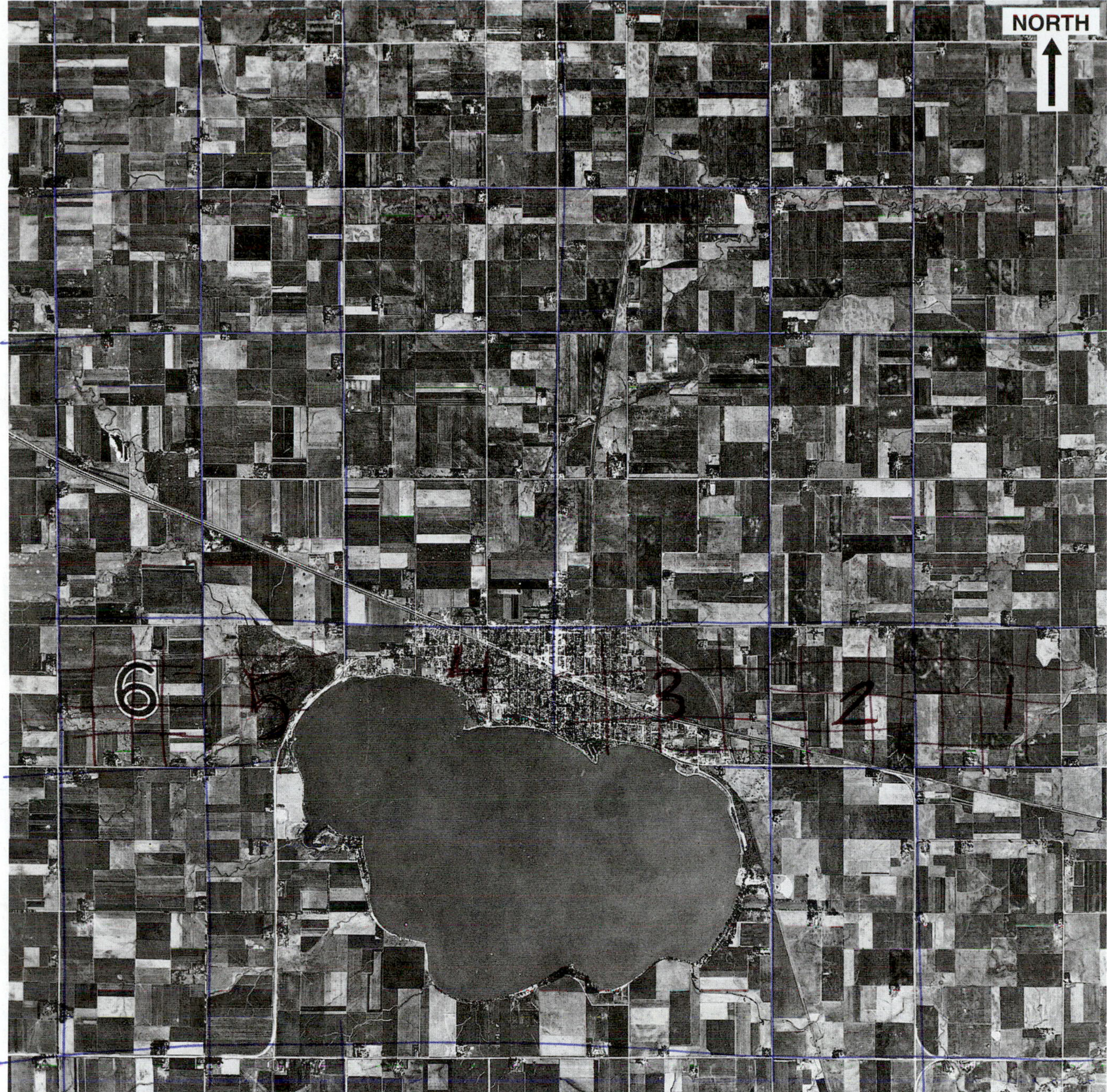

Figure 3.4

Aerial photograph of an area in northwest Iowa. The town on the north edge of the lake is Storm Lake, Iowa. The pattern of squares formed by the north–south and east–west roads are sections, 1 mile square. Section 6, T. 90 N., R. 37 W. is labeled on the photo.

Courtesy of the U.S.G.S. Photo was taken on September 23, 1950.

(1mile x 1mile) (36 sections)
Section → Township

Topographic Maps

Definition

A **topographic map** is a graphic representation of the three-dimensional configuration of the earth's surface. Most topographic maps also show land boundaries and other human-made features. The United States Geological Survey (U.S.G.S.), a unit of the Department of the Interior, has been actively engaged in the making of a series of standard topographic maps of the United States and its possessions since 1882.

Features of Topographic Maps

The features shown on topographic maps may be divided into three major groups: (1) **topography** or relief (printed in brown), depicted by the configuration of **contour lines** that show hills, valleys, mountains, plains, and the like. Shaded relief may also be used either with or without contour lines to emphasize relief (fig. 4.24 is an example); (2) **water features** (printed in blue), including oceans, lakes, ponds, rivers, canals, swamps, intermittent streams, and the like; (3) **culture features,** (printed in black), representing human-made works such as roads, railroads, prominent buildings, land boundaries, and similar features (fig. 3.5). Geographical names are also printed in black.

On some topographic maps, woodland cover (forests, orchards, vineyards, and scrub) is shown in green, important roads and public land surveys in red (see fig. 4.14), and where existing maps have been corrected through the use of aerial photos without field checks, the added features are shown in purple.

Appendix B provides detailed information about the standard symbols used on topographic maps published by the U.S.G.S. Additional information about U.S.G.S. maps is also provided in Appendix B.

Standard topographic maps of the U.S.G.S. cover a **quadrangle** of area that is bounded by **parallels of latitude** (forming the northern and southern margins of the map) and by **meridians of longitude** (forming the eastern and western margins of the map). The published maps have different scales. **Map scale** is a means of showing the relationship between the size of an object or feature indicated on a map and the corresponding actual size of the same object or feature on the ground.

Geologists make use of topographic maps because they provide them with a means to observe earth features in *three dimensions.* Unlike other maps, topographic maps show natural features to a fair degree of accuracy in terms of length, width, and vertical height or depth. Thus, by examination of a topographic map and through an understanding of the symbols shown thereon, geologists can interpret earth features and draw conclusions as to their origin in the light of geologic processes.

In the United States, Canada, and other English-speaking countries of the world, most maps produced to date have used the English system of measurement. That is to say, distances are measured in feet, yards, or miles; elevations are shown in feet; and water depths are recorded in feet or fathoms (1 fathom = 6 feet). In 1977, in accordance with national policy, the U.S.G.S. formally announced its intent to convert all of its maps to the metric system. As resources and circumstances permit, new maps published by the U.S.G.S. will show distances in kilometers and elevations in meters. The conversion from English to metric units will take many decades in the United States. In this manual, most maps used will be those published by the U.S.G.S. *prior* to the adoption of the metric system because the metric maps are still insufficient in number to portray the great diversity of geologic features presented in this manual. However, some examples of map scales in metric units will be given to acquaint you with the system.

To help you familiarize yourself with the metric system and its relationship to the English system, a conversion table is provided on the **inside front cover** as a convenient reference.

Elements of a Topographic Map

Topography

Topography is the configuration of the land surface and is shown by means of contour lines (fig. 3.5). A contour line is an imaginary line on the surface of the earth connecting points of equal elevation. The **contour interval** (C.I.) is the difference in elevation of any two adjacent contour lines. Elevations are given in feet or meters above mean sea level. The shore of a lake is, in effect, a contour line because every point on it is at the same level (elevation).

Topographic Profiles

A **topographic profile** is a diagram that shows the change in elevation of the land surface along any given line. It represents graphically the "skyline" as viewed from a distance. Features shown in profile are viewed along a horizontal line of sight, whereas features shown on a map or in **plan view** are viewed along a vertical line of sight. Topographic profiles can be constructed from a topographic map along any given line.

The **vertical scale** of a profile is arbitrarily selected and is usually, but not always, larger than the horizontal scale of the map from which the profile is drawn. Only when the horizontal and vertical scales are the same is the profile **true,** but in order to facilitate the drawing of the profile and to emphasize differences in relief, a larger vertical scale is used. Such profiles are **exaggerated profiles.**

Ideally, both the horizontal and vertical scales would be the same. This is impractical in most cases. On a section with a horizontal scale of 1:1,000,000, the topographic features would be almost impossible to see if the same vertical scale were used. Both horizontal and vertical scales are usually provided with each profile. The **vertical exaggeration** is determined by comparing the inches on the profile with feet in nature. Thus, on a profile with a horizontal scale of 1:24,000 and a vertical scale of 1/10 inch to every 100 feet of elevation,

Horizontally, 1 inch represents 24,000 divided by 12 = 2,000 ft

Vertically, 1 inch represents 10 times 100 = 1,000 ft

Vertical exaggeration is 2.0 times

Be aware that vertical exaggeration not only increases but also changes the character of the profile. A volcano such as that shown in figure 3.14 would appear as a sharp peak at 10 times vertical exaggeration.

Instructions for Drawing a Topographic Profile

Figure 3.10 shows the relationship of a topographic profile to a topographic map. It should be examined in connection with the following instructions:

1. The line along which a cross section is to be constructed may be defined by an actual line drawn on the map or by two points on the map that determine the terminals of the line of cross section.
2. Examine the line along which the profile is to be drawn and note the difference between the highest and lowest contours crossed by it. The difference between them is the **maximum relief** of the profile. Cross-sectional paper divided into 0.1 inch or 2.0 millimeter squares makes a good base on which to draw a profile. Use a vertical scale as small as possible so as to keep the amount of vertical exaggeration to a minimum. For example, for a profile along which the maximum relief is less than 100 feet, a vertical scale of 0.1 inch or 2.0 millimeter = 5, 10, 20, or 25 feet is appropriate. For a profile with 100 to 500 feet of maximum relief, a vertical scale of 0.1 inch or 2.0 millimeter = 40 or 50 feet is proper. If the maximum relief along the profile is between 500 and 1,000 feet, a vertical scale of 0.1 inch or 2.0 millimeter = 80 or 100 feet is adequate. When the maximum relief is greater than 1,000 feet, a vertical scale of 0.1 inch or 2.0 millimeter = 200 feet is appropriate. The general rule for guidance in the selection of a vertical scale is **the greater the maximum relief, the smaller the vertical scale.** Label the horizontal lines of the profile grid with appropriate elevations from the contours crossed by the line of the profile. Every other line on a 0.1 inch or 2.0 millimeter grid is sufficient.
3. Place the edge of the cross-sectional paper along the line of profile. Opposite each intersection of a contour line with the line of profile, mark a short dash at the edge of the cross-sectional paper. If the contour lines are closely spaced, only the heavy, or **index,** contours need to be marked. Also mark the positions of streams, lakes, hilltops, and significant cultural features on the line of profile. At the edge of the paper, label the elevation of each dash.
4. From the labeled elevations on the edge of the paper, draw vertical lines down to the corresponding elevations represented by the horizontal lines on the crosssectional paper. The intersection of these sets of lines will produce a series of points across your paper.
5. Connect the points by a smooth line, and label significant features such as streams and the summit of hills. Add the horizontal scale, and write a title on the profile.

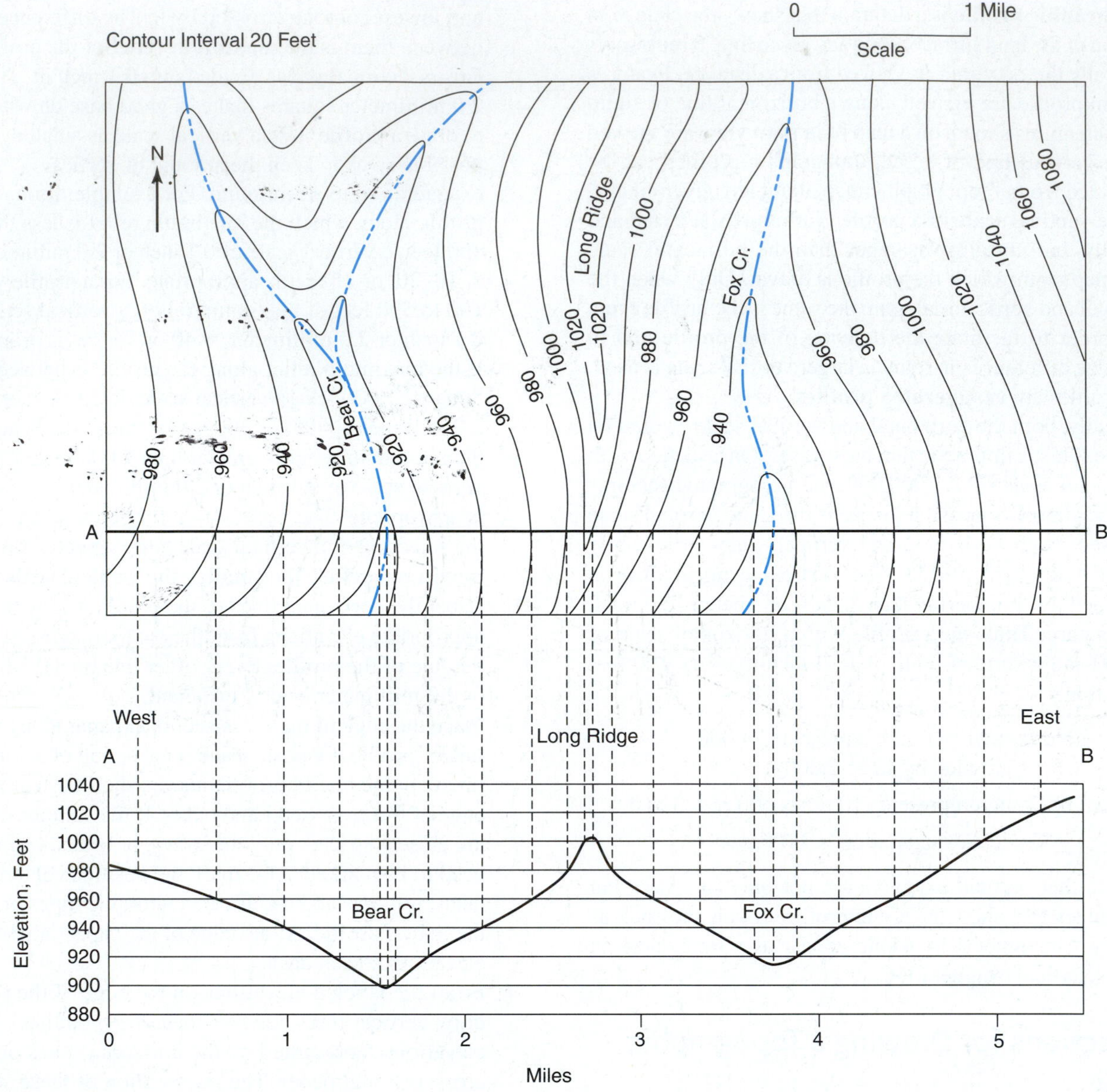

Figure 3.10

A topographic profile drawn along line A–B on the map of the hypothetical Bear Creek–Fox Creek area. See text for step-by-step instructions.

Name Elizabeth Sowers

EXERCISE 11

Section Date

Topographic Map Reading

The following questions are based on the Delaware map, figure 3.12. (NOTE: Figure 3.13 is an aerial photograph of part of the map in figure 3.12. It is shown here for comparative purposes and will be used later in Exercise 12.)

1. What is the R.F. of the map? 1:24000
2. What is the C.I. of the map? 20 ft.
3. If the C.I. were twice what it actually is, would there be a greater or lesser number of contour lines? lesser
4. What is the map distance in feet between the highest point on Mt. Lookout (1,335 feet) and the nearest point on the south shore of Lake Bailey?
   roughly 3000 ft.
5. If you walked along the line described in question 4, would the actual distance walked be greater, less, or the same as the map distance? Explain.
   Greater b/c the Scale is not talking about the change in elevation.
6. What is the maximum relief of the map area?
   700 ft. difference
7. Give the location of the following features using the township, range, section, etc. (the northernmost east–west red line separates T. 58 N. from T. 59 N., and the entire map area falls in R. 30 W.):
   (a) North Pond. Section 11 Twn 16
   (b) Mt. Lookout. Section 3 Twn 29
8. Refer to figure 3.3. Notice that each square mile of land (one section) is comprised of 640 acres. Each quarter-section contains 160 acres (640 × 1/4 = 160). Knowing that 160 acres = 1 quarter-section and that a quarter-section of land is a tract 1/2 mile (2,640 feet) on a side, determine
   (a) the number of square feet in a quarter-section (160 acres). 297
   (b) the number of square feet in one acre.
   1188
9. Approximately how many acres are covered by Lake Bailey? (*Suggestion:* First determine the area of the lake in square feet by multiplying the length of the long axis by the length of the short axis. Do the same for the island. Subtract the area of the island from the area of the lake and convert the difference to acres.) 2 miles
   640 acres
10. What is the direction of flow of the intermittent stream located in Sections 10 and 11?
    west to east
11. What is the elevation of the contour line that surrounds the small pond between Silver River and Highway 26 just south of the shore of Lake Superior? (Lake Superior is the large body of water across the northern part of the map.)
    630 ft.

SHOW YOUR CALCULATIONS:

Figure 3.12 Delaware Map

Part of the U.S.G.S. Delaware quadrangle, Michigan, 1997. Scale: 1:24,000; contour interaval: 20 feet.

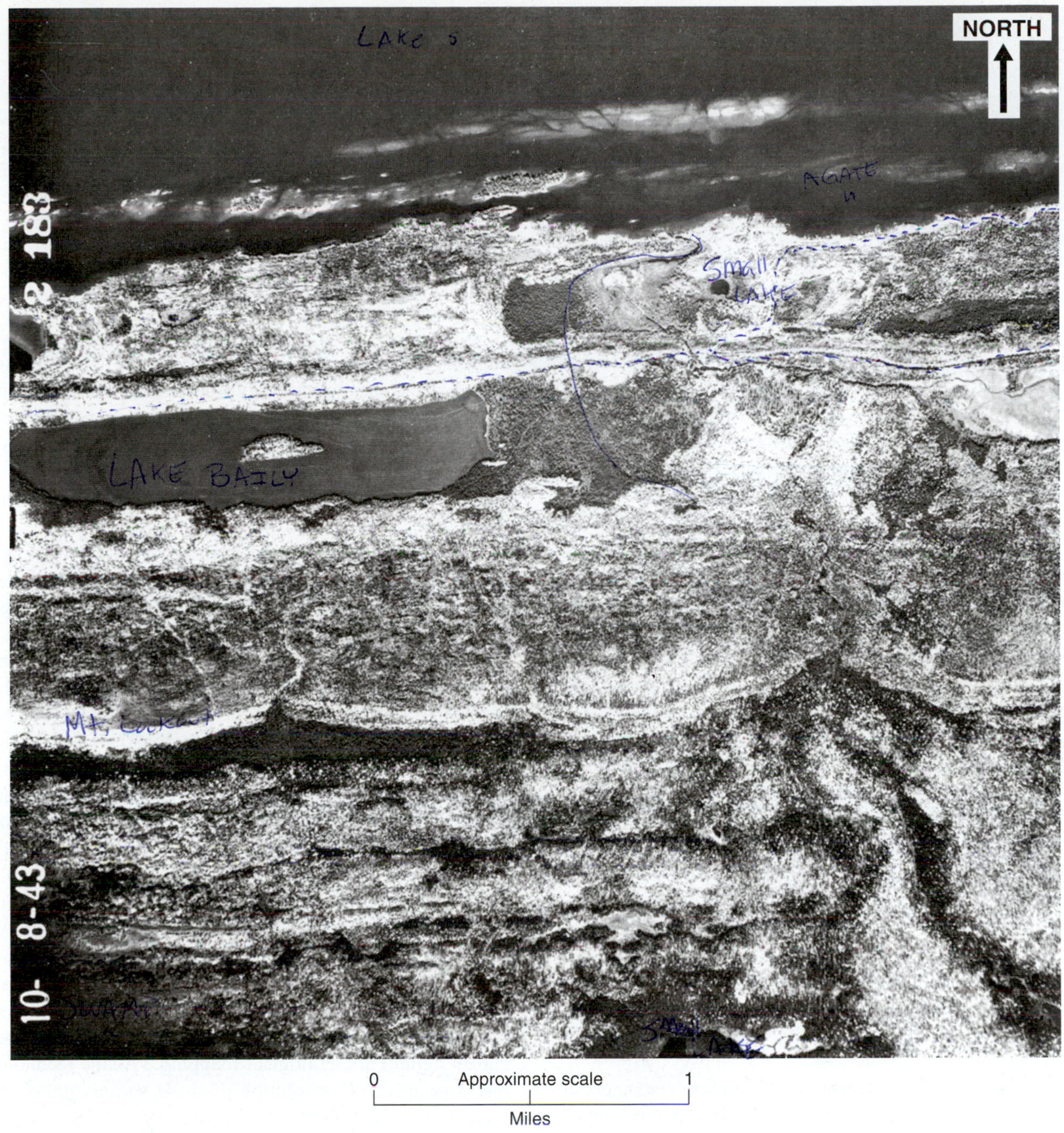

Figure 3.13

Aerial photograph of part of the Delaware map in figure 3.12. The photo is shown here for comparative purposes and will be used in a later exercise.

Courtesy of the U.S.G.S. Photo was taken on October 8, 1943.

Imagery from Remote Sensing

Background

Remote sensing is defined as the noncontact recording of information from the ultraviolet, near-infrared, thermal infrared, or microwave regions of the electromagnetic spectrum. Cameras, scanners, lasers, and linear arrays are the usual recording instruments, and the acquired information is available for analysis by digital and visual image processing. Remote sensing data are routinely acquired from a variety of platforms, including aircraft, satellites, space shuttles, and space stations such as Mir.

The most widely used systems are photographic cameras and electronic devices that record electromagnetic energy. Based on the portion of the electromagnetic spectrum that is recorded, remote sensing systems have been divided into ultraviolet (UV), visual near-infrared (VNIR), shortwave infrared (SWIR), and radio detecting and ranging (radar). The first three systems measure the reflections of the sun's electromagnetic energy; thus, they are passive systems. Radar differs from the other systems because it is an active system that sends out an energy beam and then records the signal reflected back from the surface. Remote sensing data can be collected with a single band (panchromatic) such as black-and-white or color photographs, multispectral (a few bands) such as Landsat Thematic Mapper (TM) where seven bands are recorded by the sensor, or hyperspectral (many bands) such as Hyperion where 240 bands are recorded by the sensor.

A recent advance in remote sensing is the development of Interferometric Synthetic Aperture Radar (InSAR) technique. This technique uses the phase information in the Radio Detection And Ranging (RADAR) remote sensing data to generate Digital Elevation Models (DEMs) as well as millimeter-scale imaging of surface change such as pre-, syn- and post-seismic deformation, the inflation or deflation of volcanoes, other surface elevation changes such as isostatic adjustments or changes resulting from the extraction of groundwater or petroleum, and the movement of glaciers. DEMs are usually generated from single-pass InSAR (data collected at one time but different vantage points) whereas surface change requires the use of multi-pass InSAR (data collected for the same terrance at different times).

Remote sensing has been used extensively for studies of the earth's surface for many years. The standard black-and-white or color aerial photos have been supplemented by data collected by digital cameras, line scanners, thermal infrared sensors, and multispectral line scanners that simultaneously record electromagnetic energy from different portions of the spectrum. The multispectral scanners can record visible and infrared images during the day and thermal images at night, a real advantage when dealing with weather forecasting and storms. Radar has the advantage that it can "see" through cloud cover.

The portions of the spectrum recorded are actually in shades of black and white but represent different wavelengths in the electromagnetic spectrum. The "false color images" that we see in magazines and on the television result from the assignment of specific colors (red, green, and blue) to the bands of the spectrum, with the result being a colored picture on which the various surface features appear in "false" colors resulting from their differences in the reflection of various wavelengths.

A wide variety of applications for remote sensing have developed, and the science has become a specialty that is growing in popularity among students, not only in the physical and biological sciences, but also in the social sciences.

In the geosciences, the variety of tools have been applied to mineral exploration, including metallic minerals, oil and gas, water, and environmental studies. In recent years, the use of radar interferometry to study glacial movement and surface changes associated with earthquake and volcanic activity has been of great interest. Hyperspectral remote sensing has developed techniques that allow what has been called "remote mineralogical mapping." As the technology has developed, the resolution of these various images has increased to the point where a great deal of detail may be seen in some of the images.

Aerial Photographs

An aerial photograph is a picture taken from an airplane flying at altitudes as high as 60,000 feet. Practically all of the United States has been photographed aerially by federal, state, or other agencies.

Photographs taken with the axis of the camera pointed vertically down are **vertical photos;** those taken with the axis of the camera at an angle from the vertical are **oblique photos.** Vertical photos are most useful to the geologist. Aerial photos are also used in making topographic maps, locating highways, military operations, mapping soils, city planning, and many other operations.

Determining Scale on Aerial Photographs

The scale of an aerial photo depends on both the focal length of the camera and the height of the airplane above ground surface. If these two factors are known, the R.F. of the photo can be determined (*R.F. = focal length divided by altitude. Use same units*). The photo scale can also be determined by measuring known ground distances on the photo. In many parts of the contiguous United States, section lines are ideally suited for this purpose since they are normally 1 mile apart and are usually visible on the photo as roads, fences, or other human-made features. Conversion to an R.F. or graphic scale can be made in the same manner as was described under the section on topographic map scales.

Stereoscopic Use of Aerial Photographs

If two vertical photos are taken from a slightly different position and viewed through a **stereoscope,** the relief of the land becomes visible. A **stereopair** consists of two vertical photos viewed in such a way that each of the eyes sees only one of the two photos. The brain combines the two images to form a three-dimensional view of the objects shown on the photograph.

Figure 3.14 is a stereopair for practice in using the simple lens stereoscope. Figure 3.15 shows one being used. The stereoscope is positioned over the stereopair so that the viewer's nose is directly over the line separating the two adjacent photos. If stereovision is not immediately

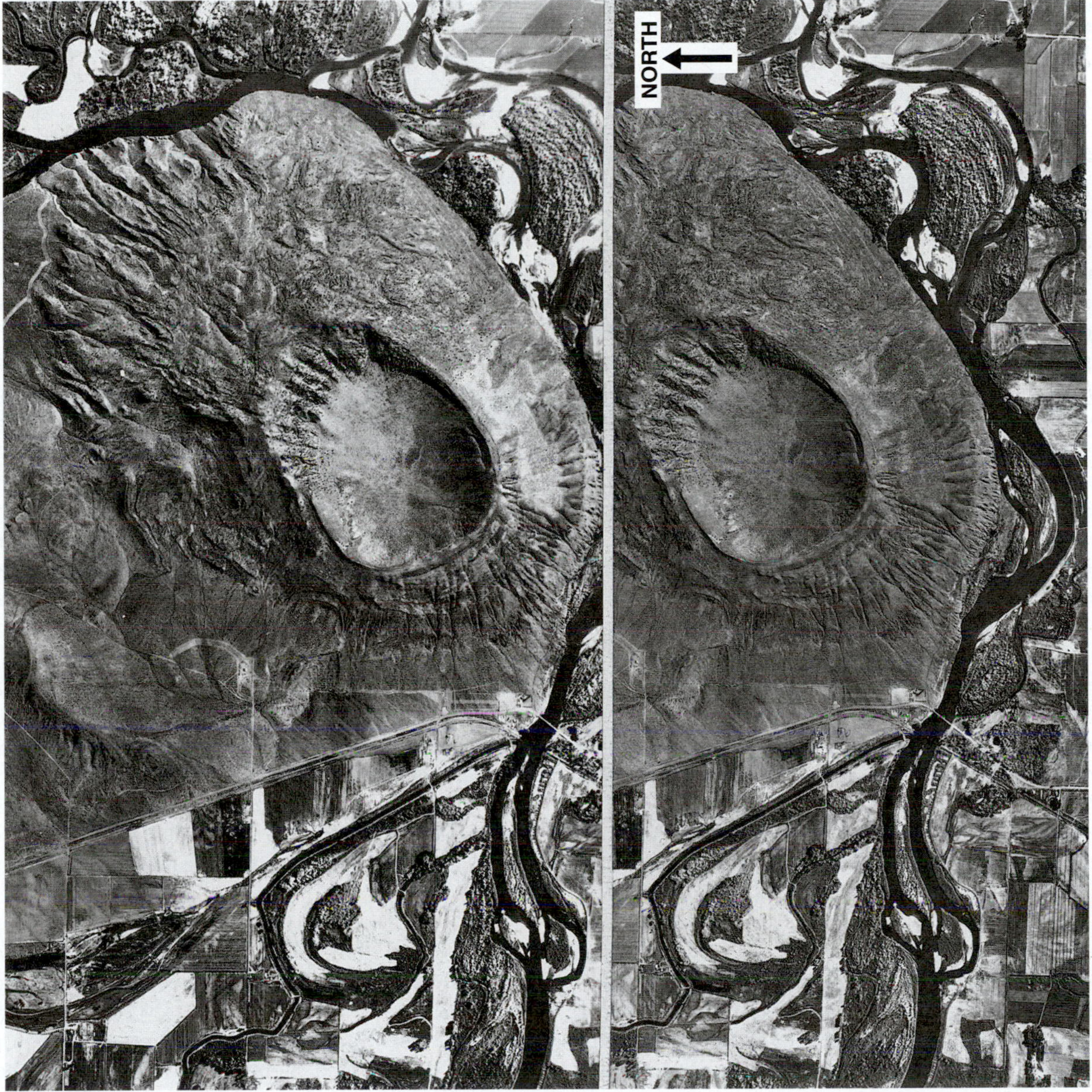

Figure 3.14

Stereopair of Menan Buttes, Idaho. The Snake River flows around the south side of the crater.

Courtesy of the U.S.G.S. Photos were taken on October 8, 1950.

Figure 3.15

Student using a lens stereoscope to view a stereopair. In actual practice, the nose is positioned in the slot between the two lenses in order to bring the eyes closer to the lense.

achieved, the stereoscope should be rotated slightly around an imaginary vertical axis passing through the midpoint between the two lenses until the image viewed appears in relief.

On photographs in which steep slopes occur (e.g., the walls of deep canyons or the flanks of high mountains), the stereoscopic image of these features is exaggerated. That is, steeply inclined canyon walls may appear to be nearly vertical when, in fact, they are not. This distortion, however, does not present a problem under normal conditions of viewing by the beginning student. Photos should be oriented with image shadows toward the observer. If photos are not oriented correctly, the topography may appear inverted. For example, view figure 3.20, then turn it upside down.

In both Parts IV and V of this manual, several exercises will require the interpretation of single aerial photos and stereopairs with reference to the terrain features and geologic phenomena shown on them. In preparation for those exercises, the paragraphs that follow will provide you with some basic information on the interpretation of aerial photographs in terms of the recognition of common natural and humanmade features.

Interpretation of Aerial Photographs

Aerial photos may be used to great advantage by geologists. Stereopairs are preferable, but single photos taken under good conditions of lighting and from proper altitudes reveal exactly what the human eye sees except for the third dimension, and even this can, with practice, be approximated from single photos.

The interpretation of aerial photos is an art acquired only after considerable experience in working with photos from many different areas. Beginning students can comprehend aerial photos amazingly well if they have a few simple instructions to follow and some basic principles to guide them.

The greatest difficulty confronting an individual looking at an aerial photo for the first time is recognition of familiar features that are seen every day on the ground but become mysterious objects when seen from the air. It is necessary, therefore, to acquire the ability to recognize certain common features before one can expect to use an aerial photo as a geologic tool. Some of these common features are described in the paragraphs that follow and are illustrated in figure 3.16.

Vegetation

Vegetational cover accounts for a great many differences in pattern and shades of gray tone on aerial photos. Heavily forested areas are usually medium to dark gray, whereas grasslands show up in the lighter tones of gray. Planted field crops are extremely varied in tone, depending not only on the kind of crop but also on the stage of growth. Cultivated fields are usually rectangular in shape and appear either in dark gray or light gray, depending upon whether the fields have just been plowed or whether the crop is already in and growing.

Soil and Rock

Soil texture controls the soil moisture, which, in turn, controls the appearance of the soil on aerial photos. Wet, clayey soils have a much darker shade of gray than do the dry, sandy soils, which usually show up as light gray to almost white in arid regions. Different tones of gray that show up in the same field are due to different degrees of wetness of the soil, a condition usually related to topographically low (wet) and high (dry) areas.

Where bedrock crops out at the surface, aerial photos reveal differences due to lithology, texture, mineral composition, and structure of the rocks (review fig. 2.13). Students cannot always distinguish a sandstone from a limestone, for example, but should have less trouble recognizing the differences among the major rock groups (i.e., igneous, sedimentary, and metamorphic). Vegetational patterns on aerial photos commonly reflect the underlying bedrock, and this is helpful in tracing out a single rock unit of the photo. On the other hand, if a thick cover of residual material, such as soil or talus, or a cover of transported material, such as glacial drift, alluvium, or eolian sands, cover the area, then all bedrock features may be partially or totally obscured.

Aerial photos used in connection with field investigations on the ground are among the most valuable tools of the professional geologist. When direct field examination is not feasible, aerial photos provide even better information than one could acquire by flying over the area in person, especially if the photographs are available for stereoscopic study. For the student of elementary physical geology, aerial photographs are useful in that they show geologic features of the earth's surface that otherwise would be difficult to describe or impossible to illustrate.

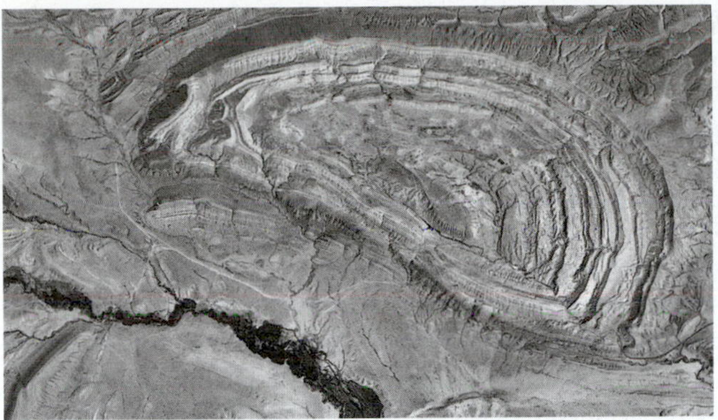

A. Sedimentary rocks uplifted to form a dome (elliptical-shaped feature) in a semi-arid climatic zone. Stream valley with deciduous trees in valley bottom (black). Wyoming. Scale: 1:80,000.

B. Deeply incised river in flat-lying sedimentary rocks. Badlands topography on either side of the river. Colorado. Scale: 1:80,000.

C. Small town in Midwest. Various cultural features shown. Deciduous vegetation in town and along river. Missouri. Scale: 1:20,000.

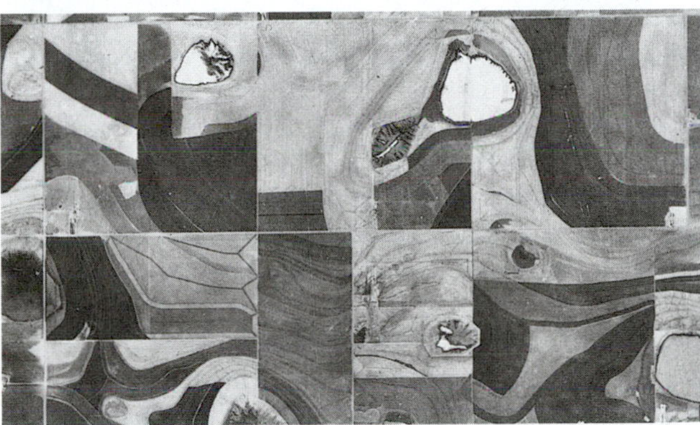

D. Agricultural field patterns (light and dark variegated areas) with undrained depressions forming ponds (white due to reflection of sun off water surface). Contour plowing evident. Texas. Scale: 1:63,360.

E. Crystalline rocks with sparse coniferous vegetation. River in deeply cut valley. Lake (black) at margin of photo. Joint pattern in rocks evident. Wyoming. Scale: 1:60,300.

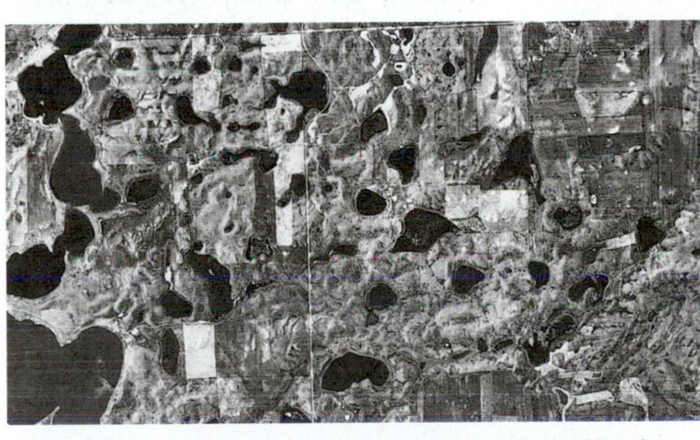

F. Glacial moraine from continental glacier. Kame and kettle topography with numerous kettle lakes (black). Rectangular field patterns in variegated colors. North Dakota. Scale: 1:60,000.

Figure 3.16

Some examples of features visible on aerial photographs.

Landsat False Color Images

The use of false color images is to emphasize one or more of the features shown on the images. The most common instruments used record the electromagnetic energy that has been reflected or radiated from the surface of the earth or from the atmosphere. This energy is classified by its wavelength, and the range of the various wavelengths are recorded as a continuous spectrum, from very short wavelengths such as gamma rays (0.001 μm), to the much longer wavelengths in radar (0.1–1.0 cm), and TV or radio (10 cm to more than 100 meters). The intervals of the various wavelengths are known as "bands."

The human eye can perceive a continuous spectrum of visible light in the range of 0.4 to 0.7 μm. The visible spectrum starts with the color violet at 0.4 μm and grades upward through blue, green, yellow, orange, to red at the upper end of the visible spectrum at 0.7 μm. Ultraviolet radiation and near-ultraviolet radiation have shorter wavelengths, and infrared radiation has a longer wavelength, but these are not in the visible spectrum. However, these wavelengths are often more useful in the study of the earth's surface.

Near-infrared radiation (0.7–1.2 μm) is very similar to visible light but cannot be seen by the human eye. Healthy vegetation is a strong reflector of near-infrared light. Shortwave infrared radiation (1.2–3.0 μm) is useful in distinguishing rock types that have a different electromagnetic "signature" on the image. Middle-infrared radiation (3–6 μm) comes mainly from the sun but is also emitted by forest fires or burning gas wells. Thermal infrared radiation wavelengths lie between 6–300 μm, the majority of which is between 8–12 μm. Heat radiation occurs during both night and day, and thermal infrared radiation images can show areas of higher and lower temperatures. Such images are used at a variety of scales and may be used to locate areas of heat loss in buildings and so forth.

The data are recorded by the instruments in various shades of gray. In the processing of the data the operator of the computer can assign a color to the various bands of wavelengths. The three primary colors of light are red, green, and blue, and the computer can display three different bands at the same time by using a different primary color for each band; thus, the images are known as "false color" images. Landsat 7 images are color composites. The color assignments most commonly used for the three bands are: red for near infrared (band 4), green for visual red (band 3), and blue for visual green (band 2). This band combination makes vegetation appear red, soils with little or no vegetation range from white to greens and browns, water is dark blue or black, and urban areas appear as bluegray. Any combinations of the primary colors and the various bands are possible and may be used, depending on the purpose of the generated image.

Landsat views do not cover a rectangular area due to the fact that the earth rotates as the satellite passes overhead. This phenomenon results in a rhombohedral "picture" such as the image shown in figure 3.17. North is generally toward the top margin of the image, but a true north–south line must be determined independently of the image margins. Where visible, agricultural fields are useful for this purpose because their boundaries are commonly oriented in north–south and east–west directions. If these are not present, other features such as a coastline or a major river can be compared with a map showing the same features to determine true north.

Figure 3.17 is a false color image of part of the Atlantic Coast of the United States showing Long Island, the Hudson River, and a part of New Jersey. The Atlantic Ocean and Long Island Sound appear black, and the light-colored patches in New Jersey are cultivated fields. The brownish area in the southern part of the image is the forested coastal plain of New Jersey. New York City and Newark, New Jersey, at the mouth of the Hudson River, are dark green. The light green of the narrow strips of longshore bars off the southern shore of Long Island and off the coast of New Jersey reflects the fact that these features are sandy and generally lack heavy vegetation. In New York City, areas of heavy vegetation appear as red on the image in the midst of the green color of the buildings and other structures.

As an example of the use of other color combinations, figure 3.18A is a "near-normal" color photograph of Wadi Umm Nar in the east central Egyptian Desert. Figure 3.18B is a false color image of the same area. On this image, red has been assigned to the ratio of band 3 to band 4 (3/4), green to the ratio of band 2 to band 1 (2/1), and blue to the ratio of band 4 to band 6 (4/6). The decision to use other than the common color assignment was done to emphasize the various rock types shown on the image. Here the red color does not show vegetation but volcanic rocks, the brown is a banded iron formation, and the upper greenish area is granite outcrops separated by a fault. A plunging anticline is clearly visible and outlined by the pattern on the banded iron formation.

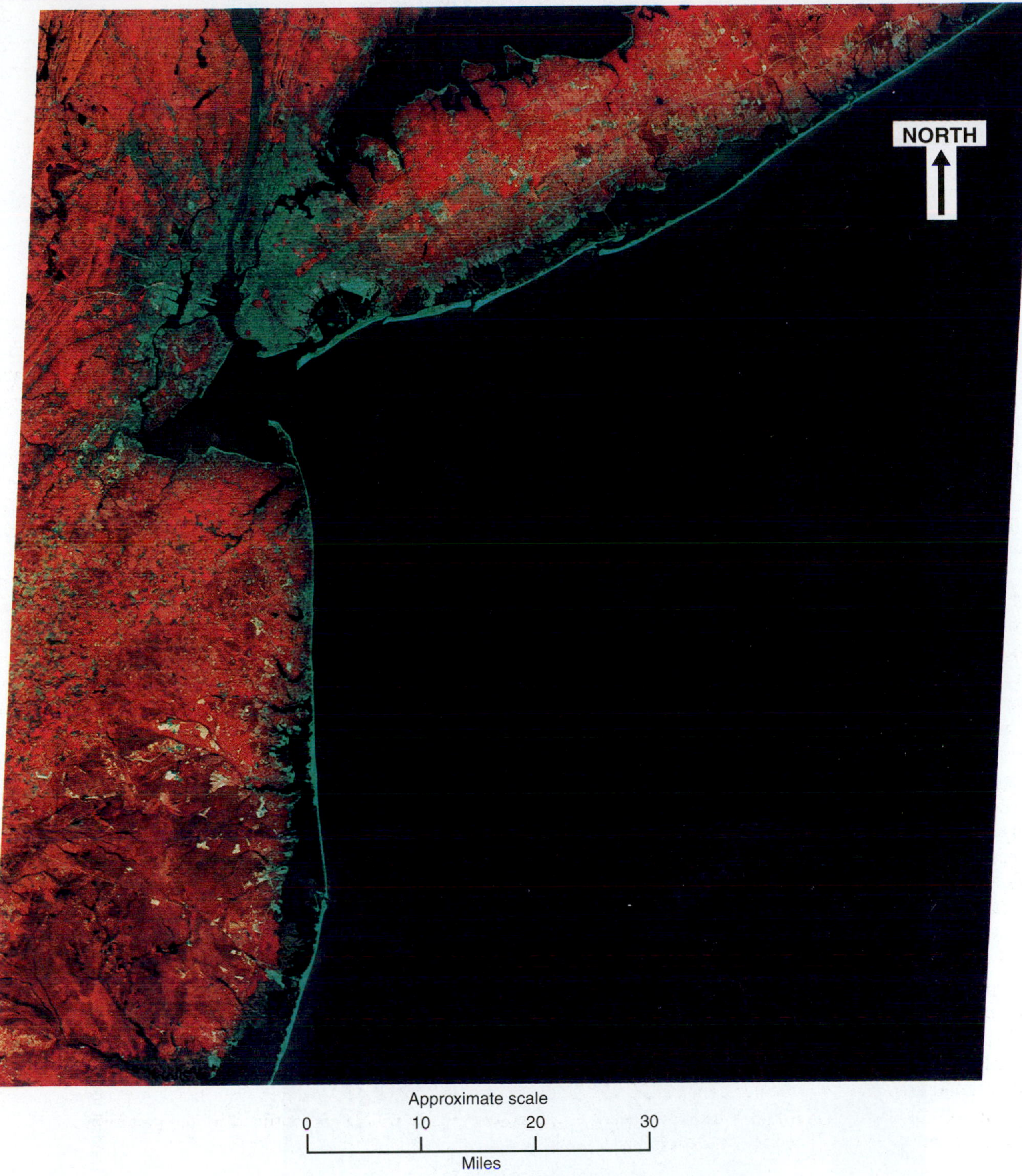

Figure 3.17

False color image of the New York City, Long Island, and New Jersey area recorded by Landsat 2 on October 21, 1975. The long narrow barrier island off the southern shore of Long Island is about 65 miles long.

NASA ERTS image E-2272–14543, U.S.G.S. EROS Data Center, Sioux Falls, South Dakota 57198.

Figure 3.18A

"Near-natural" color Landsat image of desert area, eastern Egypt. Longitude and latitude shown on image border. There is little, if any, vegetation.

EROS Center.

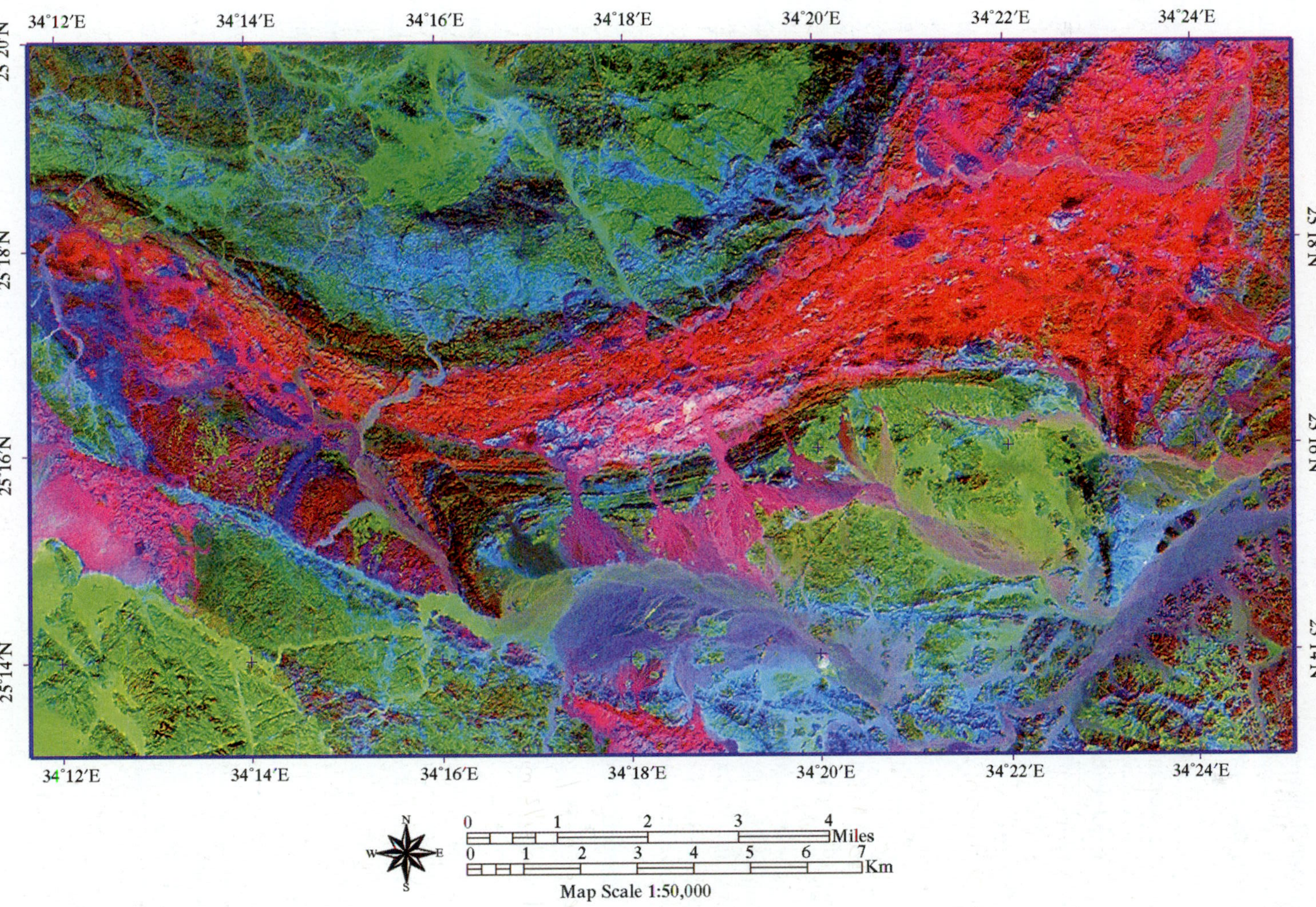

Figure 3.18B

False color image of same area. Colors assigned. Color red assigned to volcanogenic rocks, pink and white depict serpentinite, the greens at the top and lower left of the photo are granite, and the brownish rocks are banded iron formation folded in a plunging anticline. Wadi alluvium appears as blue.

EROS Center.

EXERCISE 12

Name
Section Date

Introduction to Aerial Photograph Interpretation

1. Figure 3.19 is an aerial photograph of standard size. Identify the following features on the photograph. Where a topographic map symbol is available for the feature (refer to Appendix B), draw the symbol directly on the photograph of figure 3.19 at the place where the feature occurs. If no symbol exists, draw the outline of the feature on the map and label it accordingly. Use a red pencil in all cases except for water features, which should be shown in blue.
   (a) Major road
   (b) Railroad track
   (c) River
   (d) Landing strip
   (e) Football field and track
   (f) Small stream valley
   (g) Bridge over river
   (h) Overpass
   (i) Golf course
2. The southeast corner of the golf course is also the SE corner of Section 12, T. 92 N., R. 52 W. The "T" intersection of the road at the south edge of the golf course about 4 inches to the west (left) on the photograph is at the SW corner of the SE1/4 of the SE1/4 of Section 11, T. 92 N., R. 52 W. Determine the R.F. and verbal scale of the photograph.

   R.F. distance on map / distance on ground

   Verbal scale 1inch = 1/2 mile, 4 inches = 2miles

3. Figure 3.13 covers part of the Delaware map of figure 3.12. With red pencil, label the following features with the lowercase letter that appears before each name:
   (a) Lake Superior
   (b) Lake Bailey
   (c) Agate Harbor
   (d) Small lake in Section 35
   (e) Swamp area south of Mount Lookout
   (f) Mount Lookout
   (g) Trace the roads in Sections 34, 35, and 36 that are shown by a dashed red line on the Delaware map.
4. Trace the drainage system from the Delaware map onto the aerial photo of figure 3.13. Use a blue pencil and label each creek or river that has a name.
5. Determine the scale of the aerial photo in figure 3.13.
6. Figure 3.4 is an aerial photograph of an area in central Iowa. Section 6 of T. 90 N., R. 37 W. is labeled on the photograph.
   (a) Using the information provided, determine the scale of the photograph. 1:1
   (b) Draw the township and range lines on the photograph using red pencil. Draw in the section lines with black pencil. Number all sections in black pencil. Along the photograph margins record the township and range designations in red pencil using the conventional system shown in figure 3.3.
   (c) Note the relationship between the section lines and the road pattern. Compare this relationship with that on figure 3.19. Suggest possible reasons for the differences.

       Difference in property lines / layout

   (d) A railroad track crosses the area from southeast to northwest. Use the appropriate map symbol to show this feature on the photograph. (Draw directly on the photograph with black pencil.)
   (e) The rectangular areas covering most of the area are croplands. They appear in various shades of gray to nearly black on the photo. Explain the differences.

       The crops that are darker will be more dense / lesser populated

(continued)

Figure 3.19

Aerial photograph of area in South Dakota.

Photo No. VE-1JJ; taken on June 19, 1968.

EXERCISE 12

Name

Section Date

Introduction to Aerial Photograph Interpretation *(Continued)*

7. Figure 3.20 is a stereopair. Examine it with a stereoscope and answer the following questions while completing the instructions.
   (a) Does the major stream channel contain any signs of the presence of water?
   yes, it must be spring.

   (b) Trace the drainage lines in blue pencil on the right-hand photograph using the proper symbol from the Appendix.
   (c) What is the dominant vegetation of the area?
   Sage brush

   (d) Are any human-made features visible on the stereopair? no. too far from population

   (e) The area covered by the stereopair shows two relatively flat upland surfaces away from the stream channel, each of which lies at a general elevation that differs from the other. Draw the boundaries of these areas on one photograph while viewing the stereopair with a stereoscope. Use a red pencil. Mark the lowerlying area with the number *1*, the higher area with the number *2*. Extend these boundaries on the rest of the photograph.
8. The scale of a vertical photograph depends on the focal length of the camera used and the height of the airplane taking the photo above the ground. The R.F. is the focal length of the camera divided by the height of the plane above the ground. Determine the scale of a photograph taken by a camera with a 9-inch focal length from an airplane flown at a height of 35,000 feet above the ground. (Be careful about units.)

SHOW YOUR CALCULATIONS:

focal leangth ÷ height of plane

9 inch ÷ 35,000ft.

roughly 2,916 feet.

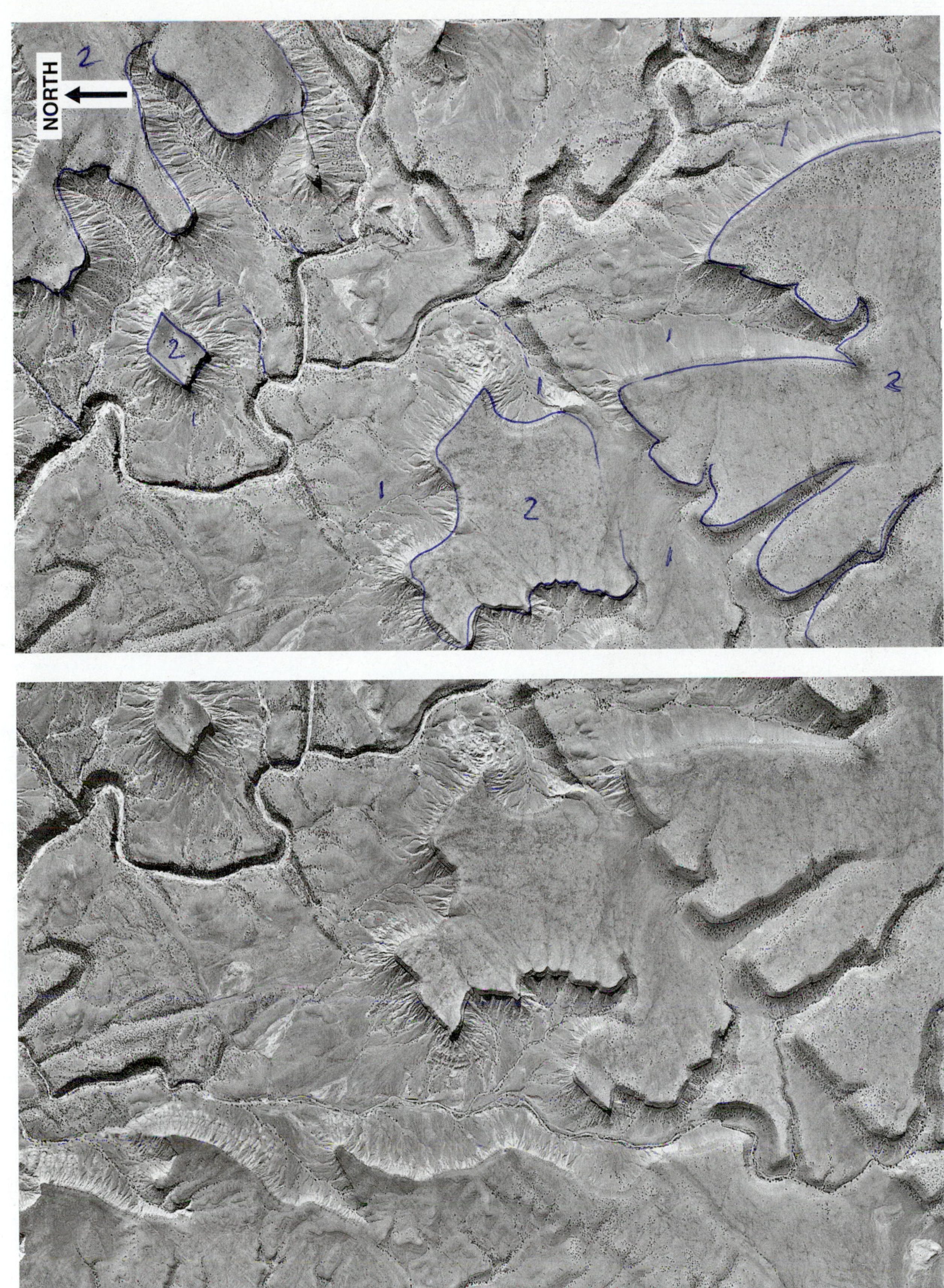

Figure 3.20

Aerial photograph stereopair, Utah, 1956. Scale, 1:20,000. (Photograph numbers GS-RR-17–43 and GS-RR-17–44.)

PART IV

Geologic Interpretation of Topographic Maps, Aerial Photographs, and Earth Satellite Images

Background

In Part III of this manual you learned that the configuration of a landform is expressed on a topographic map by contour lines and is revealed on aerial photographs when two overlapping photos are viewed stereoscopically. Topographic maps and stereopairs can thus be used for the study and analysis of terrain features in terms of the geologic processes that produced them. Images from earth-orbiting satellites are additional tools useful in the study of landforms.

Every geologic process leaves some imprint on the part of the earth's surface over which it has been operative. These processes include the work of the major geologic agents such as wind, groundwater, running water, glaciers, waves, and volcanism. Each of these agents leaves its mark on the landscape in the form of one or more characteristic landforms.

The association of geologic agents with the origin of various landforms is a subdivision of geology called **geomorphology.** Geomorphologists have systematized the relationship of geologic processes to topographic forms into a body of knowledge that can be used in deciphering the origin of topographic features shown on a topographic map or seen in a stereopair. The body of knowledge that deals with the origin of landforms is presented in all basic textbooks that deal with physical geology. Figure 4.6 is a digital shaded relief map portraying the landforms of the coterminous United States. City names have been added to assist you in finding your location on the map.

It is assumed that you will have become acquainted with the different geologic processes and their related landforms through reading appropriate chapters in a textbook on physical geology and by listening to lectures in which the origin of landforms is presented. This is prerequisite to the understanding and successful completion of the exercises that are presented in this part of the manual.

General Instructions

The purpose of Exercises 13 through 20 is to acquaint you with a variety of landforms and geologic principles associated with the geologic agents of wind, groundwater, running water, glaciers, waves, and volcanism. Topographic maps and profiles, aerial photographs (some of which are in the form of stereopairs) and other pertinent maps, satellite images, diagrams, and data are provided in Exercises 13 through 20 as the basic tools for learning the association between landform and geologic agent or to establish a specific geologic principle.

The title of each exercise identifies the geologic agent that will be under consideration for that particular exercise. It is assumed that you are thoroughly conversant with and have a good understanding of the material presented in Part III of this manual—on topographic maps, satellite images, and aerial photographs, including map and photo scales, contour lines, map symbols, and topographic profiles. The terms associated with the geologic processes covered in this part of the manual generally are defined in the background material for each exercise in which they are used and also in the Glossary. **However, it is good practice to bring your textbook to the laboratory as a reference for unfamiliar or forgotten terms that may appear in the exercises.**

All maps and photographs needed for completion of the exercises are included in the manual. Unless otherwise noted on the maps or photos, NORTH is at the top of the page. Before proceeding with the questions or problems based on maps or photographs, note the scale and contour interval on the topographic maps and the scale of the stereopairs. Some exercises require you to draw on the maps or photos with ordinary lead pencil or colored pencils. In these cases, make your initial lines very light so that if erasure is necessary, you can do so easily.

Also, use sharp pencils to ensure accuracy, especially where the drawing of a topographic profile is a requirement of the exercise. Some students who suffer from eyestrain after prolonged study of maps may find an inexpensive magnifying glass helpful for completing the map exercises. The questions for each exercise should be answered in the order given because they are arranged in a more or less logical sequence.

Geologic Work of Running Water

Stream Gradient and Base Level

Landforms produced by running water are the result of stream erosion, stream deposition, or a combination of erosion and deposition. The **gradient** of a stream is the slope of the stream bed, or surface of the stream of larger rivers, along its course and is usually expressed in feet per mile (a gradient of 5 feet per mile means the stream has a vertical drop of 5 feet for each horizontal mile of the stream course). A **longitudinal profile** of a stream shows that the gradient decreases in a downstream direction (fig. 4.1A). A stream tends to erode its channel bottom or bed in the upper portion or headwaters and deposit sediment or erode its channel walls in its lower portion.

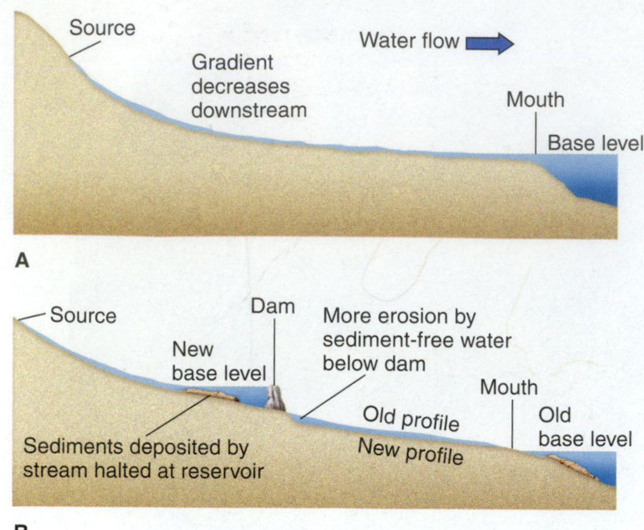

Figure 4.1

(A) Longitudinal profile of a river showing a gradual decrease in gradient downstream. Base level is the lowest level to which a stream can erode its bed. (B) Longitudinal river profile that has been interrupted by a dam. The old predam base level has been replaced by a new one.

Base level is the lowest level to which a stream can erode its bed. The base level for streams that flow to the ocean is sea level, but other temporary base levels may exist along the stream course. These include lakes resulting from natural processes, such as landslides or lava flows damming the stream, or the dam may be a human-made feature, such as Hoover Dam on the Colorado River or smaller dams on local rivers (fig. 4.1B).

The surface of a lake behind a dam constitutes a new base level for the segment of the stream above the dam. Sediment is deposited where the stream enters the lake because the stream gradient has been reduced. Below the dam, a new profile is formed because the water discharged through the gates of the dam is flowing at an increased velocity, resulting in the erosion of the stream bed to form a steeper profile.

The material in transport by a stream is the **load.** The load is made up of three types of material: the **dissolved load,** made up of material in solution; the **suspended load,** made up of materials held in suspension; and the **bed load,** made up of material moving along the bottom of the stream by traction (rolling, sliding, or dragging) or by saltation (bouncing). The **competence** of a stream refers to the size of the largest diameter particle that a stream can transport at a given velocity. The **capacity** of a stream is the total load that a stream can carry at a given discharge.

Stream Gradients and Drainage Divides

A stream together with its tributaries is called a **drainage system.** Each drainage system occupies a **drainage basin** that is separated from adjacent drainage basins by a **drainage**

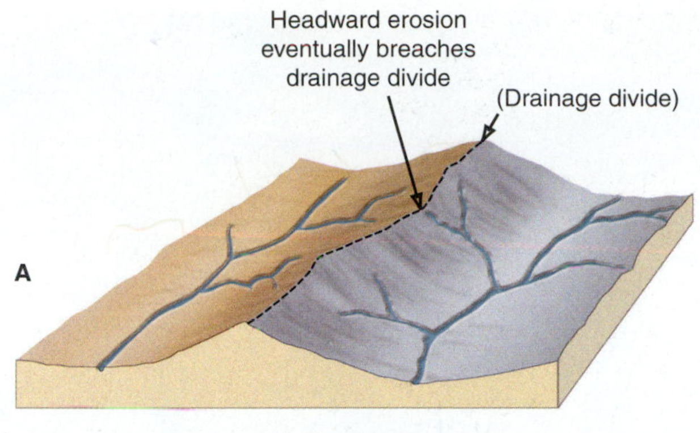

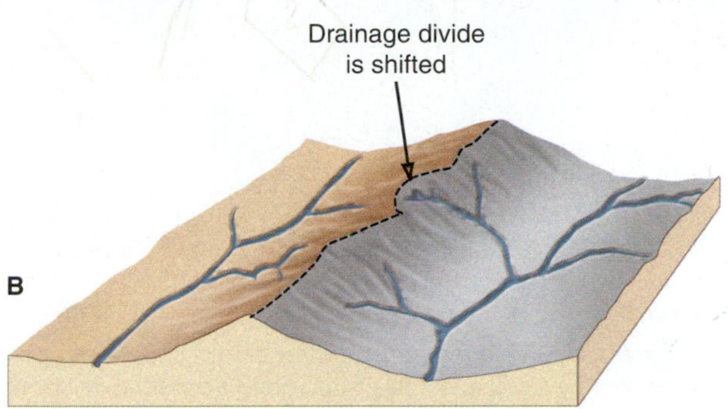

Figure 4.2

(A) A divide between two opposing drainage systems is attacked by headward erosion. (B) The divide is shifted toward the drainage system whose streams have the lower gradient of the two.

divide, an imaginary line connecting points of highest elevation between the two basins. The erosive power of a stream is a function of the stream gradient and **stream discharge.** Discharge is the rate of flow, usually expressed as volume per unit time (cubic feet per second or cubic meters per second, for example). Streams with higher gradients and higher discharge are more potent in eroding their beds than streams with lower gradients and lower discharge. If adjacent drainage systems have comparable gradients and discharge in the upper (headward) reaches, the divide between the two drainage basins they occupy will remain more or less constant. If the gradients and discharge of the stream in the two stream systems differ appreciably, the divide between the two drainage basins will shift over time toward the system with the lower gradient and discharge (fig. 4.2B).

Alluvial Fans and Pediments

These two landforms have similar but not identical topographic expressions. An **alluvial fan** is a depositional feature produced where the gradient of a stream changes from steep to shallow as it emerges from a mountainous terrain (fig. 4.3). At that point, the coarse sediments carried by the fast-flowing stream in the mountain are deposited in the form of a fan-shaped apron at the mountain front where the gradient decreases abruptly. When two or more alluvial fans coalesce, a **bajada** is formed (fig. 4.3).

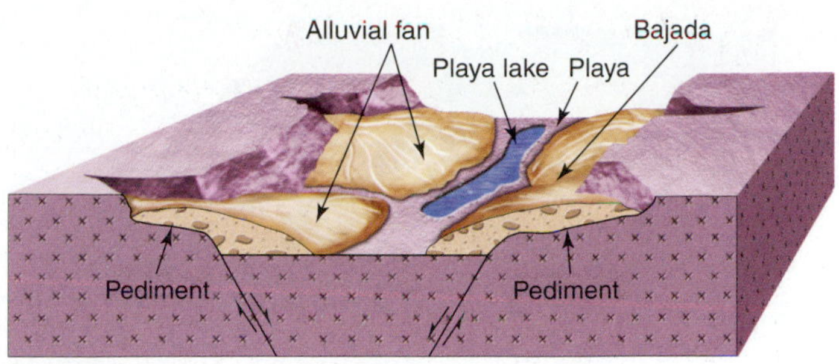

Figure 4.3

Diagram showing relationship between alluvial fans, a bajada, and pediments.

A **pediment** (fig. 4.3), on the other hand, is a gently sloping erosional surface covered with a veneer of coarse sediment. Pediments are common in arid regions. They form as the weathered mountain front recedes by attack from heavy but infrequent rainfalls. Below the coarse sedimentary cover in transit lies the planed-off bedrock that was formerly part of the mountainous terrain. One description of a pediment is that it is the end product of a mountain range consumed by weathering and stream erosion. A pediment may contain remnants of the mountainous terrain standing as isolated bedrock knobs above the pediment surface. A **playa** is a flat, undrained desert basin that contains intermittent lakes (**playa lakes**).

Meandering Rivers and Oxbow Lakes

In the lower reaches of a stream's course where the gradient is low, the stream's erosive action is directed toward the channel walls (fig. 4.4A). This process of **lateral erosion** produces a flat valley floor called a **floodplain** (fig. 4.4B). A floodplain is not produced by flood stages of the river but rather by constant shifting of the stream channel through lateral erosion across the valley floor. A floodplain does become inundated when the stream channel is unable to accommodate the volume of spring runoff or of heavy rains in the drainage basin at any time of the year.

The circuitous course of a stream flowing across its floodplain is called a **meandering course,** and the individual bends are called **meanders** (fig. 4.4C). When a meander loop is isolated from the main channel by erosion of the narrow neck of land between the upstream and downstream segments of the meander loop, an **oxbow lake** is formed (fig. 4.5). With the passage of time, an oxbow lake gradually becomes filled with sediment brought in by floodwaters and eventually is reduced to a swampy scar on the floodplain as testimony to its former existence as an active channel of the river.

During a flood stage as the river overflows its banks its velocity decreases and the river is unable to carry the material in suspension. The coarser material is deposited close to the main channel, and the finer material is deposited out on the floodplain of the river. Through time these deposits build up low ridges on either side of the channel, forming **natural levees.**

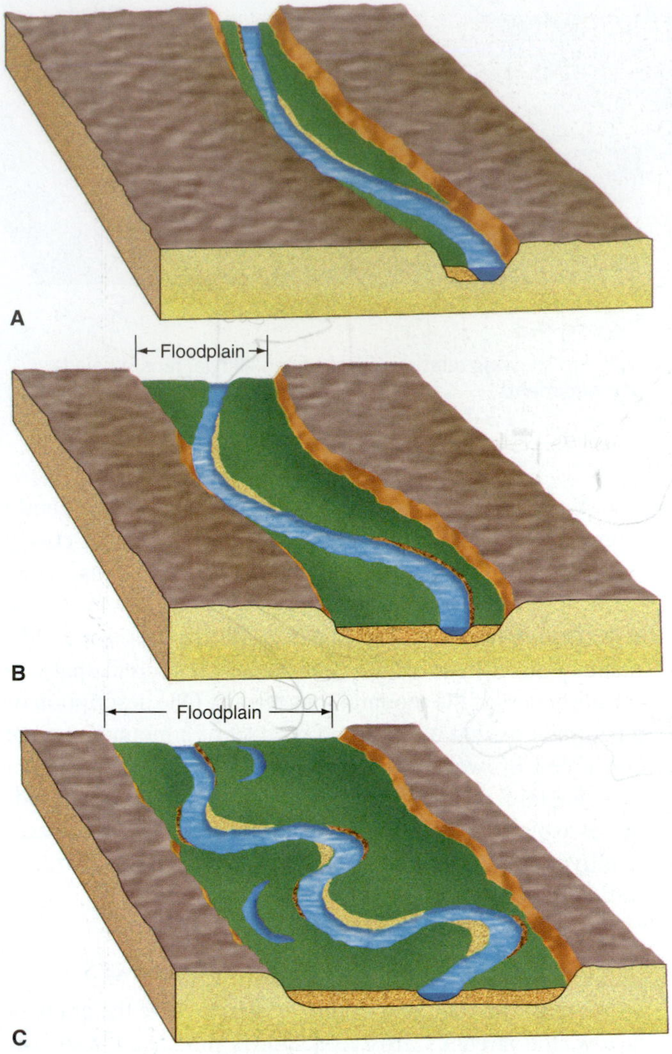

Figure 4.4

The evolution of a floodplain. (A) River widens the valley floor by lateral erosion. (B) Continued lateral erosion widens the valley floor further, and the river course becomes more meandering. (C) The river's course develops meander loops, which become oxbow lakes.

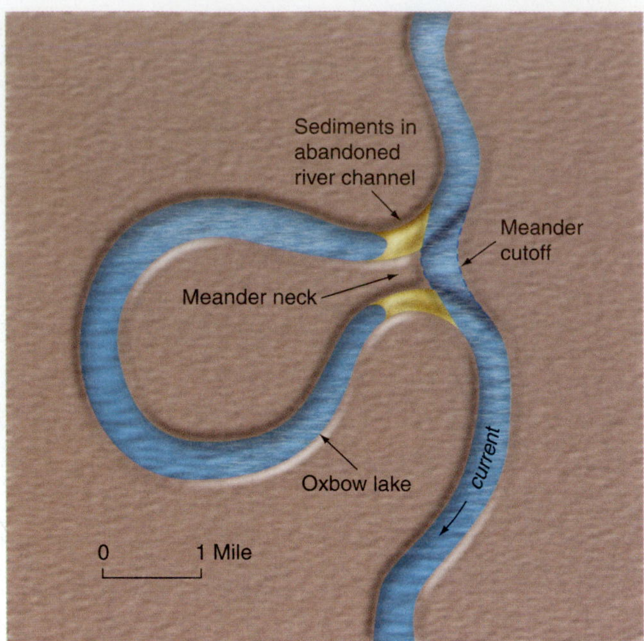

Figure 4.5

Map diagram showing a former river meander that has been severed from the river by a meander cutoff that flows across the meander neck. The ends of the old meander have been filled with sediment, thereby forming a closed depression called an oxbow lake.

These tend to be the highest points on the floodplain and may serve to contain smaller-volume floods in the channel (see cross section of natural levees in figure 2, Box 4.1, p. 173).

Historically these natural levees have been the places where towns and roads were constructed on the floodplain. Agricultural development followed, and in order to protect the structures and fields, artificial levees are often constructed on top of the natural levees. The end result of this effort is that during flood stage, the river is contained within the channel area, and when the flood recedes, the sediment is deposited in the channel rather than on the floodplain. This may result in an elevation of the stream over time, and in many places the stream bottom is above the level of the adjacent land area.

The lower Mississippi River is a classic example of a meandering stream flowing across a broad floodplain. Hundreds of miles of levees have been constructed by the U.S. Army Corps of Engineers to contain floodwaters. Additionally, the Corps of Engineers has cut off meanders from the main channel through the excavation of channels across meander necks. These artificial **cutoffs** shortened the river course in deference to the barge traffic plying the river with cargoes of various sorts.

The Mississippi River forms the boundary between many states, from Minnesota to its mouth in Louisiana where it enters the Gulf of Mexico (fig. 4.6). The states of Mississippi and Arkansas generally lie east and west of the river, respectively. Some small segments of each of these states, however, lie on the *opposite* side of the river. The reason for this is based on the meandering nature of the river and the need to fix the boundary so that the shifting course of the river does not also shift the state boundary. To reduce interstate conflicts that might arise through a constantly shifting boundary, the United States Supreme Court ruled in 1820 that where a stream forms the boundary between states and the channel of the stream changes by the "natural and gradual processes known as erosion and accretion, the boundary follows the varying course of the stream; while if the stream from any cause, natural or artificial, suddenly leaves its old bed and forms a new one, by the process known as **avulsion,** the resulting change of channel works no change of boundary, which remains in the middle of the old channel, although no water may be flowing in it."*

*Van Zandt, Franklin K. 1976. Boundaries of the United States and the Several States. U.S.G.S. Professional Paper 909, p. 4.

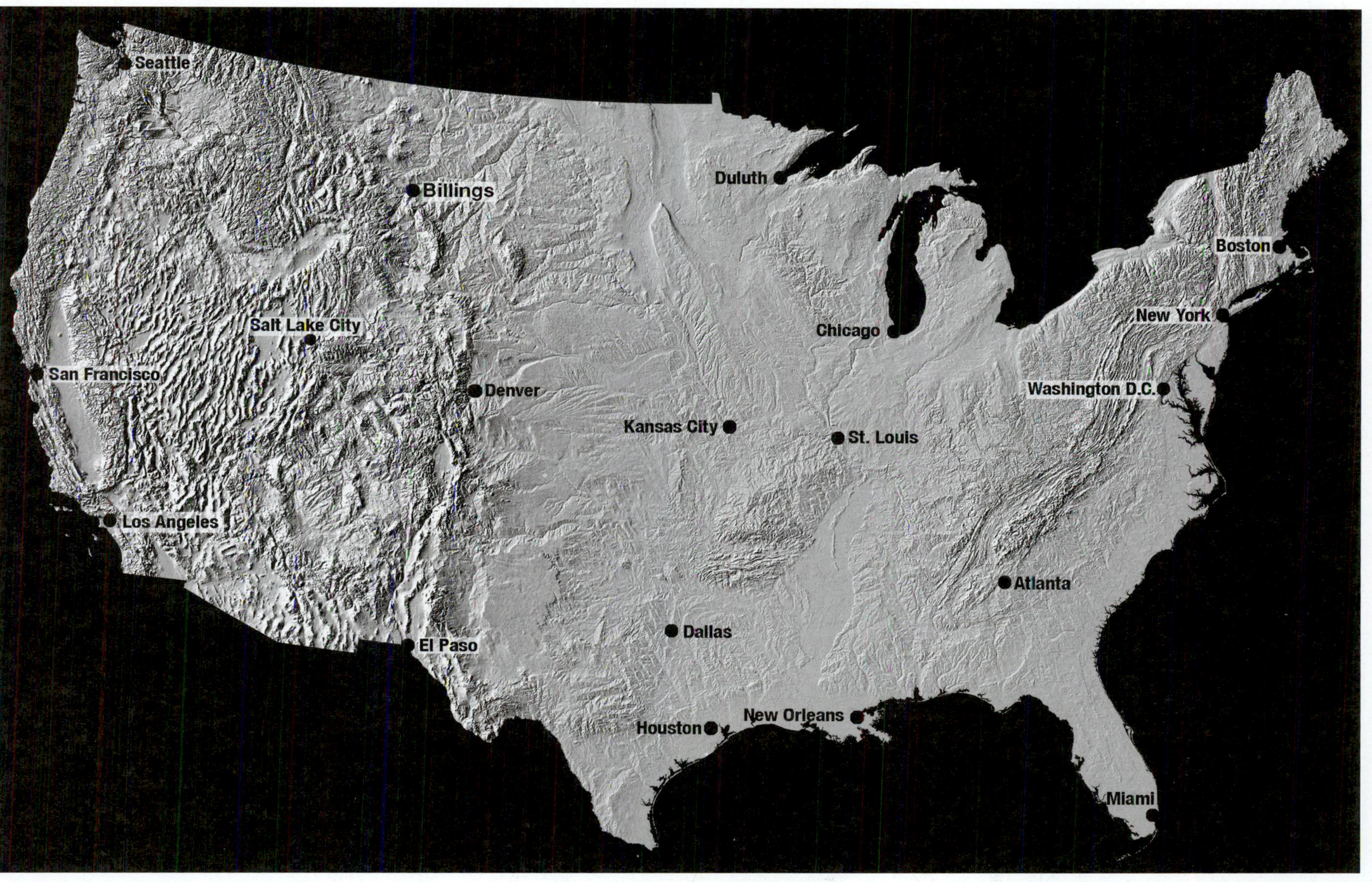

Figure 4.6

Landforms of the Conterminous United States

A digital shaded-relief portrayal. Thelin, G. P. and Pike, R. J. U.S.G.S. Misc. Inv. Series map 1-2206, 1991. (Modified by the addition of city names.)

EXERCISE 13A

Name

Section　　　　　　　　　　　　　　Date

Stream Gradients and Base Level

A segment of the Missouri River flows across the Portage map in figure 4.7.

1. What is the direction of flow of the Missouri River? west to East
2. Three dams are located along the course of the Missouri River. On the basis of human-made installations associated with each dam, what purposes do the dams serve? Trap water either for hydro electricty or other services
3. How many new base levels have been created by these dams? 6
4. In the stretch of the Missouri River between the Rainbow Dam and the Ryan Dam, is the water depth greater near the Rainbow Dam or the Ryan Dam? Ryan Dam
5. In the same stretch of river, state whether the riverbed is subject to erosion or deposition at the following sites:
   (a) Immediately downstream from the Rainbow Dam. erosion
   (b) Immediately upstream from the Ryan Dam. deposition
6. If the Ryan Dam were destroyed by an earthquake, describe the environmental effects on
   (a) The channel of the Missouri River between Ryan Dam and Rainbow Dam. The water would flow faster,
   (b) Morony Dam. The Dam might overflow
7. Beginning about 1 mile downstream from Rainbow Dam, a series of contour lines cross the Missouri River. In a distance of 1.7 river miles, the river drops 60 feet. On the basis of this information, what is the gradient of the river along this stretch? State your answer in feet per mile. 60 ft. per 1.7 miles it drops
8. The Missouri River flows a total distance of 2,500 miles from its headwaters at 14,000 feet above sea level to its confluence with the Mississippi River at 410 feet above sea level.
   (a) What is the average gradient of the Missouri River from source to mouth? State your answer in feet per mile. 5,436 feet per mile gradient
   5.5 ft./mile
   (b) What is the average gradient of the Missouri River from source to mouth? State your anwer in meters per kilometer.
   (c) Why does the average gradient differ from the gradient determined in question 7?

2,500 Miles

14,000 ft.　　　　410 ft

13 590 → 5,436 = 5.5 ft/mi

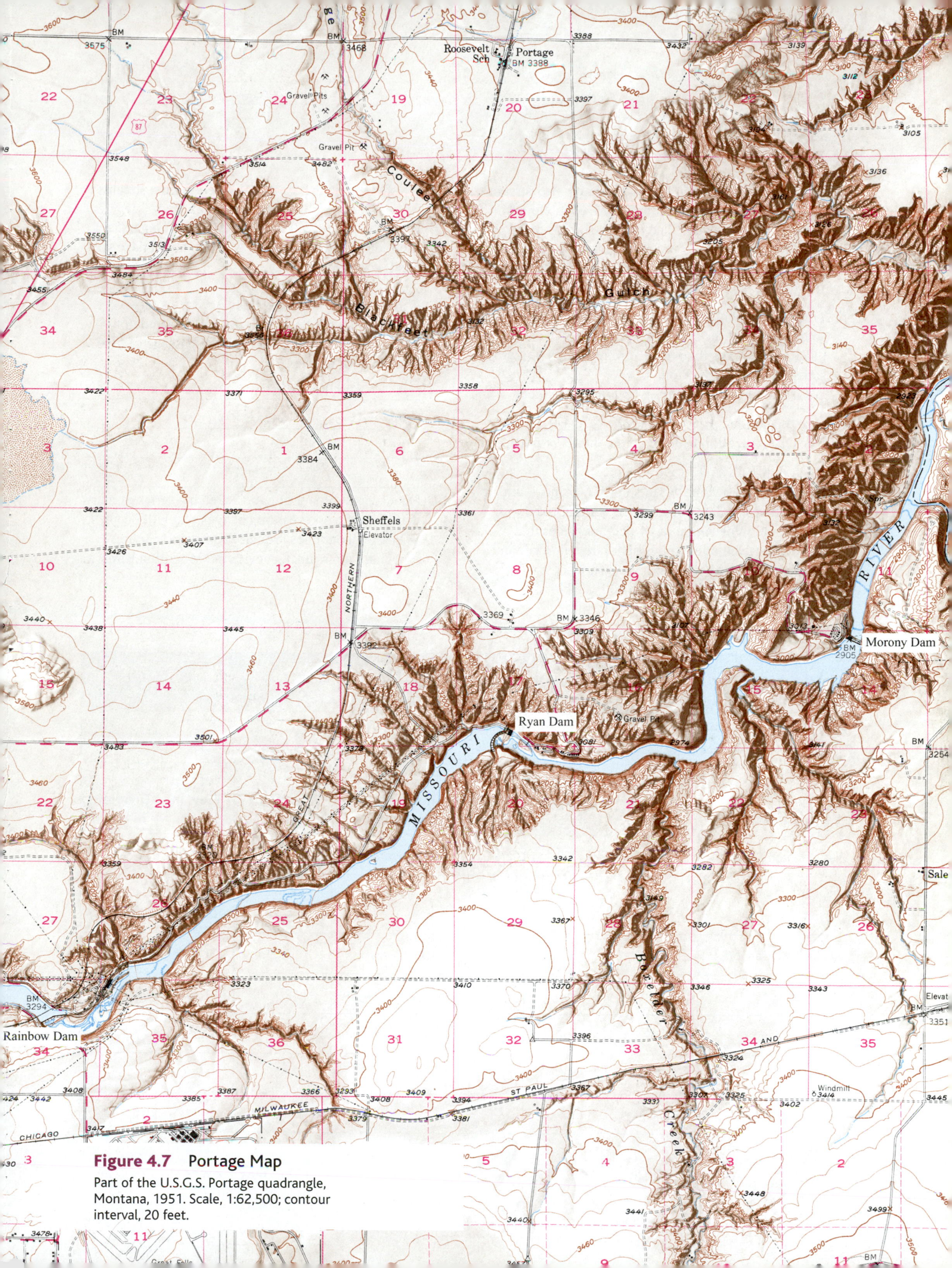

Figure 4.7 Portage Map

Part of the U.S.G.S. Portage quadrangle, Montana, 1951. Scale, 1:62,500; contour interval, 20 feet.

EXERCISE 13B

Name

Section Date

Stream Gradients and Drainage Divides

The Promontory Butte map (fig. 4.8) shows two stream systems flowing north and south of the Mogollon Rim (pronounced *Mo-gee-yone*). The topography produced by these two systems is more rugged south of the rim than north of it. In this exercise, we will examine the reason why this difference exists.

1. Trace in red pencil the drainage divide between the source areas of the two stream systems. Use a soft black pencil first, then trace it in red when you are satisfied with its position.
2. What human-made feature is *roughly* coincident with the divide? fish hatchery
3. Does the Mogollon Rim lie on the divide or north or south of the divide? South
4. Table 4.1 lists several streams north (A) and south (B) of the Mogollon Rim. For each stream, the differences in elevation of two contour lines that cross the stream a few miles apart are recorded, and the map distance between them is shown.
   (a) For each stream segment listed in table 4.1, calculate its gradient in feet per mile.
   (b) Determine the average gradient for the streams north and south of the Mogollon Rim.
5. If the erosive potential of a stream is proportional to its gradient, which system, A or B, should have the greatest erosive power? System B
6. Will the divide between the two systems move generally toward the north, toward the south, or remain more or less fixed with the passage of time? Explain your answer.
   South B/C thats the waterflow + elevation
7. On a separate sheet of paper, sketch a series of maps that show the progressive changes that will occur as erosion continues around Promontory Butte, especially at its juncture with the Mogollon Rim.

Table 4.1 Matrix for Determining Stream Gradients (A) North of the Mogollon Rim and (B) South of the Mogollon Rim

A. Stream Gradients, North of Mogollon Rim				B. Stream Gradients, South of Mogollon Rim			
Name of Stream	Length in Miles	Diff. in Elev. in Feet	Gradient in ft/mi	Name of Stream	Length in Miles	Diff. in Elev. in Feet	Gradient in ft/mi
West Leonard Canyon	2.6	250	96 ft/m	Big Canyon Creek	3.7	1,500	405 ft/mi
Middle Leonard Canyon	2.4	400	166 ft/mi	Dick Williams Creek	1.8	1,150	638 ft/mi
East Leonard Canyon	1.9	250	131.5 ft/m	Horton Creek	3.0	1,000	333 ft/mi
Turkey Canyon	1.9	250	131.5 ft/mi	Doubtful Canyon	2.4	1,250	520 ft/mi
Beaver Canyon	3.7	300	81 ft/mi	Spring Creek	2.2	1,250	568 ft/m
Bear Canyon	1.6	200	125 ft/mi	See Canyon	2.2	700	318 ft/mi
Average Gradient			121.6 ft/mi	Average Gradient			483 ft/mi

NOW LATER WAY LATER

Figure 4.8 Promontory Butte Map

Part of the U.S.G.S. Promontory Butte quadrangle, Arizona, 1951. Scale, 1:62,500; contour interval, 50 feet.

Name

Section Date

EXERCISE 13C

Pediments and Alluvial Fans

The Antelope Peak map (fig. 4.9) shows a well-developed pediment, and the Ennis map (fig. 4.11) depicts an excellent alluvial fan.

1. Draw a topographic profile of each of these features on the grids provided (fig. 4.10A, B), with the beginning and end points of the profile as described in the following. Begin each profile by plotting the *highest* elevation on the vertical axis at mile-zero.
   (a) Line of profile for the Antelope Peak pediment (fig. 4.10A): the starting point is at the SW section corner of Section 35, T. 6 S., R. 2 E., at an elevation of 1,694 feet. The profile extends as a straight line for a little more than 6 miles in a northeasterly direction, passing through the SW section corner of Section 6, T. 6 S., R. 3 E., at an elevation of 1,320 feet, and ends at the 1,300-foot contour line about one-half mile northeast of the previously described point. Draw the line of profile on the Antelope Peak map with a sharp pencil.
   (b) Line of profile for the Cedar Creek Alluvial Fan, Ennis, Montana (fig. 4.10B): the starting point is at the northeast section corner of Section 21 in the southeast part of the Ennis map area (Section 21 is the one in which the Lawton Ranch is located). The elevation at the starting point (i.e., mile-zero) is 6,080 feet above sea level. From this point, the line of profile extends as a straight line in a northwesterly direction for about 4 miles, passing through the northwest section corner of Section 7 (elevation 5,245 ft.), and ending at the 5,200-foot contour line. (In drawing this profile, you should start with the highest elevation, 6,080 feet, at the left side of your topographic profile.) Draw the line of profile on the Ennis map with a sharp pencil.

(continued)

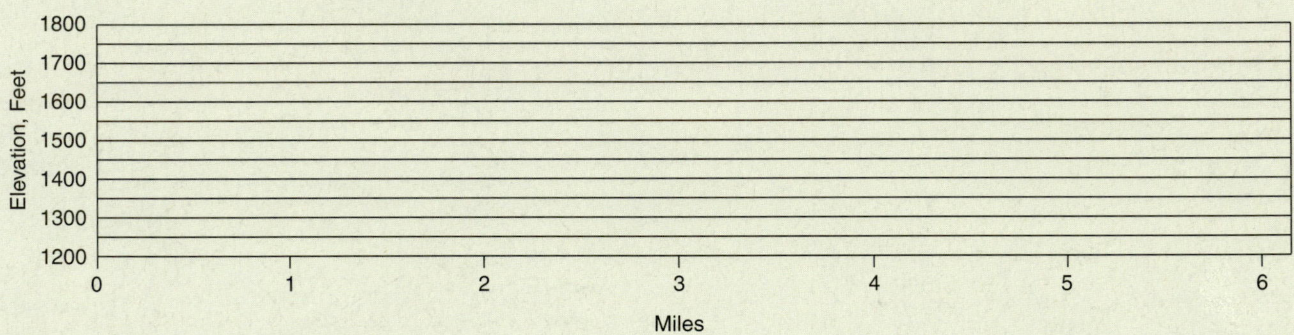

Figure 4.10A Antelope Peak Profile

Grid for drawing pediment profiles based on Antelope Peak map.

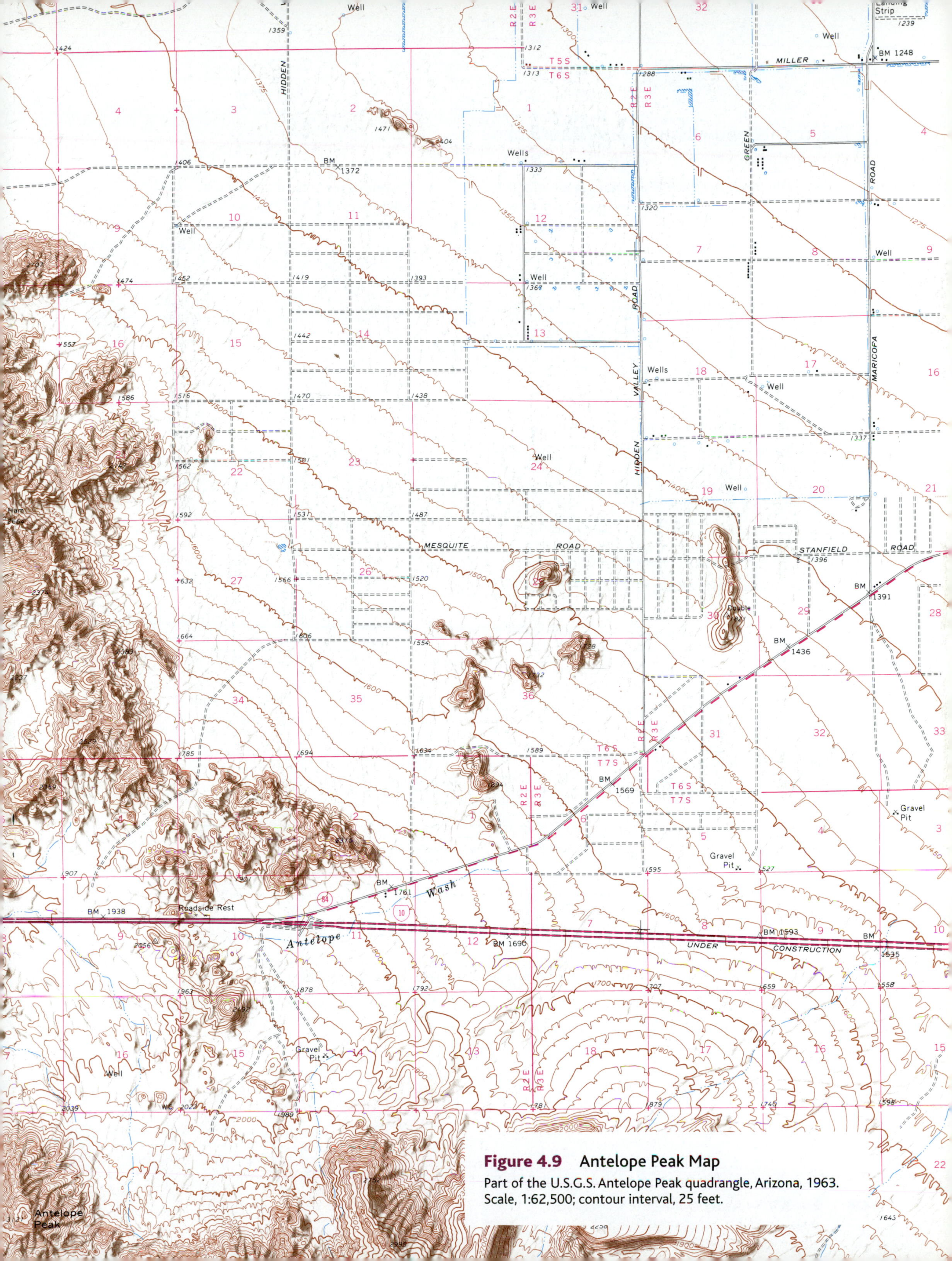

Figure 4.9 Antelope Peak Map
Part of the U.S.G.S. Antelope Peak quadrangle, Arizona, 1963. Scale, 1:62,500; contour interval, 25 feet.

EXERCISE 13C

Name ______________________

Section ______________ Date ______________

Pediments and Alluvial Fans *(Continued)*

2. Both the pediment and the alluvial fan are produced by running water. Describe each profile in terms of its shape (i.e., concave, convex, or straight) and gradient in feet per mile.

____

____

____

____

3. Inasmuch as both the pediment and the alluvial fan are produced by the action of running water, suggest a reason or reasons why the two profiles are different.

____

____

____

____

4. The uniform regularity of contour spacing on the pediment of the Antelope Peak map is interrupted by several isolated hills that rise 100 to 200 feet above the level of the pediment between Mesquite Road and Highway 84. What is the origin of these features?

____

____

____

____

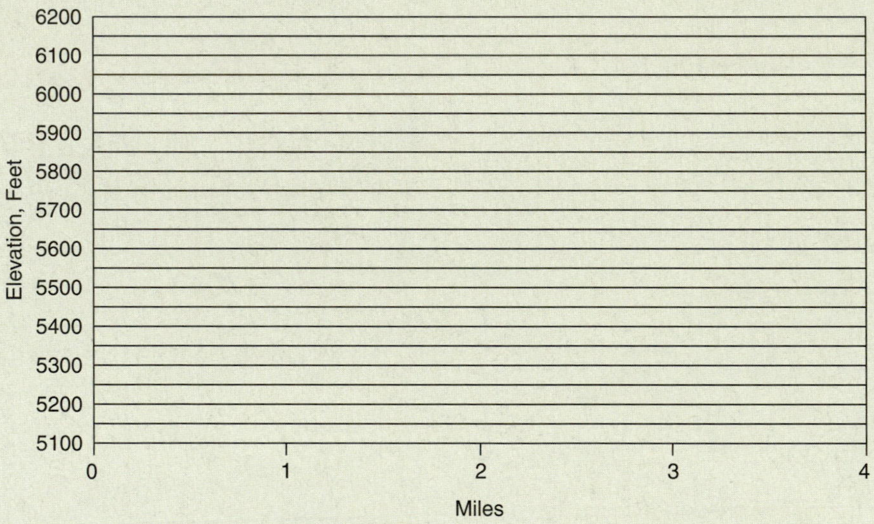

Elevation, Feet: 6200, 6100, 6000, 5900, 5800, 5700, 5600, 5500, 5400, 5300, 5200, 5100

Miles: 0, 1, 2, 3, 4

Figure 4.10B Ennis Profile

Grid for drawing pediment profiles based on Ennis Topographic map. (See Exercise 13C, question 1b for location.)

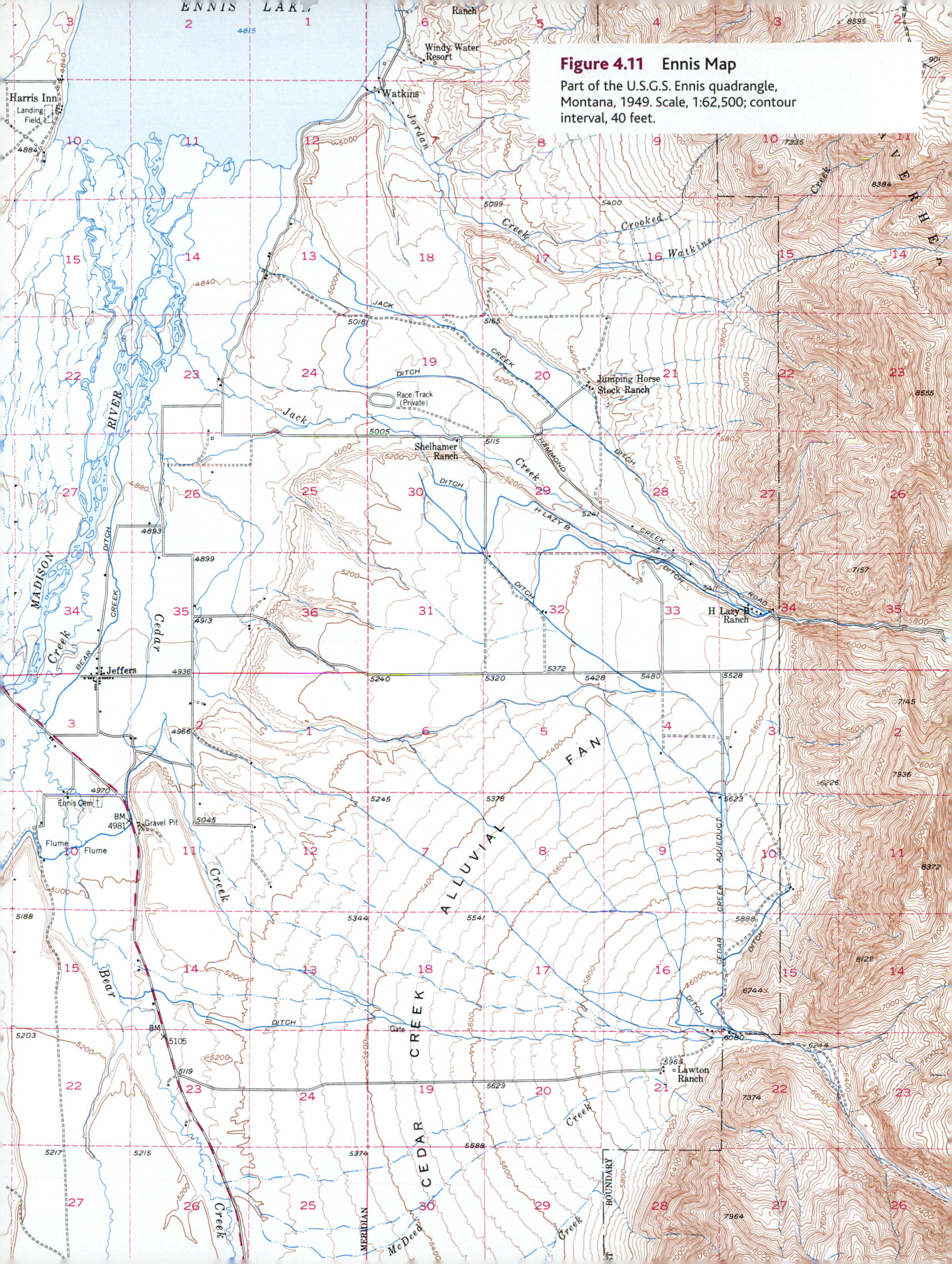

Figure 4.11 Ennis Map

Part of the U.S.G.S. Ennis quadrangle, Montana, 1949. Scale, 1:62,500; contour interval, 40 feet.

EXERCISE 13D

Name

Section Date

Charles Creek, Canada

The aerial photograph (fig. 4.12) shows a meandering main stream and a single meandering tributary. Each has shifted its channel a number of times, both by channel migration and meander cutoffs. Channel migration occurs because of erosion by the river on the outside of a meander and deposition on the inside. Crescent-shaped sandbars on the inside bank of a meander are evidence of deposition. Meander cutoffs result when two meanders migrate toward each other, causing the destruction of the narrow neck of land separating them as shown in figure 4.5. The isolation of a meander from the stream channel results in an oxbow lake. With the passage of time, the oxbow lakes become filled with sediment from flood waters and aquatic vegetation.

1. Show by a series of diagrammatic map sketches how a U-shaped meander will be transformed into an oxbow lake. Indicate on your map the points of erosion by the letter *E* and deposition by the letter *D*.
2. On the Charles Creek aerial photo (fig. 4.12), trace the present courses of the main stream and its tributary in blue pencil. Where the channel width is greater than the width of your pencil line, make your line follow the thread of maximum velocity as indicated by erosion on the outside banks of meanders and other bends in the river. Your pencil line traces the **thalweg,** a line connecting the deepest points of a river channel.
3. The point at which a tributary joins the main stream is called the **confluence.** Note the point of confluence at the present time. Indicate by a dashed red line the course of the tributary stream when its confluence with the main channel was different than at present.
4. Indicate by the letter *A* (with red pencil) the most recent meander cutoff on the main channel.
5. Indicate by the letter *B* (with red pencil) where the next meander cutoff on the main channel will occur.
6. Three former meander loops of the main channel are designated by the letters *X, Y,* and *Z* on figure 4.12. Which of these was most recently part of the river channel, and which has been an oxbow lake the longest? How is this determined?

   Z b/c most H_2O and alot of H_2O present.

7. A stream segment with many meanders has a gradient determined by the length of the stream channel and the difference in elevation of the end points of the segment. If one of the meanders in the segment becomes an oxbow lake, will the gradient remain the same, become higher, or become lower? Use figure 4.5 as a reference in determining your answer.

   lower.

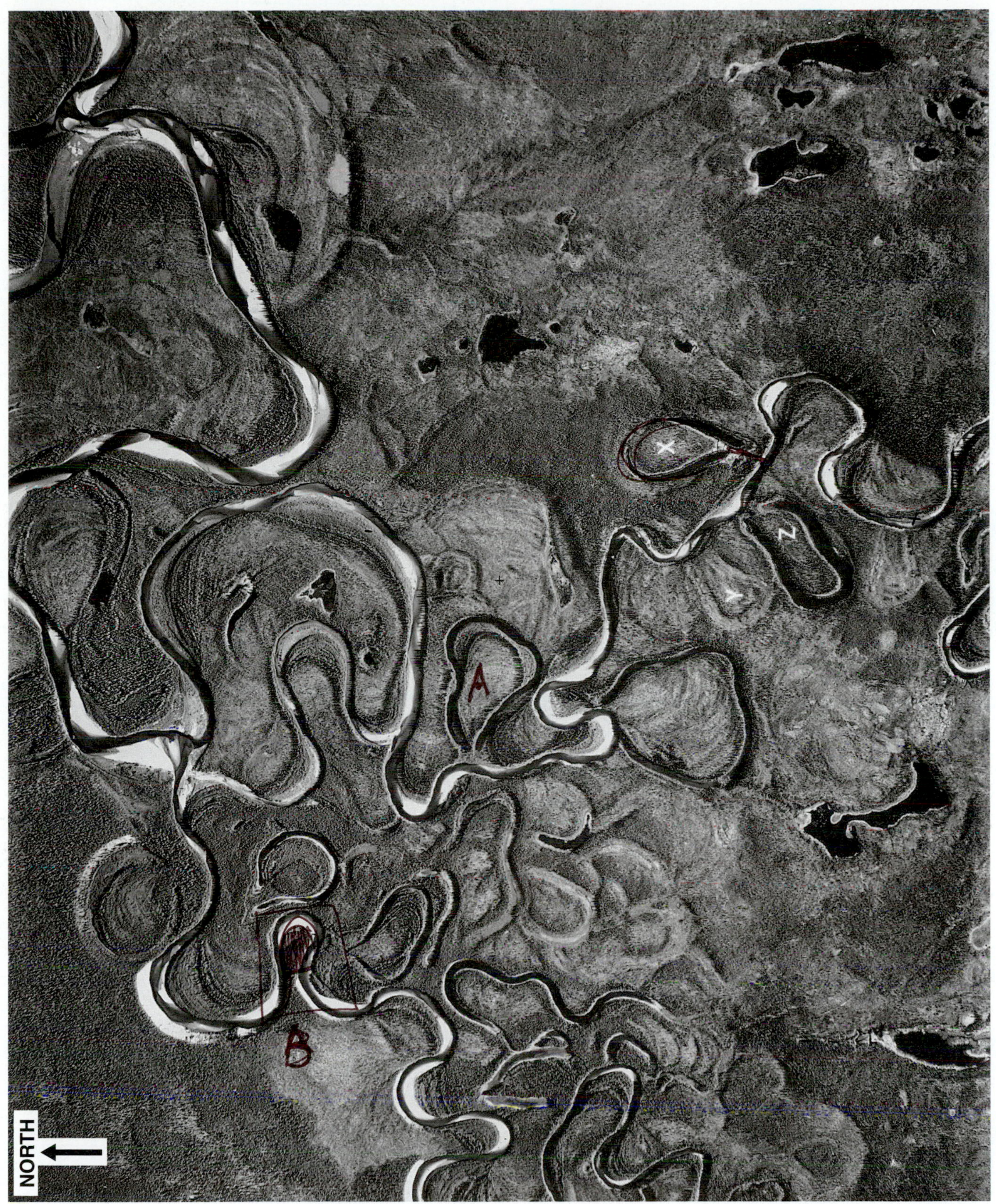

Figure 4.12

Aerial photograph of Charles Creek, south of Great Bear River, Northwest Territories, Canada. Scale, 1 inch = 1,200 feet.

By permission of the Royal Canadian Air Force.

EXERCISE 13E

Name

Section Date

Refuge Map, Arkansas–Mississippi Greenwood Map, Arkansas–Mississippi

The Refuge map (fig. 4.13) depicts a typical segment of the lower reaches of the Mississippi River. The course of the river in the map area is partly a natural one and partly an artificial one due to engineering works of the U.S. Army Corps of Engineers. These works include artificial cutoffs, which are channels dug by the Corps in an attempt to straighten the meandering course of the river, and revetments, which are concrete slabs or other materials designed to reduce erosion on river banks and impede the migration of meanders. The entire map area lies on the floodplain of the Mississippi River. The origin of a floodplain by lateral erosion of a river is shown in figure 4.4.

The modern course of the river across the map area is from Miller Bend (near the north margin of the map) southward through the Tarpley Cut-off (constructed in 1935) and the Leland Cut-off (constructed in 1933).

Previous courses of the Mississippi River are marked on the map. One is defined by the **meander line** of 1823. It is shown on the map as a dotted line with the respective date printed alongside. Another previous course of the river is coincident with the boundary line between Arkansas (west of the Mississippi River) and Mississippi that lies generally to the east of the river. Notice, however, that certain parts of the state of Mississippi lie on the *west* side of the modern course of the river. As previously noted (p. 108), this is the result of a court ruling that states where a stream or river forms the boundary between states and the channel of the stream is changed by the ". . . natural and gradual process known as **erosion** and **accretion,** the boundary follows the varying course of the stream . . . while if the stream from any cause, natural or artificial, suddenly leaves its old bed and forms a new one by the process known as avulsion, the resulting change of channel works no change of boundary, which remains in the middle of the old channel . . . although no water may be flowing in it."*

1. Draw the following lines on the Refuge map:
   (a) Using a red pencil, trace the course of the Mississippi River at the time the Arkansas–Mississippi state boundary was established. (NOTE: The thalweg is commonly used to define a boundary when a river is part of that boundary. At the time the Arkansas–Mississippi River boundary was established, the thalweg of the Mississippi River at that time was used to define the boundary.)
   (b) Using a blue pencil, trace the course of the Mississippi River during the year 1823.
2. Study the two courses of the Mississippi River that you drew to answer 1 above. Deduce whether the Arkansas–Mississippi boundary was established before or after the 1823 course. (Meanders tend to migrate in a *downstream* direction.)

(continued)

*Van Zandt, Franklin K. 1976. Boundaries of the United States and the Several States. U.S.G.S. Professional Paper 909, p. 4.

Figure 4.13 Refuge Map

Part of the U.S.G.S. Refuge quadrangle, Arkansas–Mississippi, 1939. Scale, 1:62,500; contour interval, 5 feet.

EXERCISE 13E

Name

Section Date

Refuge Map, Arkansas–Mississippi; Greenwood Map, Arkansas–Mississippi *(Continued)*

3. Figure 4.14 shows a segment of the Mississippi River along the border between the states of Arkansas and Mississippi. The course of the modern river is not coincident with the boundary line between the two states, which was fixed by Congress before the modern course of the river was established.

   We will refer to the course of the Mississippi River shown on figure 4.14 as the "Modern Course" and the course of the river when the Arkansas–Mississippi boundary was set as the "Boundary Course." From the northern to southern boundaries of the Greenwood map, the Modern Course is 65 miles and the Boundary Course is 136 miles.

   (a) Express as a percent the amount of shortening of the Modern Course compared to the Boundary Course.

   (b) What is the impact of this shortening on the *competence* and *capacity* of the river?

4. Draw the boundaries of the Refuge map on the Greenwood map. The Refuge map was published in 1939 and the Greenwood map in 1979. Compare the Refuge map with its corresponding area on the Greenwood map. Describe the changes in former or modern channels that have occurred in this 40-year period with particular reference to the following:

   (a) Areas around Linwood Neck, Luna Bar, and Point Comfort.

   (b) The course of the Tarpley Cut-off with respect to the dike on its east side.

Figure 4.14 Greenwood Map

Part of the U.S.G.S. Greenwood, Miss.–Ark.–La. map, 1953, revised in 1979. Scale: 1:250,000; contour interval, 50 feet with supplementary contours at 25-foot intervals.

Groundwater Movement, Groundwater Pollution, and Groundwater as a Geologic Agent

Groundwater

Groundwater is the water that occurs beneath the surface of the earth. As the water from various sources moves downward, it first moves through an **unsaturated zone** in which both water and air fill the pore space. The contact between the unsaturated zone and the **saturated zone** is the **water table.** This is not a fixed surface but may move up or down depending on water supply and is called the **capillary fringe.**

Sources of groundwater that are permeable and saturated with water are **aquifers.** Many cities depend on groundwater aquifers for their water supplies, and much of the agriculture in the western half of the United States is dependant on groundwater to supply the water for irrigation systems. Water moves through the aquifer, and where the water table intersects the surface of the earth, it may feed water directly to surface streams or form **springs**—places where water flows out from the aquifer.

Water Table Contours

The water table is a planar to irregular surface that tends to parallel the surface of the land above it. Its configuration can be defined by contour lines in the way that the configuration of the earth's surface can be defined on a topographic map. Contour lines on the water table can be drawn if enough points of known elevation on the water table are available for plotting on a base map. These points are assembled from water wells and other places of known elevation where the water table intersects the earth's surface, such as a stream or a lake.

Flow Lines

Groundwater moves under the influence of gravity through porous rock or unindurated sand and gravel, but the movement is very slow compared to flowing water in a river. Groundwater moves along flow lines. A **flow line** is a path followed by a water molecule from the time it enters the zone of saturation until it reaches a lake or stream where it becomes surface water.

Figure 4.15 is a map of a hypothetical area underlain by clean sand showing the relationship of the water table contours and flow lines to a permanent surface stream flowing south. For example, a water molecule at point *a* will follow the path of the flow line until it enters the stream at *a′*. The same relationship holds true for water molecules at *b, c, d,* and *e.* They follow the flow lines as they move down the slope of the water table at right angles to the water table contours until they enter the stream at points *b′, c′, d′,* and *e′* respectively.

Flow lines can converge or diverge, but they cannot cross each other. Moreover, as can be deduced from figure 4.15, groundwater from the west side of the stream cannot move across the stream to be commingled with groundwater on the east side of the stream, and vice versa, because the stream intercepts the flow of groundwater from both sides. From this simple analysis, it follows that any pollutant introduced into the zone of saturation will be carried by the flow of groundwater along flow lines until it eventually is discharged into a lake or stream.

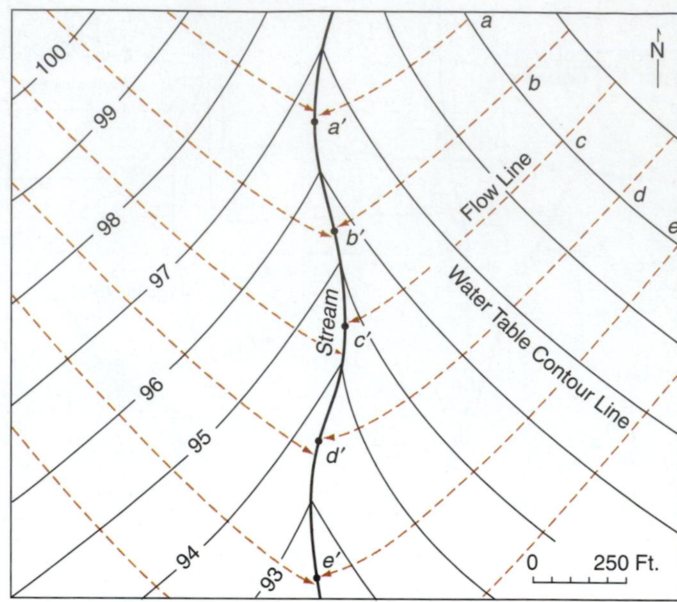

Figure 4.15

Map of a hypothetical area underlain by a well-sorted coarse sand showing a south-flowing permanent stream and contours on the water table. The ground surface is roughly the same shape as the water table but about 5 to 10 feet higher. The contour interval of the water table contours is 1 foot. Flow lines are shown in dashed colored lines. (See text for further explanation.)

Conversely, if a pollutant is introduced into the stream of figure 4.15 at point *a′*, it would not enter the water table but would flow down the stream channel in the direction of *b′*. This relationship between groundwater flow and stream flow holds only for the case of an **effluent stream,** which is **a stream fed by groundwater;** that is, the groundwater is discharged into the stream channel. An **influent stream,** on the other hand, is one in which the **groundwater flows away from the stream channel.** In such cases, pollutants dumped directly into the stream will move into the zone of saturation. Influent streams also are commonly **intermittent streams,** streams that flow only during certain times of the year when rainfall is sufficient to supply surface runoff directly to them. With these basic principles in mind, we will apply them to the following exercises.

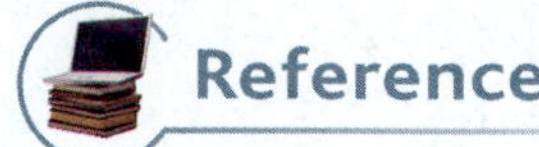

Reference

Sniegocki, R. T. 1959. Geologic and groundwater reconnaissance of the Loup River Drainage Basin Nebraska, *U.S.G.S. Water-Supply Paper 1493.* Washington, D.C.: U.S. Government Printing Office, 106 pp.

NOTES

EXERCISE 14A

Name ______________________

Section ______________ Date ______________

Water Table Contours

The Ashby map (fig. 4.16) covers part of the Sand Hills of western Nebraska. The Sand Hills are formed from ancient sand dunes that are now more or less stabilized by surface vegetation consisting mainly of native grasses, an environment that makes this an excellent area for the grazing of cattle.

Sand dunes are composed of windblown sand that is very porous. Rain striking the surface in the Sand Hills very quickly percolates to the zone of saturation where it becomes groundwater. The water table in the area covered by the Ashby map is quite shallow (i.e., close to the surface) in the interdune "valleys," which explains the many lakes and marshes that occur there.

The numbers imprinted on some of the lakes in this map area are elevations of their water surfaces. For example, the water surface elevation of Castle Lake is 3,767 feet above sea level. Assuming that these elevations are points on the water table, it is possible to draw a rough approximation of the water table contours. Some lakes have no elevations marked on the map. In these cases, the lake elevation can be estimated by noting the elevation of the contour nearest to the lakeshore. Careful inspection of the map reveals that the lowest known lake elevation is North Twin Lake in the northeast corner of the map (3,737 feet above sea level) and that the highest lake elevation is Melvin Lake in the northwest corner of the map area (3,798 feet above sea level). From this relationship, we can deduce the water table must be sloping downward in an easterly direction. Therefore, the contour lines on the water table must extend at right angles to the direction of slope of the water table, or in a general north–south direction.

1. With a soft pencil (easily erasable), sketch in the water table contours for the entire map area. Use a C.I. of 10 feet. Start with the 3,800-foot contour line that begins just west of Melvin Lake and extends in a southerly direction until it passes just west of Hibbler Lake (about 2 miles northeast of the town of Ashby), with an elevation of 3,796 feet above sea level. Draw all water table contours on the map. For best results work first on the contours in the south-central part of the map area where more lakes are present, hence more control points are available. (Be sure that each water table contour is labeled.)
2. Locate the point in Section 20 (south half of the map) marked with a brown × and the numerals 3,862, which is the surface elevation at that point. On the basis of the water table contours that you have drawn on the map, estimate the depth of the water table at this point.

   ____

3. Why are there so few surface streams in this area?

   ____

   ____

   ____

   ____

   ____

Figure 4.16 Ashby Map

Part of the U.S.G.S. Ashby quadrangle, Nebraska, 1948. Scale, 1:62,500; contour interval, 20 feet.

EXERCISE 14B

Name

Section Date

Groundwater Pollution

Figure 4.17 is the map of a hypothetical area underlain by well-sorted coarse sand crossed by a permanent stream, Clear Creek, that flows in a southeasterly direction. The ground surface is gently sloping to the southeast, and the shallow water table is defined by the water table contours. The location of a dump is shown on the map. The dump is privately owned and operated by a small company that hauls trash and garbage for residents of a nearby small town. The dump is an excavated pit, the bottom of which lies just above the water table.

The Jones property lies southeast of the dump, and its west property line borders on Clear Creek. Mr. Jones owns horses that he keeps in the barn and corral part of the time. The rest of the time the horses graze on the property and occasionally drink the water from Clear Creek.

The Smith estate lies west of Clear Creek and also has frontage on the creek. Both Jones and Smith derive their domestic water supplies from wells that penetrate the shallow water table. To assure themselves that the water was suitable for human consumption, Jones and Smith had well-water samples analyzed by the county health department at the time their wells were completed. Both Smith and Jones owned their respective properties for many years prior to the establishment of the dump and have enjoyed potable water until recently.

Recently, Jones' well water began to deteriorate in quality. This was verified when Jones had his water tested again at the county health department. Jones attributed this to pollution of the groundwater from leaching of domestic waste deposited in the dump. In talking with his neighbor, Smith, Jones suggested that the two of them should file suit against the owner of the dump and obtain a court injunction that would require cessation of all further dumping.

Smith had water from his well tested again and found that it had not changed in quality since the tests conducted prior to the creation of the dump. Water samples along the stretch of the stream that forms the boundary between the Jones and Smith properties were also analyzed by the county health department and were found to be contaminated by materials similar to those found in the recent water samples from the Jones well. This was sufficient evidence to convince Smith that he ought to join in the suit with Jones against the dump operator. Smith reasoned that if the creek were contaminated with the same materials found in the Jones well, it would be only a matter of time until his well would also be polluted.

At that point, Jones and Smith hired an attorney to file the suit. The attorney sought the advice of a geologist at a nearby university who had access to publications of the U.S.G.S. in the school's library. There the geologist discovered a report of the geology of the area. A section of the report on groundwater contained a map showing water table contours based on static water levels in other wells not shown in figure 4.17. The geologist transferred those contours to a map he was preparing for the lawyer. This map is figure 4.17.

1. On the map of figure 4.17, several dots are shown along the 800-foot water table contour line. Assume that these are the points of intersection of flow lines with the water table contours. Using the relationship of flow lines to water table contours as shown in figure 4.15, sketch a network of flow lines on figure 4.17. One flow line should pass through each of the dots on the 800-foot contour line. (Use a soft black pencil because you may have to erase several times before you are satisfied with your results.) Extend the flow lines across the entire map area so that the movement of groundwater can be ascertained.
2. On the basis of the flow line network that you have constructed on figure 4.17, answer the following questions:
   (a) Is there reasonable evidence to conclude that seepage from the dump has contaminated the Jones well? Explain your answer.

   ______

   ______

   ______

   ______

   ______

   (b) Is there reasonable evidence that the stretch of Clear Creek adjoining the Jones and Smith properties has been contaminated by seepage from the dump? Explain your answer.

   ______

   ______

   ______

   ______

   ______

(continued)

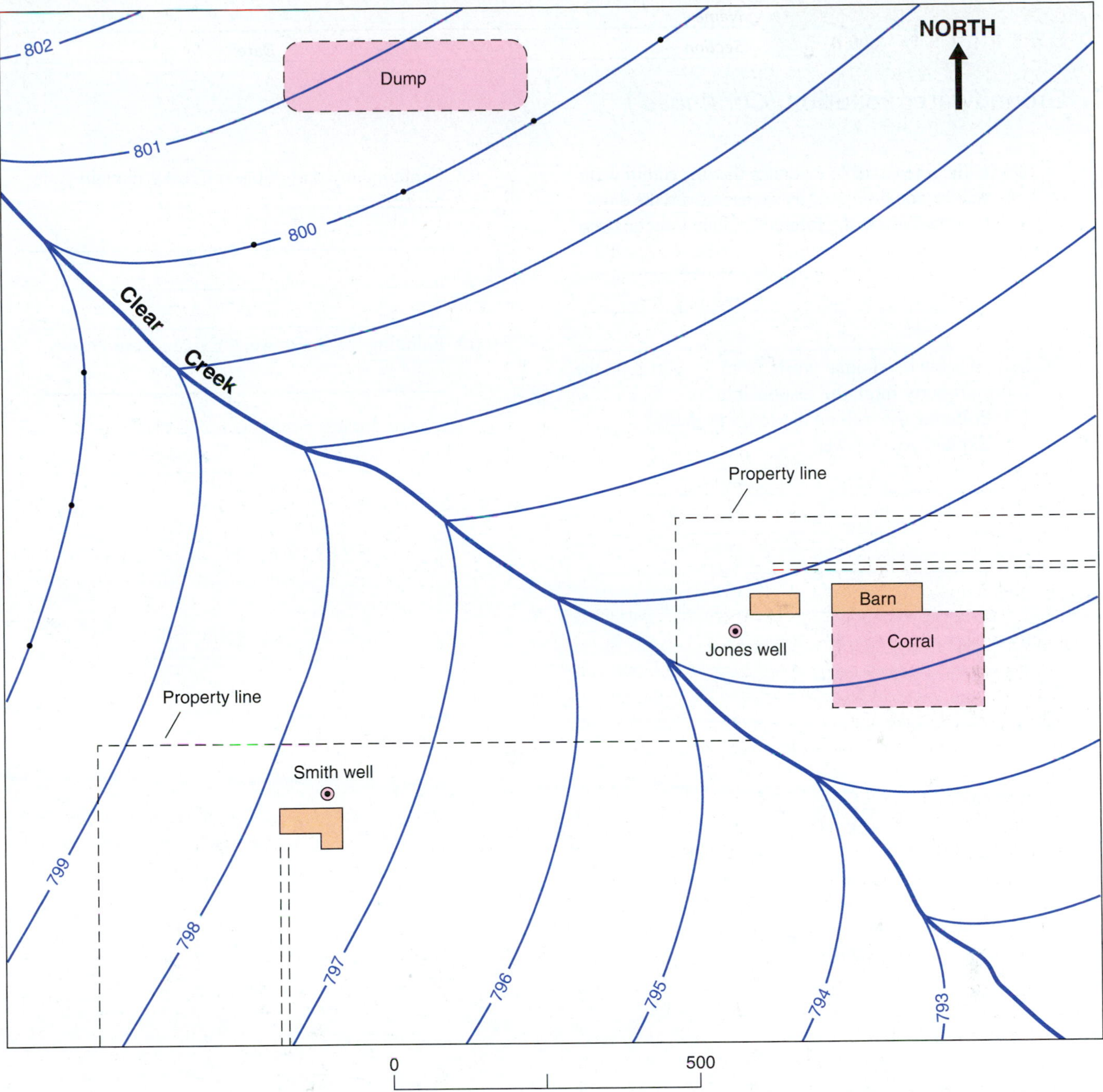

Figure 4.17

Map of a hypothetical area underlain by a well-sorted coarse sand about 50 feet thick. Clear Creek flows to the southeast. The water table contour interval is 1 foot. The water table lies about 8 to 10 feet below the ground surface, except near Clear Creek, where the water table becomes shallower until it intersects the creek. The dump is an excavated pit, the bottom of which does not quite reach the water table.

EXERCISE 14B

Name ______________________

Section ______________ Date ______________

Groundwater Pollution *(Continued)*

(c) Is there reasonable evidence that the Smith well will be contaminated by seepage from the dump at some time in the future? Explain your answer.

____

____

____

3. Is it possible the animal waste from the corral on the Jones property might be responsible for:
   (a) Polluting the well on the Jones property? Explain your answer.

____

____

____

____

(b) Polluting any part of Clear Creek? Explain your answer.

____

____

____

(c) Polluting the Smith well? Explain your answer.

____

____

____

Karst Topography

Karst, a unique geomorphic landscape, is formed by the dissolution of soluble rocks (fig. 4.18). In most areas the development of landforms is a result of mechanical erosion and tectonic uplift. In karst areas mechanical erosion plays a much smaller role, and it is the dissolution of soluble rocks—primarily carbonate rocks such as limestone and dolostone but in restricted areas also gypsum and rock salt—that is the most important agent of erosion. The type locality for karst is a limestone plateau region of southwestern Slovenia and northeastern Italy.

Karst topography develops by the dissolution of soluble rocks with a resulting land surface characterized by **sinks** or **sinkholes,** disappearing streams (with the result that there is little or no surface drainage system), underground passageways (caves) and stream systems, and residual hills that represent the remnants of the soluble rocks in which the dissolution process is not yet complete. Areas in Puerto Rico and South China are noted for the residual towers and cones that have formed during the karstification in those countries. The end result of the karst process is the removal of the soluble rocks and the reestablishment of a surface drainage system.

The dissolution of carbonate rocks in pure water is a very slow process, but the rate of dissolution increases with increased acidity in the water. Carbonic acid, H_2CO_3, the most common acid in nature, forms when CO_2 dissolves in water. The dissolution reaction is

$$CaCO_3 + H_2CO_3 \rightarrow Ca^{2+} + 2HCO_3^-$$

with both the calcium ion and the bicarbonate ions carried away in solution. This type of dissolution is dependent upon a source of carbonic acid and a fairly abundant water supply. Thus, the areas where karst is best developed have carbonate rocks and an ample supply of CO_2 and water. The result is that karst is uncommon in arid and semiarid regions.

The formation of caves and caverns in the subsurface is the result of the movement of acidic waters through the carbonate bedrock and the dissolution of the bedrock. The source of the water is either groundwater percolating downward into the bedrock or surface water carried into the underground drainage systems by streams that enter the system, often in the form of disappearing streams. An analysis of the complex hydraulic systems involved in cave formation is beyond the scope of this discussion.

Sulfuric acid produced in the subsurface apparently formed Carlsbad Caverns and associated caves in west Texas. The acid was produced by the mixing of H_2S that had moved into the area from the adjacent Delaware Basin oil field with groundwater in the Permian reef limestone of the Guadalupe Mountains. The acid dissolved the limestone and the materials were carried away in solution. This is a case in which the formation of the caves had little or no connection with surface processes.

It is estimated that 20% of the earth's land surface is karst and that 40% of the United States east of 96°W longitude is karst. The continued growth and development of urban areas in karst brings special problems to planners, builders, and regulators. Many areas are dependent on groundwater from karst, and the hidden subsurface voids present interesting problems in the construction of roads, bridges, and large buildings. The collapse of arches in the subsurface may result in differential settling or, in the extreme case, the formation of large sinkholes and the destruction of houses, roads, and other structures.

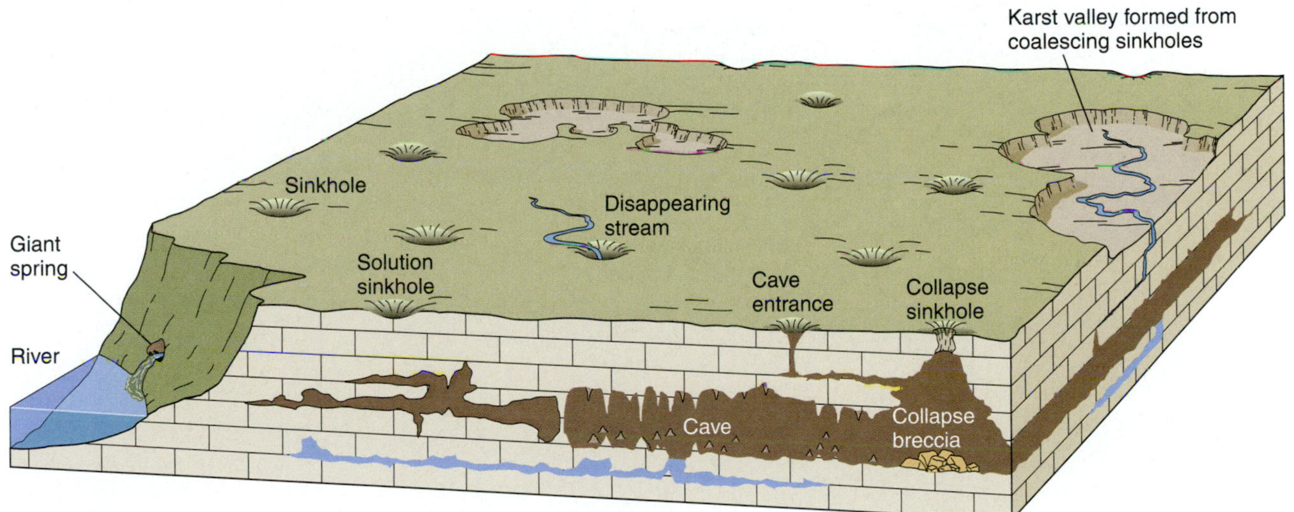

Figure 4.18

Karst topography is characterized by surface sinkholes, disappearing streams, possible cave entrances, and collapse features. Caves, collapse breccia, and streams may be present in the subsurface.

EXERCISE 15A

Name

Section

Date

Karst Topography

Many of the sinks in the area covered by the Interlachen map (fig. 4.19) are identified by concentric depression contours. Some of these sinks contain lakes.

1. A pair of sinks straddles the boundary between Sections 5 and 8 about 1.5 miles north of the east–west highway that crosses the central part of the map area. What is the approximate elevation of the bottoms of these two sinks?

2. In the same general area, an unnamed lake occupies the sink in the NE 1/4 of Sec. 7, and another occurs in the SW 1/4 of Sec. 4. Assuming the water surface of these two lakes lies 5 feet below the elevation of the contour nearest the lake shores, what is the elevation of these two lakes? (Refer to them as Section 7 Lake and Section 4 Lake.)

3. Why are the two sinks in Sections 5 and 8 dry—that is, they contain no lakes—when Section 7 Lake and Section 4 Lake both lie less than a mile away?

4. Generally speaking, the northern half of the area has greater relief than the southern half. The northern half has no streams while the southern half has Gun Creek, Cabbage Creek, and Little Cabbage Creek. Assuming that these two areas reflect different stages in the evolution of karst topography, will the northern half become more like the southern half during the passage of time, or vice versa?

Figure 4.19 Interlachen Map

Part of the U.S.G.S. Interlachen quadrangle, Florida, 1949. Scale, 1:62,500; contour interval, 10 feet.

Sinkhole Formation

A Description of the Process of Sinkhole Formation in Unconsolidated Sediments Overlying Cavernous Limestone

(From Foose, R.M., 1981, Sinking can be slow or fast. *Geotimes,* v. 26, no. 8, August 1981, p. 22.)

"The progressive downward movement of unconsolidated materials into underground limestone openings naturally would affect the surface of the land and produce funnelshape sinkholes at the land surface.

". . . Where the unconsolidated overburden is thick, say in the range of 30 to 500 ft., the likelihood of seeing funnel sinkholes developed at the surface in response to the downward migration of unconsolidated particles in limestones at great depth is much less likely. That is not to say that there is no downward migration of particles into underlying cavernous openings within the limestone. In fact, drilling into such limestones often establishes the presence of large amounts of sand, clay, and silt within cavernous openings.

". . . The downward movement of particles into cavernous openings frequently results in another cavity developing entirely within the unconsolidated materials above the bedrock, with its lower point connected to an opening into the limestone. Doubtless it begins as a small cavity. As material falls from the roof of this cavity it may migrate downward into the limestone, and hence the upper cavity enlarges. If the upper cavity, in unconsolidated materials, is within the zone of saturation, movement of material from the roof to the bottom of the cavity and on down into the limestone may be very slow. At the surface, one would not know that an open cavity existed within totally unconsolidated material **above** the bedrock. However, with the continued slow enlargement of the cavity due to particles occasionally falling from the cavity roof, it may grow to a size so large that the overlying lithostatic load (the weight of all the material between the top of the cavity and the land surface) can no longer be supported. At that moment the cavity collapses—a catastrophic event! . . ."

EXERCISE 15B

Name

Section Date

Sinkholes

Geologic events of a catastrophic nature are usually attributed to earthquakes, landslides, volcanic eruptions, and the like. The subsidence of a part of the earth's surface, while not as devastating as the foregoing, is nonetheless catastrophic when it occurs in the course of a few hours or days.

Such an event occurred on May 8–9, 1981, in Winter Park, a town in central Florida. Figure 4.20 shows a large collapsed zone in suburban Winter Park. The collapsed feature is a sinkhole about 115 × 100 meters in horizontal dimensions. The water surface in the bottom of the depression is about 13 meters below the surrounding undisturbed land surface.

The geology of the area consists of limestone bedrock overlain by uncemented sediments (i.e., sand, silt, clay, or a mixture of these). The limestone contains large openings or caverns formed by the gradual dissolving action of groundwater percolating along joints and bedding planes. As the initial voids grow larger with the passage of time, they coalesce into a large, cavernous opening. Gradually, the material from the *base* of the overlying sediments moves downward into the cavern, thereby leaving a large cavity in the lower part of the unconsolidated layer, but the upper part of this layer remains undisturbed. Eventually, however, the overlying sediments collapse suddenly to fill the empty space at the bottom of the sedimentary layer, and a conical sinkhole such as the one in figure 4.20 is formed.

When such an event occurs in an urbanized area, considerable property damage results. Human-made structures collapse, sewer lines are ruptured, and underground utility lines (gas, pipelines, electric cables, telephone lines, etc.) are left hanging or are severed. Study figure 4.20 and answer the following questions.

1. Identify the following human-made features that have been damaged or destroyed by the collapse. (Write the letter of the feature directly on the photo; e.g., "a" for paved street, "b" for swimming pool, etc.)
   (a) Paved street
   (b) Swimming pool
   (c) Parking lot
   (d) Unbroken utility line
   (e) Broken utility line
   (f) A van or recreation-type vehicle
2. Figure 4.21 is a sketch of the probable cross section of the geologic conditions that prevailed long before the Winter Park Sinkhole was formed. It depicts a network (not to scale) of voids (shown in black) in the limestone caused by groundwater solution along joints and bedding planes. Figure 4.22 shows the geologic conditions that prevailed immediately after the collapse. During the geologic time that lapsed between the conditions depicted on figures 4.21 and 4.22, the limestone continued to dissolve until a large cavern formed beneath the site of the future sinkhole. When certain conditions were achieved, the collapse occurred.

A description of the process of sinkhole formation in unconsolidated sediments resting on a cavernous limestone bedrock is described on page 134. Read it carefully and then complete the geologic cross section of figure 4.23, showing the conditions that existed *immediately prior* to the collapse that formed the Winter Park Sinkhole (assume that the base of the large limestone cavern is below the bottom of the diagram).

Figure 4.20
Photograph of Winter Park Sinkhole taken on May 13, 1981, by Rick Duerling.
Photo: Courtesy of Florida Bureau of Geology, Tallahassee, Florida.

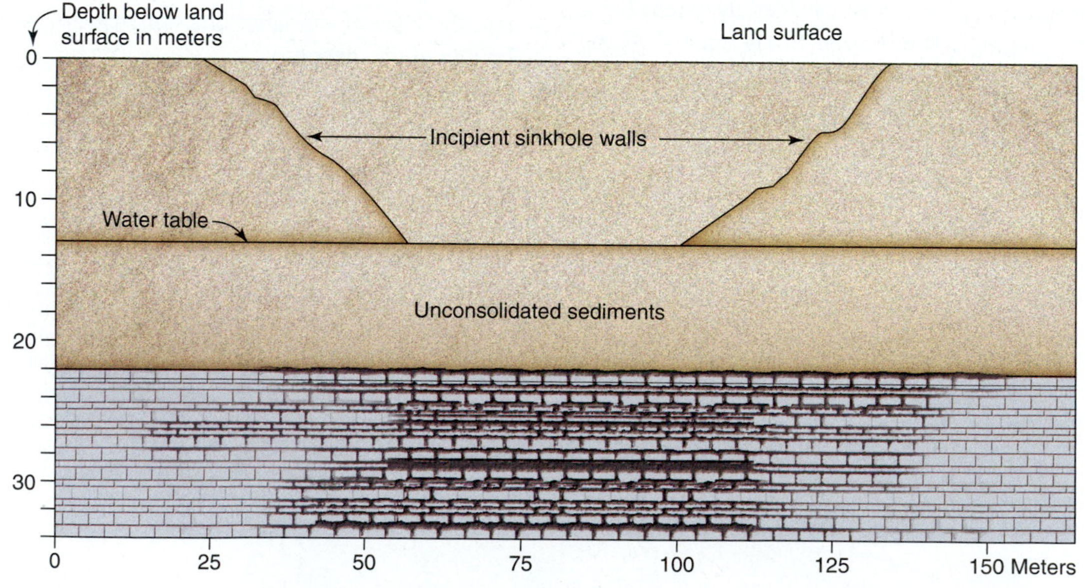

Figure 4.21
Geologic cross section of Winter Park Sinkhole area long before collapse occurred.

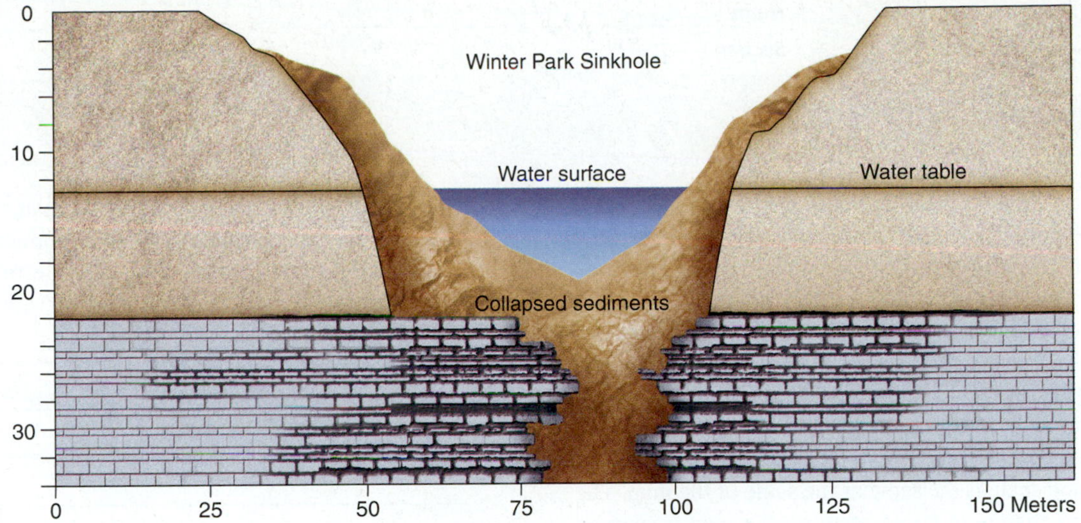

Figure 4.22
Geologic cross section of Winter Park Sinkhole *immediately after* sinkhole was formed.

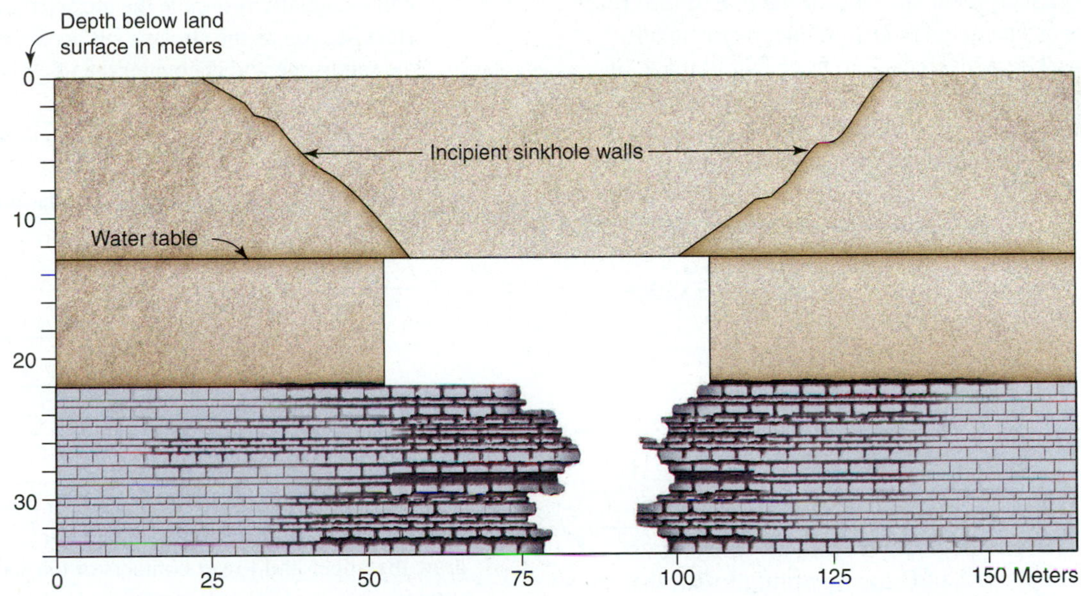

Figure 4.23
Incomplete geologic cross section of Winter Park Sinkhole *immediately before* collapse. (See Exercise 15B, question 2, for instructions.)

EXERCISE 15C

Name

Section Date

Evolution of a Karst Terrain

The area covered by the Mammoth Cave map (fig. 4.24) is a good example of the effect of underlying strata on the topography. A cursory inspection of the map reveals that, from a topographic point of view, the southern one-third of the map area is markedly different from the northern two-thirds. The reason for this is readily apparent from the geologic cross section in figure 4.25. This north–south cross section passes through the main road intersection in the town of Cedar Spring near the center of the map, and its horizontal scale is the same as the scale of the map. The pronounced change in topography on the map is marked by the Dripping Spring Escarpment lying just north of and roughly parallel to the Louisville and Nashville Road.

1. Draw the line of the geologic cross section on the map in black pencil.
2. Generally speaking, the area north of the Dripping Spring Escarpment and *west* of the line of the cross section is characterized by an integrated stream system. Use a blue pencil to trace the drainage lines occupied by permanent and intermittent streams. Use the correct map symbol for each. Take care not to extend your blue lines beyond the limits of the streams as they are shown on the map. What is the dominant bedrock in the area drained by the stream system?
3. Why does the stream flowing north into Double Sink end so abruptly there?

____

____

____

____

4. Examine the topography *east* of the line of cross section and north of Dripping Spring Escarpment. A number of valleys such as Cedar Spring Valley, Woolsey Hollow, and Owens Valley resemble stream-cut valleys, but there are no streams flowing in them. Account for the fact that the topography of this area resembles a stream-dissected terrain even though no streams occupy the existing valleys.

____

____

____

____

____

5. In making a comparison of the terrains east and west of the line of the cross section (fig. 4.24), which of the following statements is most likely the correct one?
   (a) The area *west* of the line of cross section will eventually resemble the area *east* of the cross section as the streams cut deeper into the sandstone and encounter the underlying limestone.
   (b) The area *east* of the line of cross section will eventually resemble the area *west* of the cross section as the overlying sandstone is further eroded.

____

____

____

____

____

6. Assume that the sandstone formation in figure 4.25 had an original thickness of 100 feet. On figure 4.25, draw the upper and lower contacts of the sandstone formation as it existed at some time in the geologic past before stream erosion began stripping it away.

Figure 4.24 Mammoth Cave Map

Part of the U.S.G.S. Mammoth Cave quadrangle, Kentucky, 1922. Scale, 1:62,500; contour interval, 20 feet.

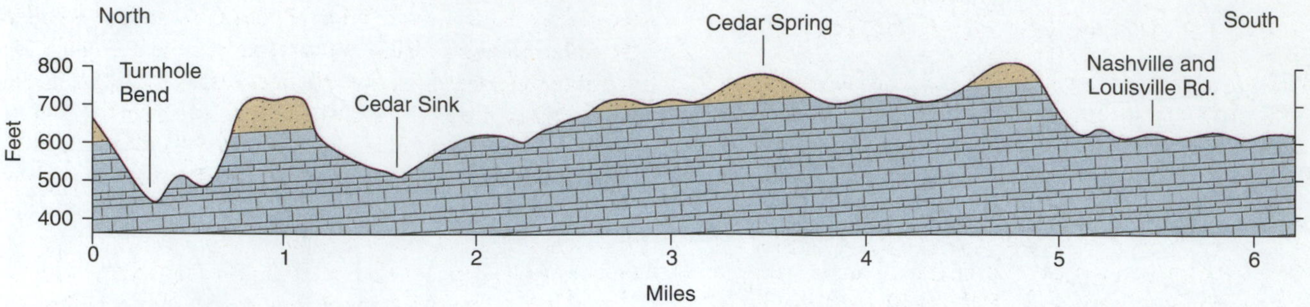

Figure 4.25

North–south geologic cross section from Turnhole Bend on the Green River to a point about three-fourths of a mile south of the Louisville and Nashville Road, Mammoth Cave map. Vertical scale exaggerated about 10 times.

Glaciers and Glacial Geology

A **glacier** is a body of ice and snow formed on land that shows downslope or outward movement under the influence of gravity. **Glaciology** is the study of existing glaciers and of ice and snow. **Alpine glaciers** (known also as valley or mountain glaciers) are confined to valleys and are literally rivers of ice. An **ice sheet** or **continental glacier** covers an area of continental proportions and is not confined to a single valley or valley system. **Glacial geology** is the study of the distinctive landforms produced by both alpine and continental glaciers, landforms that can be recognized on topographic maps and remote sensing images.

Accumulation and Ablation

The **budget year of a glacier** is the 12-month period of snow accumulation and summer ablation. In either hemisphere, the budget year begins at the end of the melt season just before the first snows of the winter and ends about 12 months later at the end of the ablation season.

During the ablation months of the budget year, the previous winter's accumulation is partly removed from the upper reaches of the glacier. The snow that remains there lies in the **zone of accumulation.** Over the lower reaches of the glacier, the previous winter's snowfall is completely removed, and some of the underlying glacier ice is also melted. The area of the glacier that suffers ablation of both snow and ice is called the **ablation zone.** The line that separates the accumulation and ablation zones for a given budget year is the **equilibrium line.** This is shown by a dashed line on the photograph of the Snow Glacier in figure 4.26 that was taken at the end of the ablation season. The white area above the snow line of the Snow Glacier is the residue of the previous winter's snowfall that survived the summer melting. On the lower portions of the Snow Glacier, all of the winter snow was melted so bare glacier ice is exposed. Some of this ice has also been lost by summer melting. The streams flowing from the glacier terminus in the lower right-hand corner of the photograph are fed by melted snow and ice from the glacier.

Glacier Mass Balance

It is possible to measure both accumulation and ablation for a given glacier during a budget year. The relationship between the annual accumulation and ablation over a period of years described the general "health" of a glacier. If accumulation exceeds ablation for several years, the glacier thickens and the terminus advances. If ablation exceeds accumulation for a period, the total mass of the glacier will decrease, a condition usually reflected by the thinning of the glacier and the retreat of the terminus.

A comparison between accumulation and ablation for a given budget year yields the **net mass balance** of the glacier. Net mass balance is expressed numerically in feet or meters

Figure 4.26

Oblique aerial photo of Snow Glacier, Kenai Peninsula, Alaska. Equilibrium line shown as dashed line.

Photo by Austin Post, U.S.G.S.

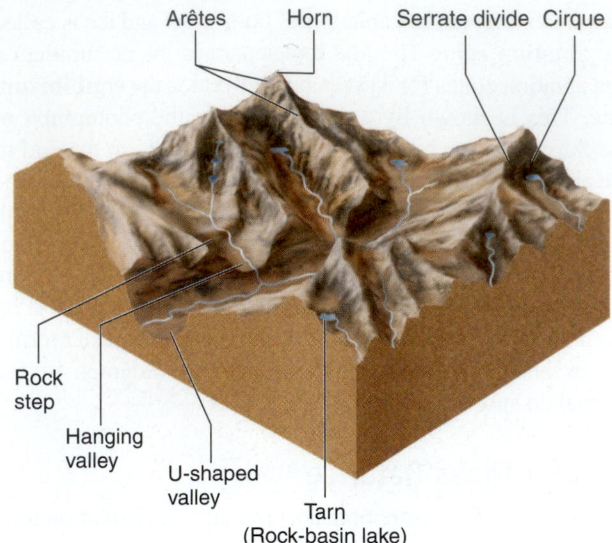

Figure 4.27
Erosional landforms produced by alpine glaciers.

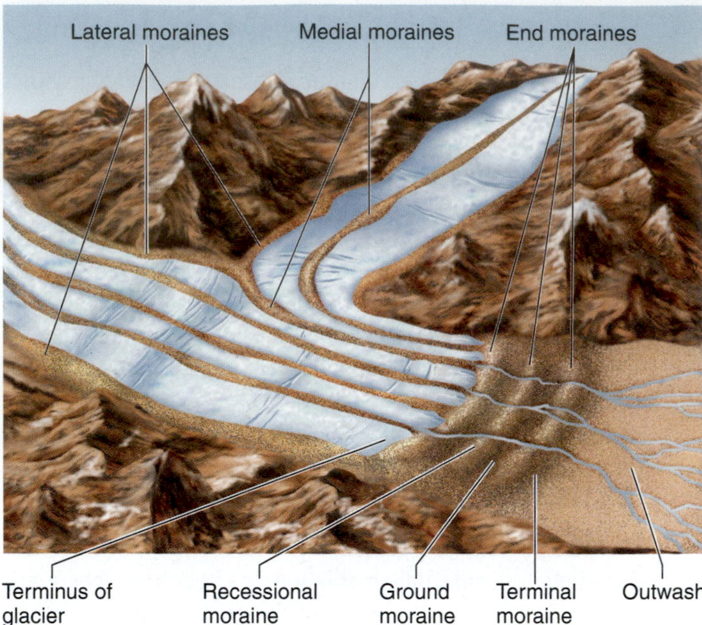

Figure 4.28
Moraines associated with alpine glaciers.

of water equivalent and represents a hypothetical layer of water determined by many measured thicknesses of columns of snow or ice on the glacier's surface. If a glacier gains in mass during a budget year (i.e., accumulation exceeds ablation), the glacier is said to have a **positive net mass balance.** If the glacier loses mass during the budget year, it has a **negative net mass balance.**

Landforms Produced by Alpine Glaciers

Erosional features usually dominate a terrain that was shaped by alpine glaciers (see fig. 4.27). A glacially eroded valley is **U-shaped** in cross section, and its headward part may contain a **cirque,** an amphitheaterlike feature. The small lakes or ponds that often form in the bottom of abandoned cirques or glacial valleys are **tarns** or **tarn lakes** (rock-basin lakes). **Hanging valleys** form where tributary glaciers once joined the trunk glacier, and waterfalls cascade from the hanging valleys to the main valley floor.

A narrow, rugged divide between two parallel glacial valleys is an **arête,** and the divide between the headward regions of oppositely sloping glacial valleys is a **serrate divide.** A pyramid-shaped mountain peak near the heads of valley glaciers or glaciated valleys is a **horn,** the most famous of which is the Matterhorn near Zermatt, Switzerland.

An alpine glacier erodes the floor and walls of the valley through which it flows. Furthermore, it functions as a conveyor belt that transports debris from the valley floor and walls to the glacier terminus where it is deposited. Debris carried and deposited directly by a glacier is **till,** an unsorted mixture of particles ranging in size from clay to boulders.

Till accumulates in various topographic forms called **moraines.** Debris eroded from the walls of an alpine glacier is transported along the margin of the glacier as a **lateral moraine.** The joining of two lateral moraines at the confluence of a tributary glacier and the main glacier forms a **medial moraine** (fig. 4.28). Both lateral and medial moraines ultimately reach the glacier terminus or snout to form an **end moraine.**

Moraines are identified on photos of glaciers as dark bands in the ablation zone (see fig. 4.26). Moraines are not visible in the accumulation zone because they are covered by the perennial snow that exists there. Lateral and medial moraines define the flow lines of an alpine glacier and cannot cross each other, but they may converge toward each other near the glacier terminus. An end moraine can remain long after the terminus where it was formed has retreated. Successive end moraines lying beyond the snout of a retreating glacier are called **recessional moraines.** Moraines left by former alpine glaciers are rarely visible on standard topographic maps because the contour interval is usually larger than the height of most moraines.

Glaciers that terminate in a lake or the ocean produce **icebergs,** masses of glacier ice that become detached from the glacier terminus, a process called **calving.** Icebergs float freely but can become grounded when they are carried to shallow water by wind and currents. Icebergs eventually melt.

EXERCISE 16A

Name

Section Date

Mass Balance of an Alpine Glacier

Figures 4.29A–C show three aerial photos of the South Cascade Glacier in the state of Washington. Photo A was taken on September 27, 1960, photo B was taken in early fall 1983, and photo C was taken on September 24, 2006. A comparison of the features shown in these three photographs reveals much about the recent history of this glacier. A longitudinal profile along the centerline of the glacier is shown in figure 4.30.

1. On each of the photographs, draw a dashed pencil line showing the boundary between the accumulation and ablation zones (ignore small patches of snow surrounded by bare ice). Label the line on photo A "equilibrium line 1960," the one on photo B, "equilibrium line 1983," and the one on photo C, "equilibrium line 2006."
2. During the time that has elapsed between the three dates of the photographs, has the equilibrium line generally remained stationary, moved to a lower elevation, or moved to a higher elevation?

3. In comparing the three photographs, what is the evidence that the glacier has thinned in the ablation zone during the period 1960 to 2006?

4. Why are there no icebergs in the lake on photo B?

5. The annual net mass balances of the South Cascade Glacier for each of the years 1959–2009 are presented in graphic form in figure 4.31, and the actual values for annual net accumulation or net ablation, expressed in meters of water equivalent over the entire glacier surface are given in table 4.2. Determine the algebraic sum of the net balance values in table 4.2 and use the result to refute or verify the evidence of glacial retreat based on the visual inspection of the three photographs.

6. Based on the data in table 4.2, in which year did the South Cascade Glacier come closest to being in equilibrium—that is, when was the net mass balance almost zero?

7. In figure 4.31 and table 4.2 you have available a 50-year record of the mass balance of South Cascade Glacier. If the trend of that record continues, what is the probable future of the glacier?

Figure 4.29A
Oblique aerial photographs of the South Cascade Glacier, Washington, September 27, 1960; Neg. No. FR6025-50.

Figure 4.29B

Oblique aerial photograph of the South Cascade Glacier, Washington, October 10, 1983; Neg. No. 83R1-188.

Photo: Courtesy of Andrew G. Fountain, U.S. Geological Survey, Tacoma, Washington.

Figure 4.29C

Oblique aerial photograph of South Cascade Glacier, Washington, September 24, 2006.

© James Carter

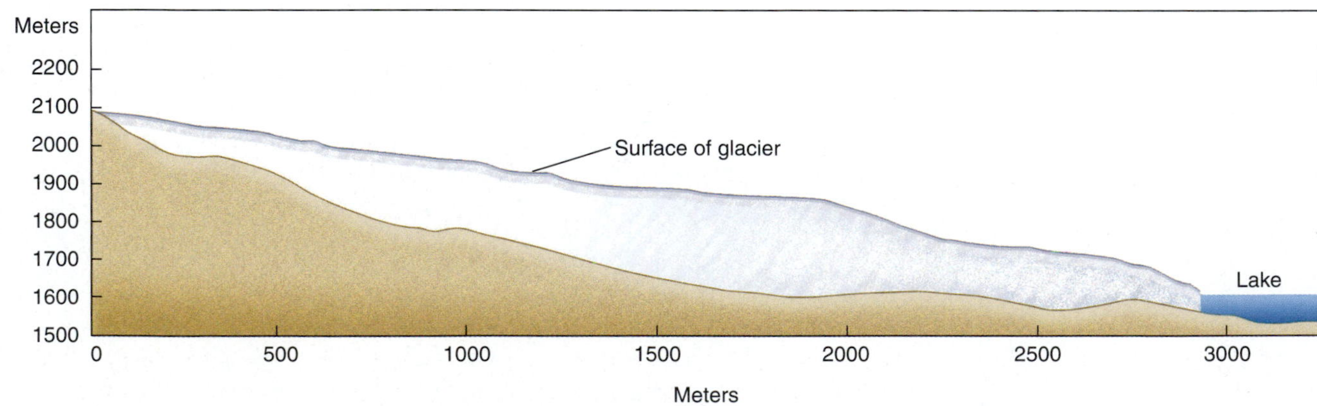

Figure 4.30

Longitudinal profile of the South Cascade Glacier during the 1965–66 budget year. No vertical exaggeration.

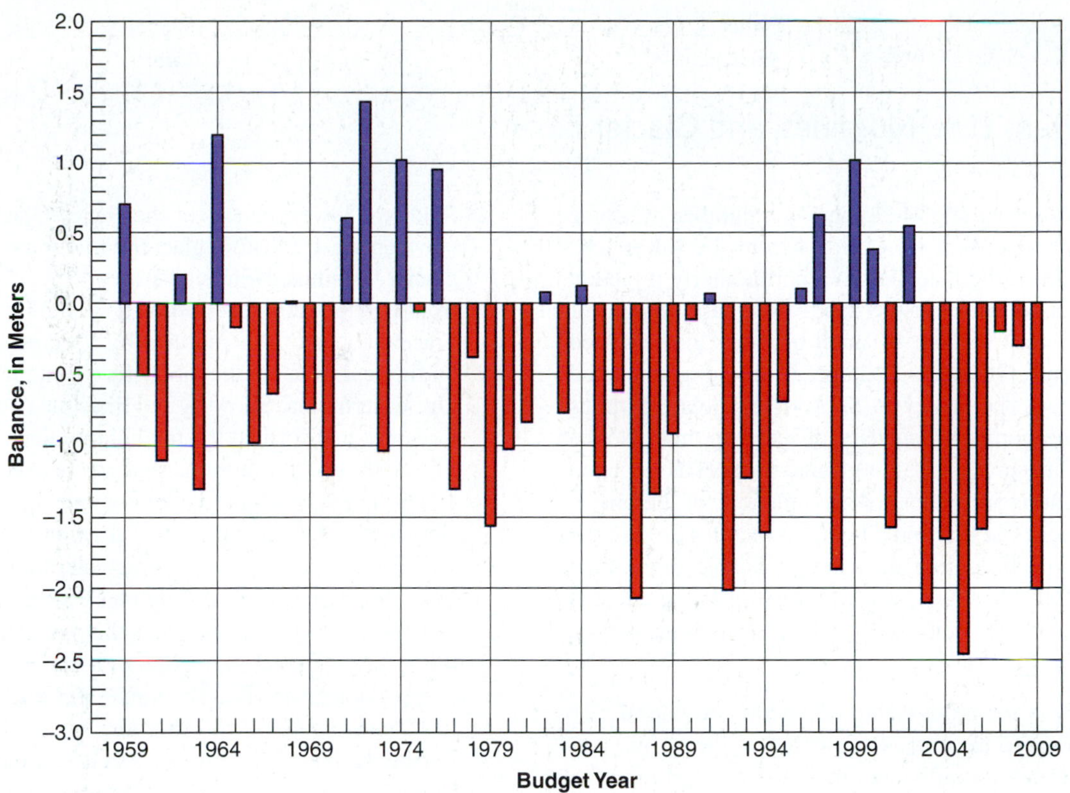

Figure 4.31
South Cascade Glacier balance, 1959–2009, based on data from table 4.2.

Table 4.2 Annual Net Mass Balances for the South Cascade Glacier, Washington, for the Budget Years 1959 to 2009

Budget Year	Net Balance*	Budget Year	Net Balance*	Budget Year	Net Balance*
1959	+0.70	1976	+0.95	1993	−1.23
1960	−0.50	1977	−1.31	1994	−1.60
1961	−1.10	1978	−0.38	1995	−0.69
1962	+0.20	1979	−1.56	1996	+0.10
1963	−1.30	1980	−1.02	1997	+0.63
1964	+1.20	1981	−0.84	1998	−1.86
1965	−0.17	1982	+0.08	1999	+1.02
1966	−0.98	1983	−0.77	2000	+0.38
1967	−0.63	1984	+0.12	2001	−1.57
1968	+0.02	1985	−1.20	2002	+0.55
1969	−0.73	1986	−0.61	2003	−2.10
1970	−1.20	1987	−2.06	2004	−1.65
1971	+0.59	1988	−1.34	2005	−2.45
1972	+1.47	1989	−0.91	2006	−1.58**
1973	−1.03	1990	−0.11	2007	−0.20
1974	+1.02	1991	+0.07	2008	−0.30
1975	−0.05	1992	−2.01	2009	−2.00

*Net balance values expressed as meters of water equivalent spread over the entire glacier.
**NOTE: Values for 2006 to 2009 are provisional and subject to revision.

EXERCISE 16B

Name

Section Date

Equilibrium Line, Moraines, and Glacier Flow

The Cordova map (fig. 4.32) shows a number of alpine glaciers. The Heney Glacier flows from the lower left-hand margin of the map area in a northeasterly direction until it ends in a lake at the upper right-hand margin of the map area. A number of small tributary glaciers feed into the Heney Glacier, and a large, 3-mile-long, unnamed tributary joins the Heney in Section 15 about 3.5 miles south of the northern boundary of the map area. Notice that the configuration of the glacier surfaces is depicted by blue contour lines that are continuations of the brown contour lines. The C.I. and all other principles used in the interpretation of contour lines are applicable to the blue contours. (NOTE: The pattern of brown stippling on parts of the glaciers represents morainal material on the glacier surface.)

1. Based on the relationship of medial and lateral moraines to the annual snow line on the Snow Glacier photograph (see fig. 4.26) and assuming the Cordova map is based on glacier conditions at the *end* of the ablation season, determine the *maximum* elevation of the snow lines on the Heney Glacier and the McCune Glacier, and describe the reasoning used in arriving at your answer. (Assume the equilibrium line for each glacier is more or less coincident with a contour line across the glacier surface.)

___

___

___

___

___

___

___

___

___

2. Study the lower reaches of the Heney Glacier between the 1,700-foot glacier contour and the glacier terminus. Notice how the glacier contours from 800 to 1,200 feet outline a medial moraine. Trace the axis of this moraine by a red pencil line on figure 4.32 from the 1,600-foot contour to the glacier terminus. Show by red lines the two lateral moraines that formed the medial moraine.
3. Locate the *medial* moraine that lies between the 1,500- and 2,500-foot glacier contours on the west side of the McCune Glacier. Trace this moraine with a black pencil line up the glacier to its logical point of origin. Show also by black pencil the lower reaches of the lateral moraines that formed it. Do the same for the larger medial moraine on the center of the glacier. (NOTE: The name *McCune Glacier* lies almost on top of this moraine.)
4. Locate the point in the upper reaches of the Heney Glacier where the 4,500-foot contour line intersects the narrow rock outcrop south of the word *crevasses.* Assume that a large boulder becomes dislodged from the eastern edge of this rock island and falls onto the glacier surface. Using a blue pencil, draw a line on figure 4.32 from the point where this boulder begins its journey down-glacier to the point where it reaches the glacier terminus. (This line is a *flow line* and cannot cross a medial moraine.) If the average velocity of the glacier is 2 feet per day, and assuming the glacier terminus remains fixed for the entire period, how long will it take for the glacier to transport the boulder along the route defined by your pencil line?

___

___

___

___

___

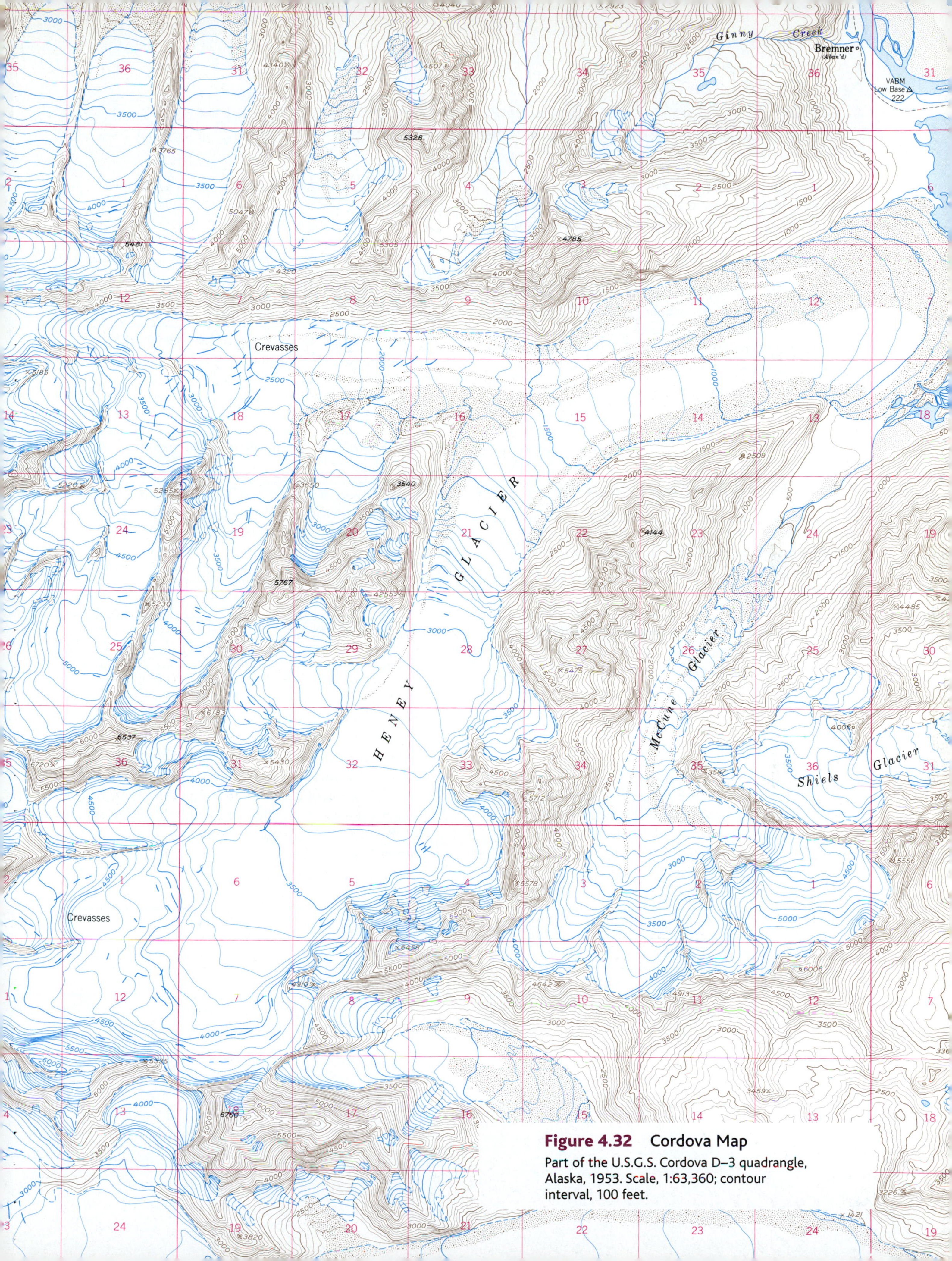

Figure 4.32 Cordova Map

Part of the U.S.G.S. Cordova D–3 quadrangle, Alaska, 1953. Scale, 1:63,360; contour interval, 100 feet.

EXERCISE 16C

Name

Section Date

Erosional Landforms Produced by Former Alpine Glaciers

The area shown in the Matterhorn Peak map (fig. 4.33) was once occupied by a system of valley glaciers. The valleys produced by these glaciers now contain streams whose courses more or less coincide with the longitudinal axes of the former glaciers.

1. Trace the existing drainage lines in blue pencil. How do the valleys containing these streams differ in cross section from the stream valleys south of the Mogollon Rim on the Promontory Butte map of figure 4.8?

2. Sawtooth Ridge on the Matterhorn map is the drainage divide between south-flowing streams such as Spiller Creek and streams flowing north to Robinson Creek.
   (a) Visualize a topographic profile along the dashed line that defines the crest of Sawtooth Ridge. Does this profile lie more or less parallel to the contour lines or does it generally cut across them?

   (b) Visualize a profile along the dashed line that approximates the divide on the Promontory Butte map of figure 4.8. Does this profile more or less parallel the contour lines or does it generally cut across them?

   (c) Why do the two profiles on the two map areas differ?

3. Refer to the paragraphs on page 142. Assign an appropriate name to each of the following topographic features shown on the Matterhorn Peak map:
   (a) Stanton Peak (SE quadrant of the map).

   (b) Spiller Lake (SE quadrant of the map).

   (c) The part of Horse Creek Valley lying up-valley from the waterfalls at about the 7,600-foot contour line (NE quadrant of the map).

   (d) The Cleaver (central part of the map).

   (e) Sawtooth Ridge.

4. The topography of the Matterhorn Peak map area is dominated by landforms produced by glacial erosion. It is not unreasonable to assume that during the time when the former valley glaciers were receding, recessional moraines might have been formed. Remnants of an end moraine are represented by the 7,120-foot contour line north of the Twin Lakes Campground in the extreme northeast corner of the map. With this one exception, no other recessional moraines are in evidence on the entire map. Suggest a reason for this.

Figure 4.33 Matterhorn Peak Map

Part of the U.S.G.S. Matterhorn Peak quadrangle, California, 1956. Scale, 1:62,500; contour interval, 80 feet.

Continental Glaciation

Bodies of glacier ice covering most of Greenland and Antarctica today are of continental proportions and are called **continental glaciers** or **ice sheets.** At times during the last 1.8 million years of the **Pleistocene epoch** or **Ice Age,** continental glaciers covered much of North America, northern Europe, western Russia, and Antarctica. The evidence of the past existence of these glaciers is the eroded areas of bedrock such as that shown in figure 2.13 and the myriad of depositional deposits and landforms formed by them.

Depositional landforms of continental glaciers are recognized on topographic maps or stereopairs, but unlike the landforms produced by alpine glaciers, the full extent of which can be shown on a single topographic quadrangle, only a part of the landscape produced by a continental glacier can be covered on a standard topographic map. Nevertheless, contour maps with a scale of 1:24,000 to 1:62,500 and a contour interval of 25 to 50 feet adequately display the vestiges of continental glaciation.

One of the major objects in the study of continental glacial landforms is to reconstruct the configuration of various lobes of the Pleistocene ice sheets. A **glacier lobe** is a lobate appendage of the main ice sheet. These lobes range in breadth from less than 50 miles to more than 100 miles. By tracing end moraines over hundreds of miles, geologists can map the former extent of the ice margin of a single lobe as it existed during the waning phases of the Ice Age.

End moraines are belts of hummocky terrain a few miles in width that consist of till and intermixed deposits of stratified sand and gravel. While a glacial lobe occupied a position indicated by an end moraine, streams flowing from the glacier terminus deposited sand and gravel beyond the front of the glacier as an **outwash plain.** Closed depressions lying in end moraines and on outwash plains are sites of former stagnant ice masses that became separated from the retreating glacier. When these ice masses melted, they left in their place depressions called **ice block pits.** An outwash plain with many pits, some of which contain lakes, is called a **pitted outwash plain.** An ice block pit in an end moraine is called a **kettle.**

A stream flowing in a stagnant ice channel deposits sand and gravel along the ice walls of the channel. When the ice melts, the stratified deposits remain in a variety of topographic forms called **ice-contact deposits.** One form of ice-contact deposit is a **kame**—a knob, hummock, or conical hill composed of sand and gravel. A chaotic assemblage of kames and kettles common to many end moraines is called **knob and kettle topography.**

Other products of continental glaciation are eskers and drumlins. An **esker** is an ice-contact deposit in the form of a ridge of sand and gravel believed to have been formed beneath the glacier surface by a sediment-laden stream flowing through an ice tunnel. A **drumlin** is an elongate and streamlined hill whose long axis is parallel to the direction of ice movement. The steep side of one of the elongate ends of a drumlin lies toward the direction from which the ice was flowing, and the more gentle side of the other end of the drumlin axis lies in the downstream direction of ice flow. Drumlins are molded at the base of the ice sheet and commonly occur in swarms called **drumlin fields.** Drumlins may consist of molded till but recent studies suggest that some are rock cored while others may result from the effects of subglacial meltwaters or simply by the erosion of the existing landscape as the ice advances.

NOTES

EXERCISE 17A

Name

Section Date

Moraines and Outwash Plains

The topography shown on the Whitewater map (fig. 4.36) was produced by the last advance and retreat of the continental ice sheet. Three general terrain types are present: (1) ground moraine, (2) end moraine, and (3) pitted outwash plain. Because the area covered by the map is so small compared to the total area covered by the continental glacier, the relationship of the Kettle moraine (an end moraine) to the position of the ice front at the time the moraine was built is not immediately clear from a casual inspection of the map.

Let us assume that the Kettle moraine on the Whitewater map is part of an end moraine that was deposited by the "Kettle lobe" as shown in the sketch map of figure 4.34. On that sketch map, two possible locations of the Whitewater map are shown as A and B. Each shows the Kettle moraine trending across the map area from southwest to northeast, as it does on the Whitewater map. Beyond the end moraine of figure 4.34, outwash associated with the "Kettle lobe" is shown. In addition, we can assume that the area formerly underlain by the "Kettle lobe" will be covered with ground moraine.

Figure 4.34

Map of the hypothetical "Kettle lobe" showing two possible locations of the Whitewater map (fig. 4.36), A or B. (See Exercise 17A for instructions.)

1. Study the Whitewater map and the topographic profile of figure 4.35 and determine which of the two locations, A or B on figure 4.34, is the one that most likely represents the Whitewater map. State in concise terms the reasons for your choice.

   ____

   ____

   ____

   ____

   ____

2. Draw the ice-surface profile of the "Kettle lobe" on figure 4.35 as it may have been when the "Kettle lobe" was building the Kettle moraine. (In drawing the ice-surface profile of the "Kettle lobe," assume that the ice slopes upward from the highest point on the end moraine until it attains a thickness of about 500 feet in a horizontal distance of about 4 miles.)

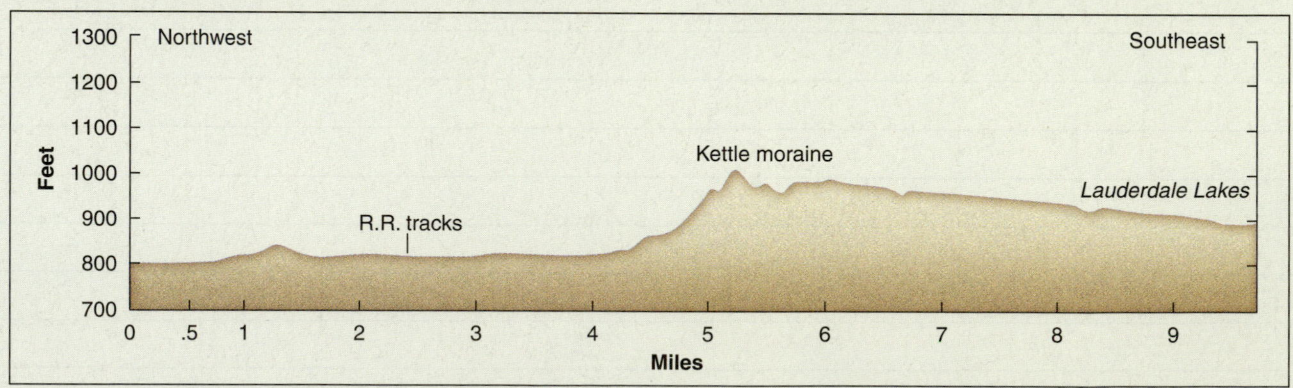

Figure 4.35

Generalized topographic profile from northwest to southeast across the Kettle moraine of the Whitewater map (fig. 4.36). (Vertical scale exaggerated.)

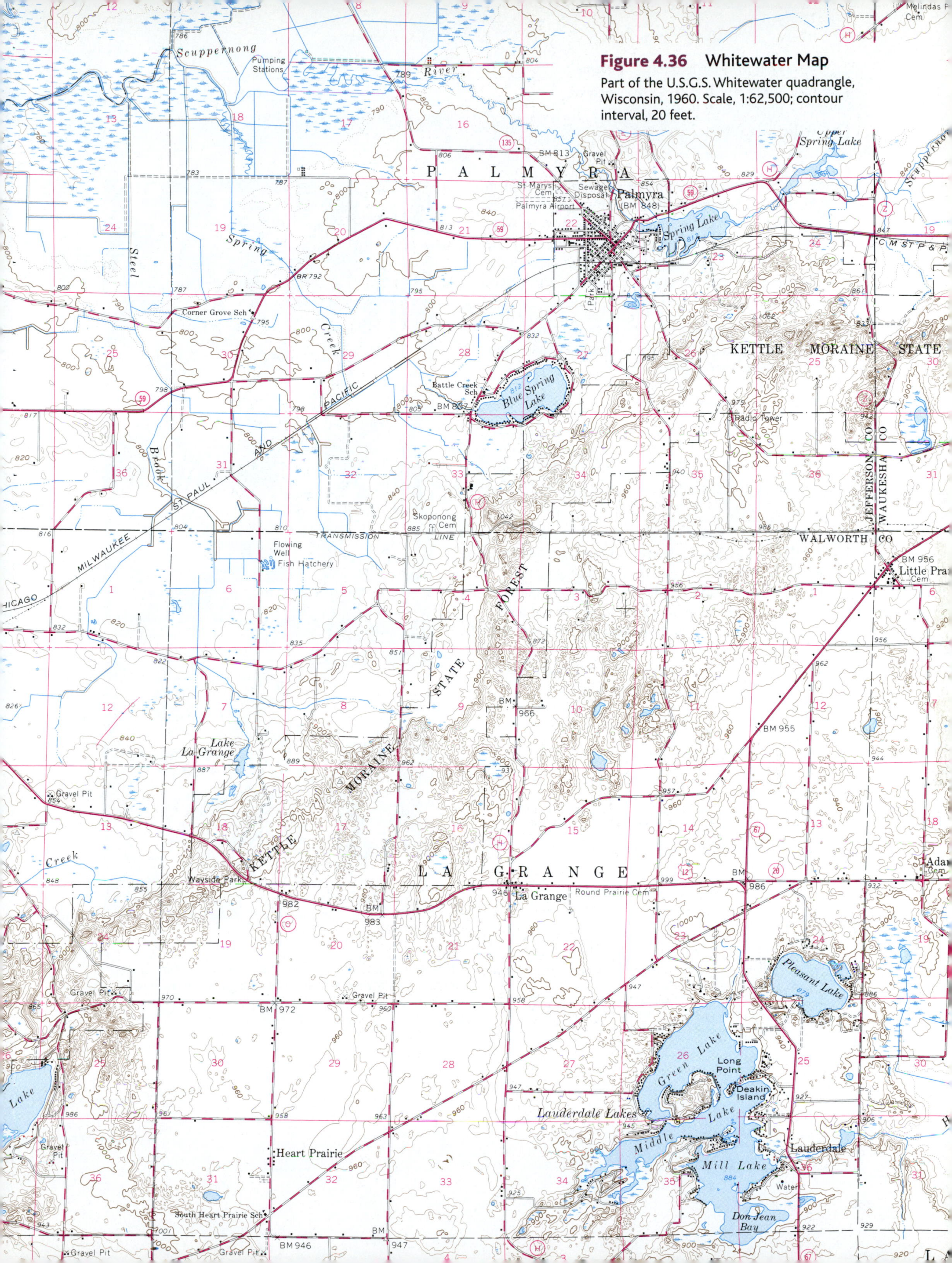

Figure 4.36 Whitewater Map

Part of the U.S.G.S. Whitewater quadrangle, Wisconsin, 1960. Scale, 1:62,500; contour interval, 20 feet.

EXERCISE 17B

Name

Section Date

Drumlins

Figure 4.37 is a stereopair of part of a drumlin field in New York state. The upper photo in the stereopair covers the area of the Sodus topographic map of figure 4.38. Mud Pond is marked with an X on the stereopair. Study the stereopair with your stereoscope in conjunction with the topographic map. (The steepness of slopes on the stereo-pair is exaggerated in stereovision.)

1. What was the general direction of flow of the continental glacier across the map area? What is the evidence to support your answer?

2. On figure 4.38 locate the main road that roughly parallels the western boundary of the map, Norris Road, and the north–south railroad tracks that pass through Zurich. On the upper photo of the stereopair, show these roads in red pencil and the railroad in black pencil. Use sharp pencils so as not to obscure your stereovision.
3. To what extent has the topography influenced the routes of the two roads and the railroad?

4. To what extent has stream erosion modified the topography of the area?

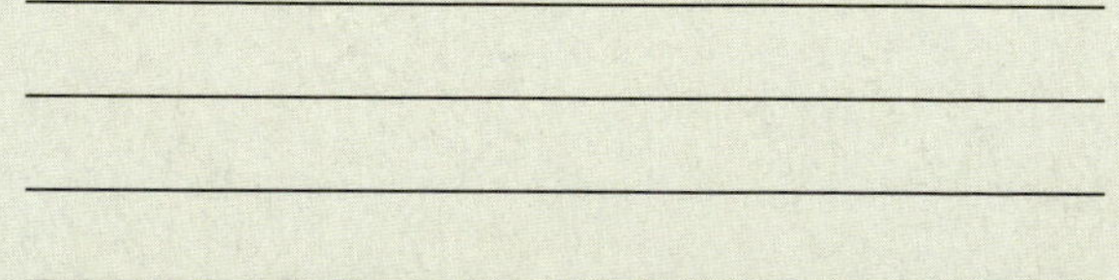

Figure 4.37

Stereopair of part of a drumlin field in New York state. The upper photo covers the same area as the topographic map in figure 4.38.

© James Carter

(continued)

EXERCISE 17B

Name

Section Date

Drumlins *(Continued)*

5. What geologic material is likely to predominate in the areas occupied by the drumlins?

____

____

6. A pair of power lines extends in an east–west direction near the southern margin of the Sodus map (fig. 4.38).
   (a) How is the route of the power lines identified on the stereopair?

   ____

   ____

   (b) Is the map route of the power lines controlled by the topography?

   ____

   ____

   ____

7. Draw the outline of Mud Pond on the upper photo. What has happened to Mud Pond during the 34 years that have elapsed between 1952, when the map was published, and 1986, when the photos were taken?

____

____

____

____

8. What is the R.F. of the stereopair? (See page 82 for discussion of R.F.)

____

____

____

____

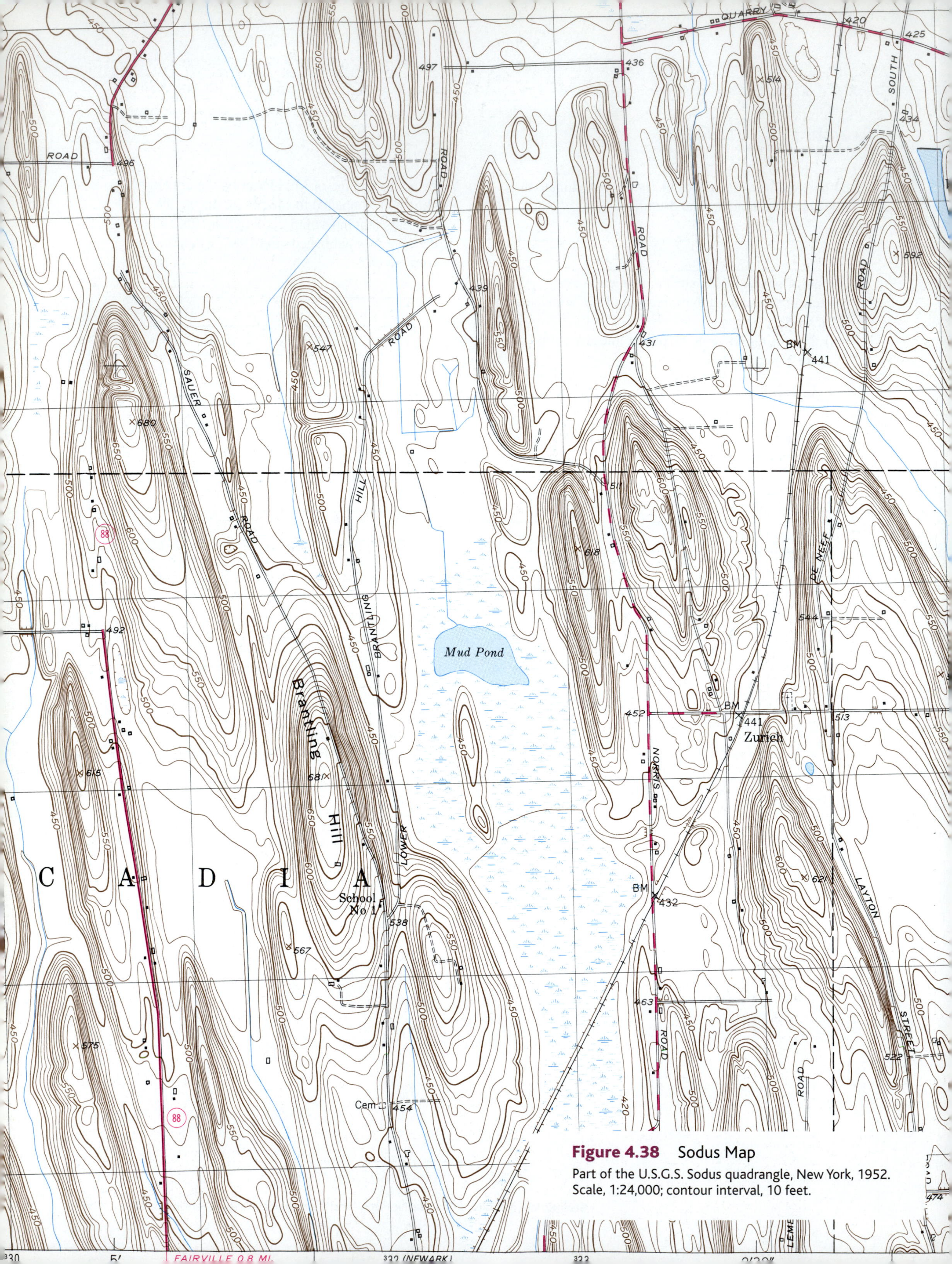

Figure 4.38 Sodus Map

Part of the U.S.G.S. Sodus quadrangle, New York, 1952. Scale, 1:24,000; contour interval, 10 feet.

NOTES

Name

Section Date

EXERCISE 17C

Ice-Contact Deposits

Figure 4.39 is a stereopair of an esker in Michigan.

1. Trace the crest of the esker in red pencil.
2. Visualize the esker crest in profile. Use a black pencil to draw a single, closed contour line showing the highest part of the esker crest. Draw the contour line on the right-hand photo while viewing the stereopair in stereovision.
3. Just east of the bend in the esker is a conical hill or knob. What is the name for this ice-contact deposit?
4. What constructional materials might be available from the esker and the ice-contact deposit?

Figure 4.40 shows two north–south trending eskers.

5. Trace the crest of each esker with a light-colored, felt-tip highlighting pen.
6. What local name is used for the eskers?
7. What map evidence verifies the assumption that eskers are ice-contact deposits?
8. The crest of the esker east of the Penobscot River does not lie at a constant elevation along its course. Given this observation, why is it necessary to invoke the presence of glacier ice to account for the origin of an esker?

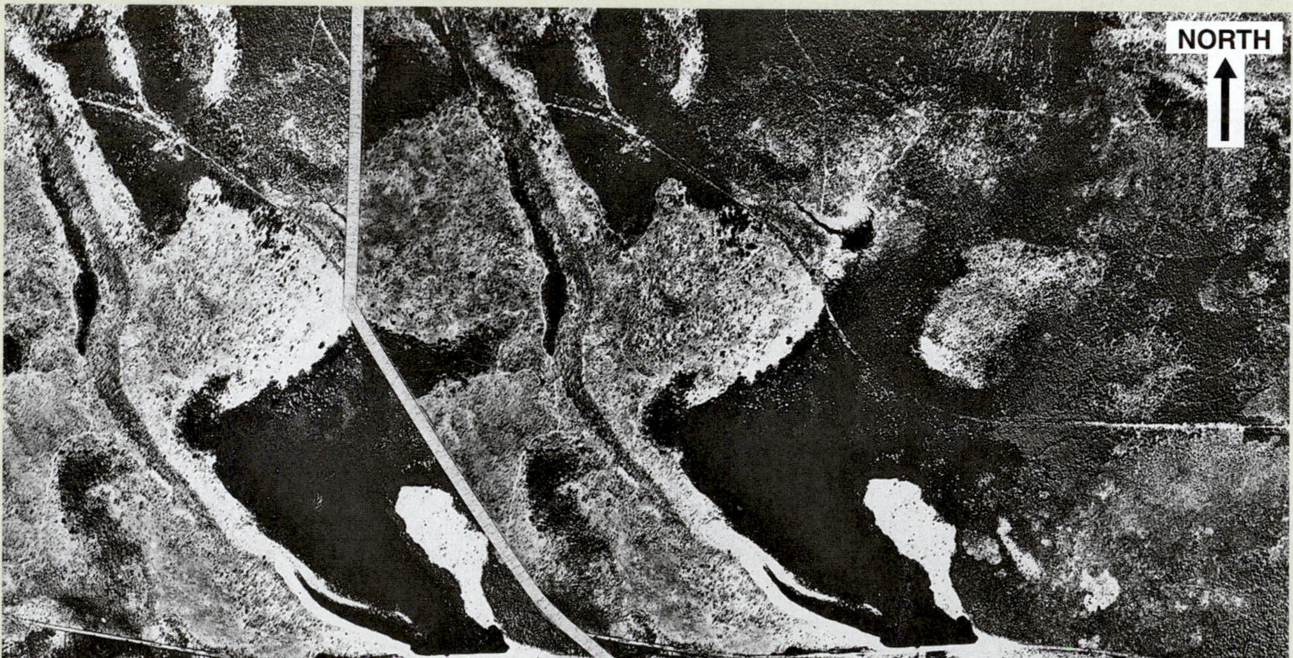

Figure 4.39

Stereopair of part of an esker in Michigan. Scale, 1:24,000.

Courtesy of U.S. Geological Survey.

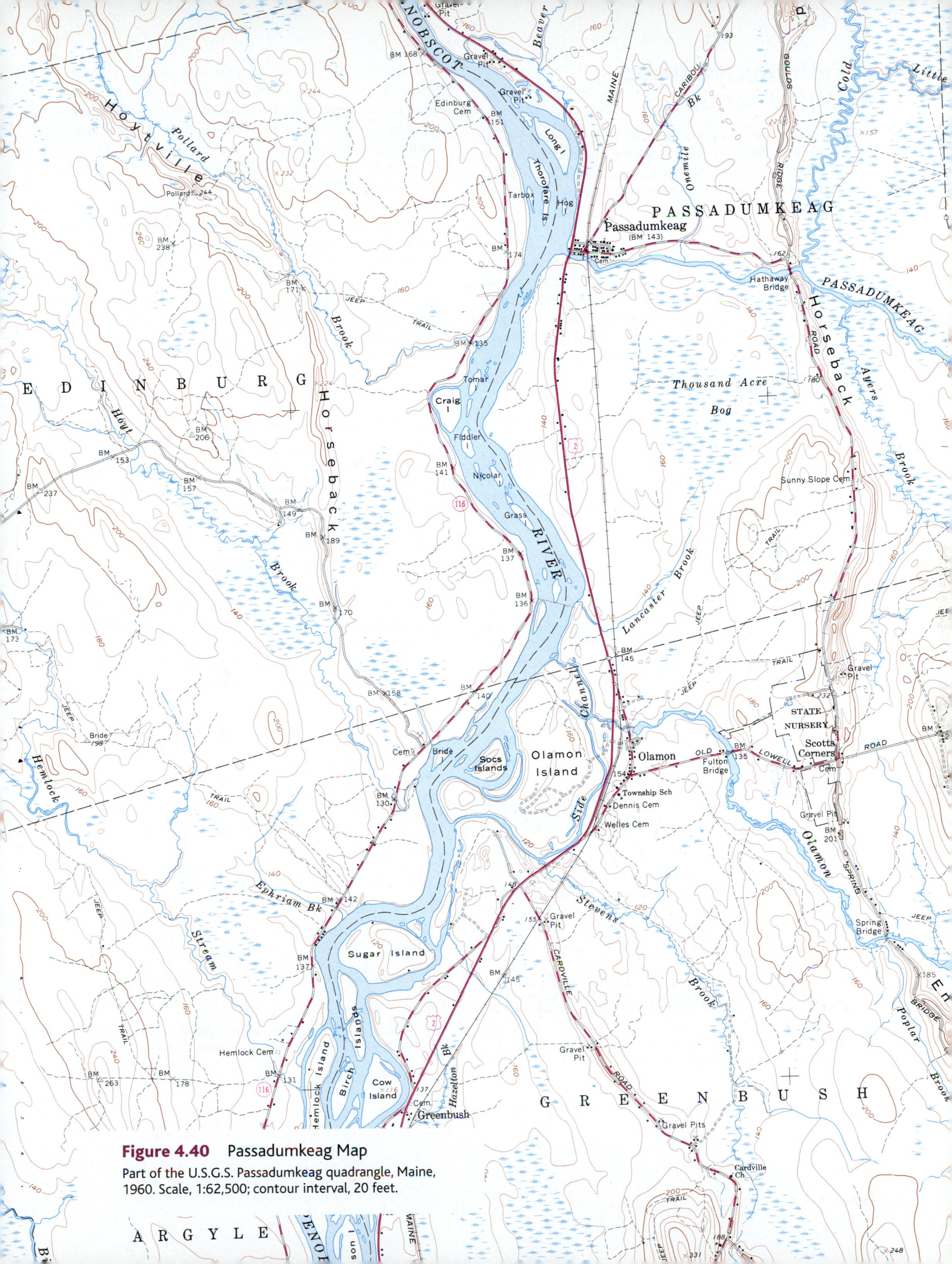

Figure 4.40 Passadumkeag Map

Part of the U.S.G.S. Passadumkeag quadrangle, Maine, 1960. Scale, 1:62,500; contour interval, 20 feet.

Landforms Produced by Wind Action

Wind is a geologic agent of erosion, transportation, and deposition. This section deals with landforms produced by wind erosion and wind deposition that are visible on standard topographic maps and aerial photographs or other images.

Blowouts

A **blowout** is a shallow depression caused by the localized removal of sand or smaller particles through the process of **deflation.** Deflation is the removal of loose surface material by wind that has a velocity sufficient to move sand or silt particles and transport them to another location. The movement of the sand grains occurs as surface **creep** or by **saltation** (bouncing) (fig. 4.41). Deflation cannot lower the ground surface below the water table.

Blowouts are irregular in map view, are formed where there is little or no vegetation, and commonly are defined by closed contour lines on a topographic map. If a blowout contains a shallow pond or lake, it can be inferred that the water table has risen after the blowout formed. A change to a more humid climate (i.e., increased rainfall) could cause the water table to rise.

Sand Dunes

A **sand dune** is a ridge or mound of sand deposited by wind. Winds that blow consistently from the same direction are called **prevailing winds.** Sand dunes formed by prevailing winds have a distinctive topographic profile. The side of a dune ridge facing the wind is called the **windward** side and is characteristically less steep than the slope on the **downwind** or **leeward** side. The leeward slope is referred to as the **slip face,** and its angle with the horizontal is about 30 degrees (fig. 4.42). The identification of the slip face of a dune allows one to infer the direction of the prevailing winds that formed it. Dunes that do not exhibit a slip face are formed by winds that blow in different directions at different times but with comparable speeds.

Dunes occur mainly in arid or semiarid regions with little or no vegetational cover. Dunes also occur along coastal regions where sand beaches provide a source of sand for the prevailing onshore winds. Dunes may occur in single isolated geometric forms, or they may exist in swarms called **dune fields.** The factors that control the geometric form of a dune are wind direction or directions and velocity, supply of sand, and the extent of vegetational cover.

Types of Sand Dunes

Sand dunes formed by prevailing winds occur in a variety of geometric shapes. Figure 4.43 shows four common forms that can be identified on maps or photos. A **barchan** is crescent-shaped in map view, and the horns of the crescent point downwind. A **transverse dune** is a linear ridge with its long dimension oriented at right angles to the prevailing wind direction. The slip face of a transverse dune provides the basis for determining the wind direction unambiguously. **Parabolic dunes** are characterized by horns that point upwind and are usually anchored by vegetation. They are common near beaches and may be associated with blowouts. The slip face is in the convex side of the dune. A **longitudinal dune (seif)** has a linear form with its long axis parallel to the wind direction.

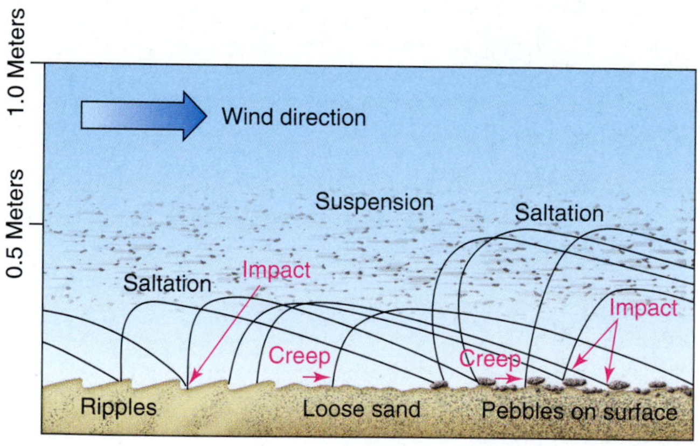

Figure 4.41

Movement ofwind-transported sediments by suspension, saltation, and creep.

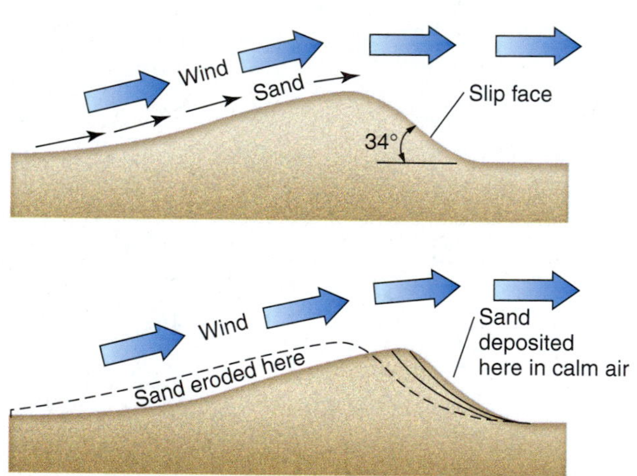

Figure 4.42

Movement of sand to form a dune. Note gentle windward slope and steeper slip slope at the angle of repose on the leeward side. The movement of the sand forms cross-beds (fore-set beds) on the slip face and results in the slow movement of the dune down-wind.

Barchans are common on flat desert surfaces with a sparse supply of sand and little or no vegetation. Transverse dunes occur where the sand supply is abundant and vegetation is sparse. Longitudinal dunes occur in deserts with strong winds that vary only slightly from a single direction. Some elongated sand ridges in coastal regions appear to be longitudinal dunes, but on close inspection, they are seen to be ridges of windblown sand anchored by vegetation and separated by wind-scoured troughs parallel to the dunes.

Active dunes are those that migrate in a down-wind direction or are constantly changing in shape in response to multiple wind directions at different times. **Inactive dunes** are those that have become stabilized by the growth of a vegetational cover to the extent that their migration ceases (remember Exercise 14A, page 126). Stabilized dunes are indicative of a climatic change to more humid conditions. When patches of vegetation on inactive dune fields die from drought or other causes, the exposed sand is deflated to form local blowouts.

Coastal dunes that invade an area occupied by stream valleys sloping toward the coast may block the valleys and cause ponding of the stream into small lakes or ponds. Such waterfilled depressions are not to be confused with blowouts.

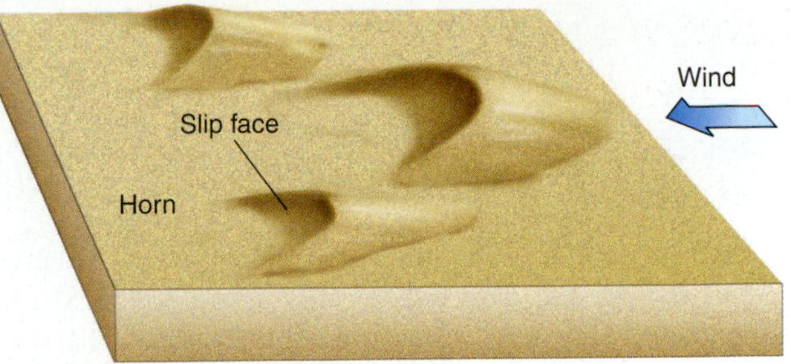

A. Barchans

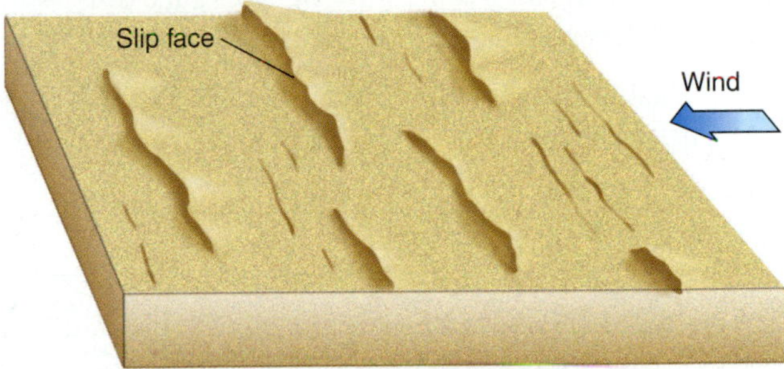

B. Transverse dunes

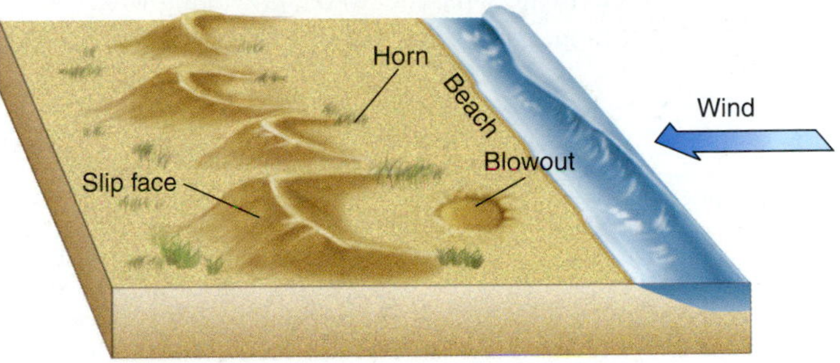

C. Parabolic dunes

D. Longitudinal dunes (seifs)

Figure 4.43

Types of sand dunes. (A) Barchans. (B) Transverse dunes. (C) Parabolic dunes. (D) Longitudinal dunes.

Name

Section Date

EXERCISE 18A

Barchans

The crescent-shaped dunes in figure 4.45 are active barchans on the desert floor west of the Salton Sea in southern California. Ground measurements show that they moved between 325 and 925 feet during a 7-year period. Study the stereopair of figure 4.45 and compare it with the topographic map of figure 4.44. Notice the difference in scales between the stereopair and the topographic map.

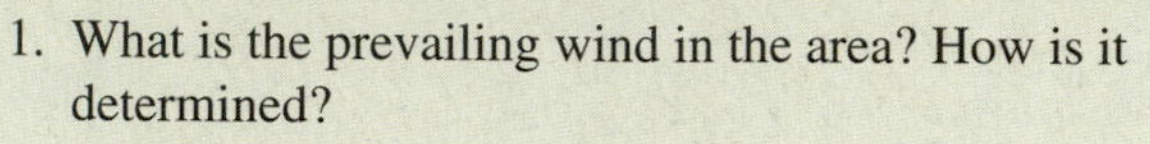

1. What is the prevailing wind in the area? How is it determined?

2. Assuming that an airstrip should be aligned approximately parallel to the prevailing wind direction, is the airstrip properly aligned?

3. The stereopair reveals many more sand dunes than the number visible on the topographic map. Account for this difference.

4. Assuming the dunes will continue their direction and rate of migration within the limits already stated, what will be the shortest and longest times, in years, for the barchan about a mile from the west end of the landing strip to reach the landing strip?

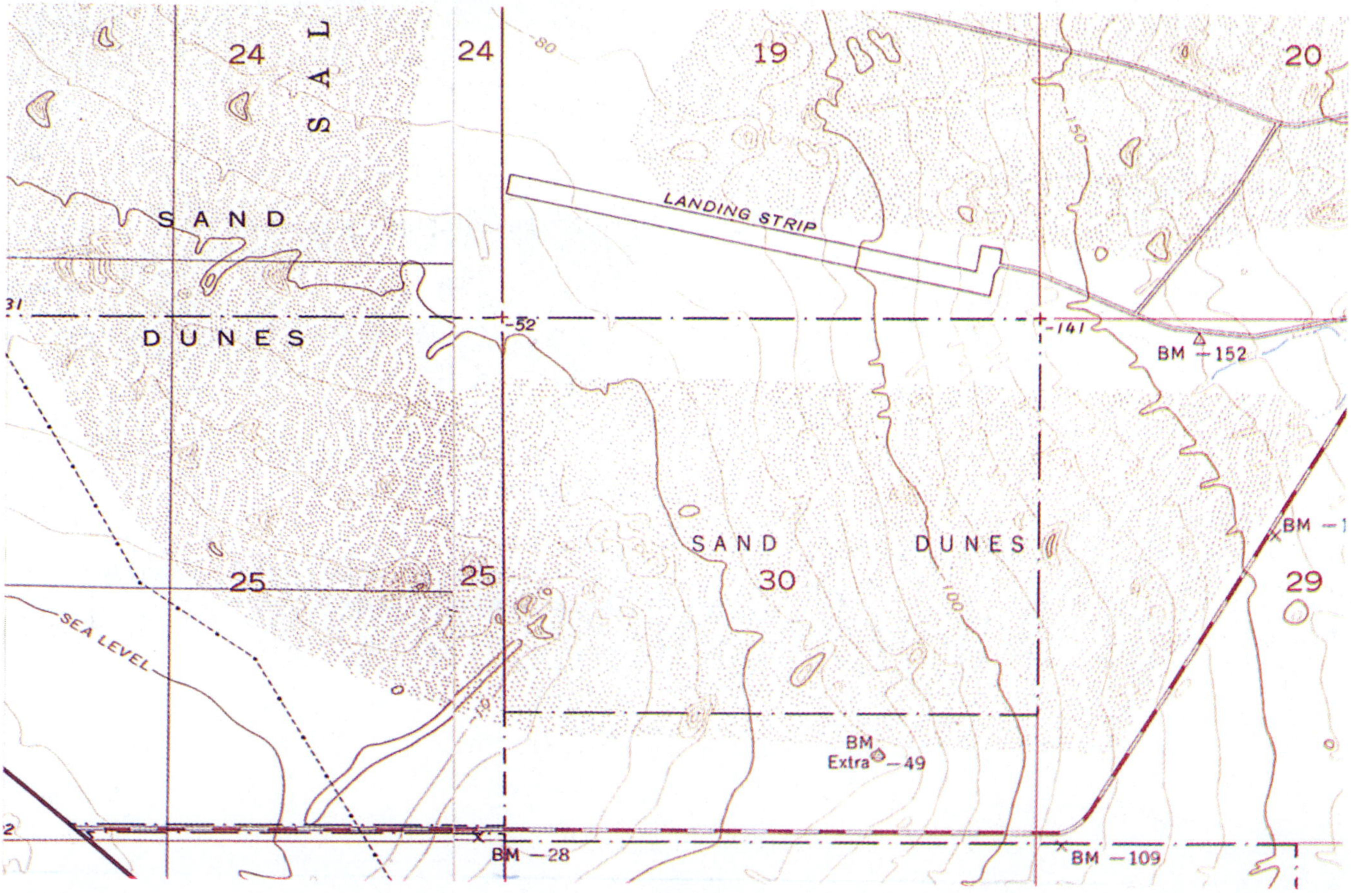

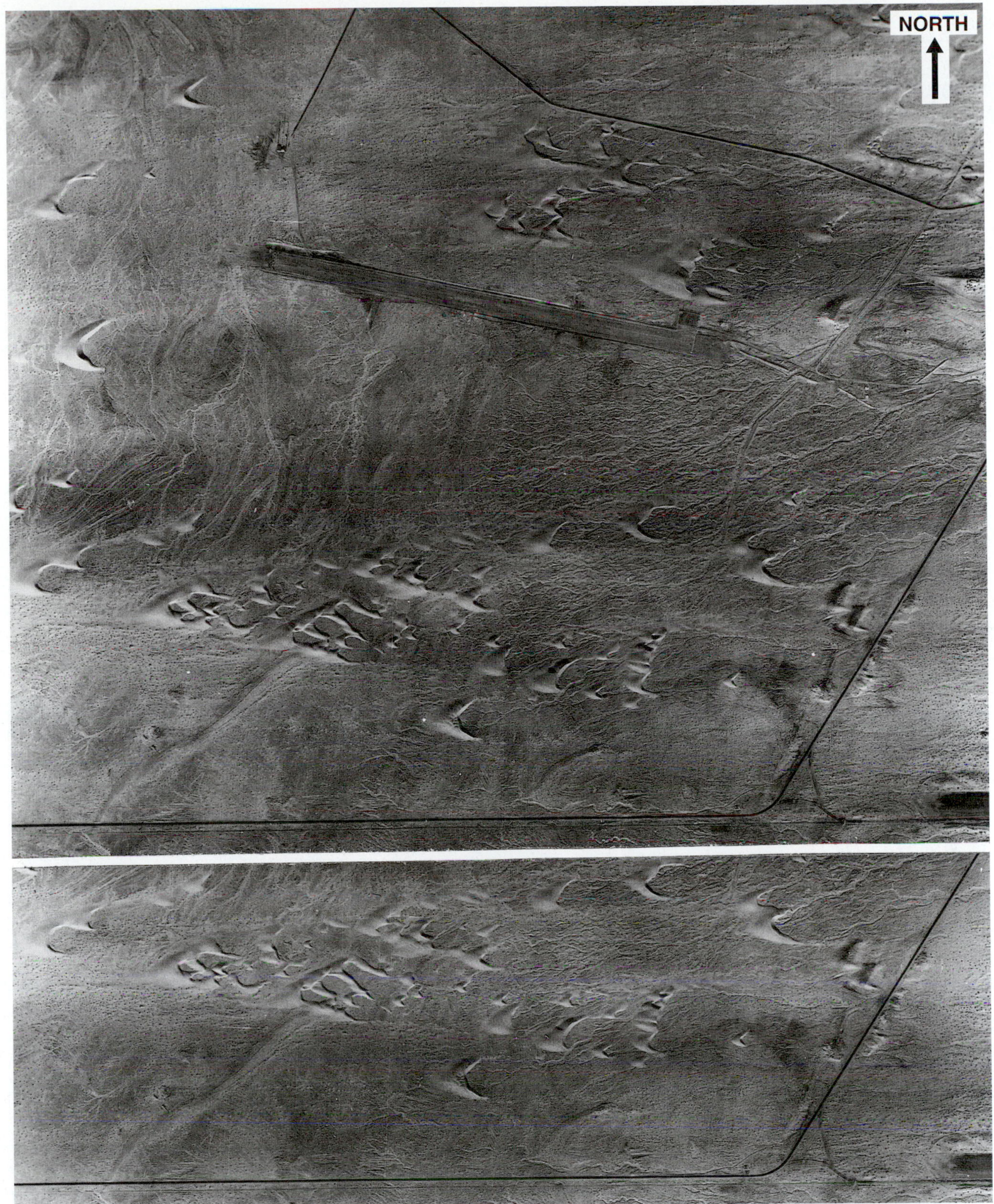

▲ **Figure 4.45**

Stereopair of Kane Spring sand dunes, California. Scale, 1:20,000; taken November 10, 1959.

◀ **Figure 4.44** Kane Spring Map

Topographic map of parts of the U.S.G.S. Kane Spring NE and Kane Spring NW quadrangles, California, 1956. Scale, 1:24,000; contour interval, 10 feet.

EXERCISE 18B

Name ______________________

Section ______________ Date ______________

Coastal Dunes

Figure 4.47 is a topographic map of coastal dunes in southern California north of Santa Barbara. Figure 4.46 is a stereopair of the southern two-thirds of the map area. Study the stereopair under stereovision and identify features on it that are labeled on the map. When you first view the stereopair under stereovision, you may see the sand dunes in *inverse topography.* In inverse topography, the hills appear as depressions, and the steep slopes appear to be facing in the direction opposite to their true orientation. In the stereopair of figure 4.46, the steep slopes on the sand dunes face *away* from the coast in true relief. You may have to "coax your mind" to bring the relief into true perspective. By turning the stereopair upside down, you may find it easier to bring the true relief into view.

1. What kind of dunes are those with their ridges lying parallel to the coast?

____

____

____

2. The narrow linear ridges next to the coast have their axes aligned at an angle of about 70 degrees to the coastline. Dark lines of vegetation are associated with the ridges. Do the vegetational lines occur generally on the tops of the ridges or between them? What is the likely origin of the ridges? Are they longitudinal dunes?

____

____

____

____

____

3. Draw an arrow in black pencil on figure 4.47 showing the inferred direction of prevailing winds.

Retain this page for use in Ex. 19A.

(continued)

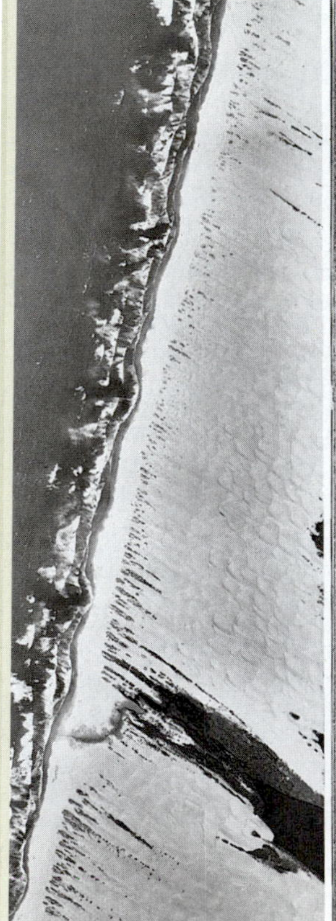

Figure 4.46

Stereopair of sand dunes on the California Coast south of San Luis Obispo. Scale, 1:28,800. (Photos taken June 27, 1964. Series GS-VASJ, numbers 1–262 and 1–265.) This stereopair will be used again in Exercise 19A. Please keep it to complete that exercise.

Figure 4.47 Oceano Map

Part of the U.S.G.S. Oceano quadrangle, California, 1965. Scale, 1:24,000; contour interval, 20 feet.

EXERCISE 18B

Name

Section Date

Coastal Dunes *(Continued)*

4. What is the source of the sand contained in the dunes?

5. What is the likely origin of Little Oso Flaco Lake and Mud Lake?

EXERCISE 18C

Name

Section Date

Inactive Dune Fields

Figure 4.48 is a Landsat image similar to a black-and-white aerial photograph taken from an altitude of 570 miles over western Nebraska in the winter of 1973. This area is known as the Sand Hills because the topography consists of a massive dune field that is now virtually inactive (stabilized) because of a lush vegetational cover of prairie grasses. The dunes were formed during a time when the rainfall was less than at present and unable to support the grasses that grow there now. A heavy cover of snow enhances the topography.

The relief of the Landsat may appear inverted. If the dune ridges appear to be depressions, turn the photo upside down so that the north arrow points toward you, and the true relief will "pop" into view.

The rectangular box in the northwestern corner of the map is the area covered by the Ashby map of figure 4.16. The two dark bands in the lower part of the image are the North and South Platte Rivers, which converge toward the town of North Platte, Nebraska, near the eastern margin of the image.

1. What is the R.F. of the Landsat image?

2. What are the only features visible on both the Landsat image and the Ashby map?

3. Some of the interdune depressions visible on the Ashby map contain lakes. Assuming that some of the interdune depressions were sites of deflation during the period of dune formation, what does the presence of the lakes imply about the elevation of the water table when the dunes were active?

4. The topography of the dunes on the Ashby map reveals that many of them are irregular in map view and contain no clear-cut distinction between gentle and steep slopes. However, those that are elongated in a general east–west direction are typical of those visible on the Landsat image. Study the contour lines on the north and south sides of the east–west trending dunes on the Ashby map, determine which slopes are the steepest, and infer the direction of the prevailing winds at the time of deposition.

5. Based on the answer to question 4, draw arrows about 1/4 inch long at several points on the Landsat image to show the direction of the prevailing winds when the dune field was formed.

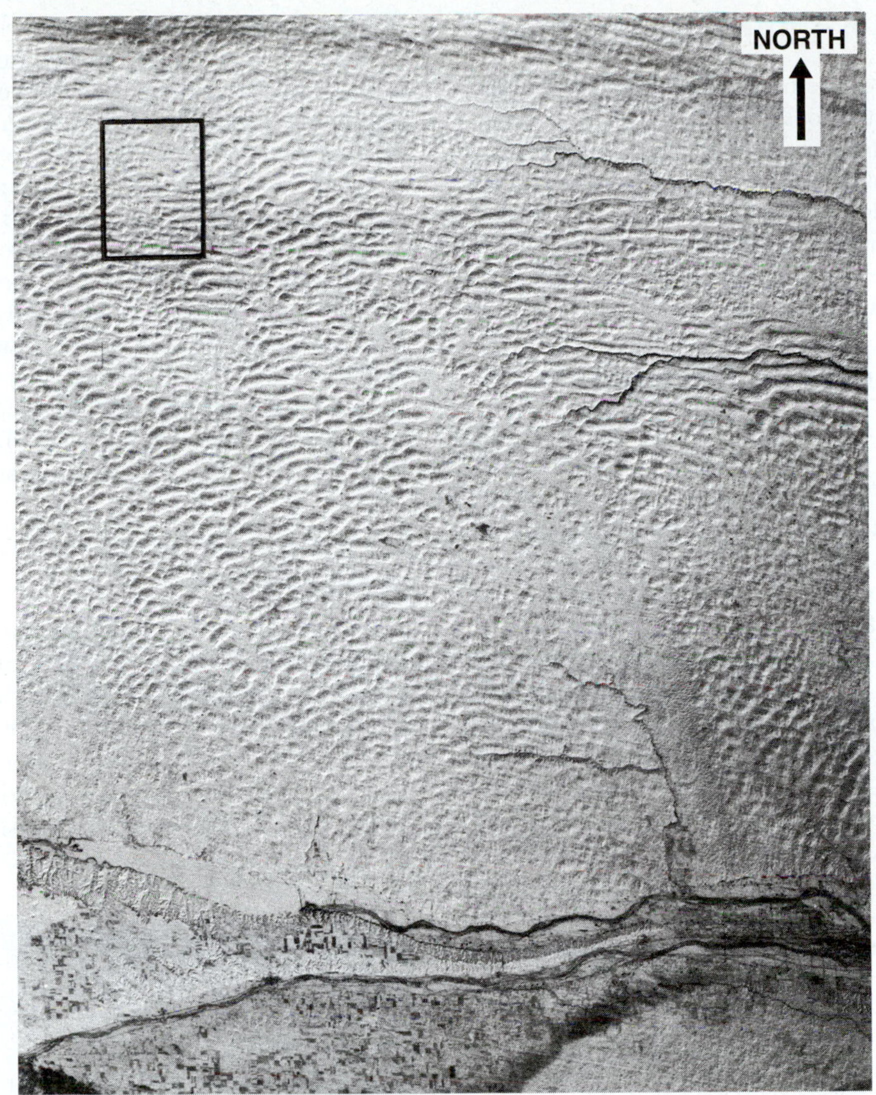

Figure 4.48

Landsat image of part of western Nebraska made on January 9, 1973, from ERTS-1 at an altitude of 570 miles. The rectangle in the northwest corner is the area covered by the Ashby map of figure 4.16.

NASA ERTS E-1170–17020, Courtesy of NASA and the U.S.G.S. EROS Data Center, Sioux Falls, South Dakota 57198.

Modern and Ancient Shorelines

The interaction between the ocean or a large lake and the land occurs at the coastline or shoreline. Waves generated by winds eventually reach the shore, where the wave energy is dissipated. Waves breaking along a shore are geologic agents of erosion, transportation, and deposition. This section deals with the modern and ancient landforms produced by one or a combination of these agents as they are portrayed on topographic maps and aerial photographs.

Depositional Landforms Produced by Wave Action

A wave crest moving toward shore is bent or **refracted** as it approaches shallow water (fig. 4.49). As the successive incoming waves break just off shore, a **longshore current** is initiated (fig. 4.50). A wave breaking on the shore carries sand particles up the beach in the direction of wave movement. When the water from the spent wave flows back down the beach, it follows a course controlled by the slope of the beach. Sand particles exposed to this alternate movement are moved along the beach, a process known as **longshore drift** (fig. 4.50). The longshore current moves fine sand, and longshore drift moves coarser sand particles parallel to the shoreline. When the longshore current slows due to deeper water associated with an indentation on the coast, such as a bay or estuary, sand is deposited in the form of a **spit.** A spit may grow across the mouth of a bay to form a **baymouth bar** (fig. 4.51). A spit that is curved shoreward is a **recurved spit.**

Other sandy features along the coastline include **beaches** and **barrier islands.** Indentations along the coastline that become isolated from the main body of water become lakes or lagoons. **Lagoons** are shallow-water bodies lying between the main shoreline and an off-shore barrier. A **barrier island** is an elongate sand island parallel to the shoreline. A break or passageway through a bar or barrier island is a **tidal inlet,** so called because it allows currents to flow into the lagoon during rising tides and out of the lagoon during falling tides. Lagoons and coastal lakes eventually become filled with sediment and are transformed into **mud flats, marshes,** and **wetlands.**

Beach sand is attacked by onshore winds that carry the sand inland to form coastal dunes. The sand lost to wind action is replenished by wave action and longshore drift.

Erosional Landforms Produced by Wave Action

An initial shoreline consists of bays and headlands. A **headland** is a part of the coast that juts out into the lake or ocean. Wave refraction concentrates the energy of waves against headlands (see fig. 4.49). The landforms produced by this intensified wave action are **wave-cut cliffs,** seawardfacing escarpments formed by wave erosion at their bases, and **stacks,** rocky pillars that are remnants of retreating wave-cut

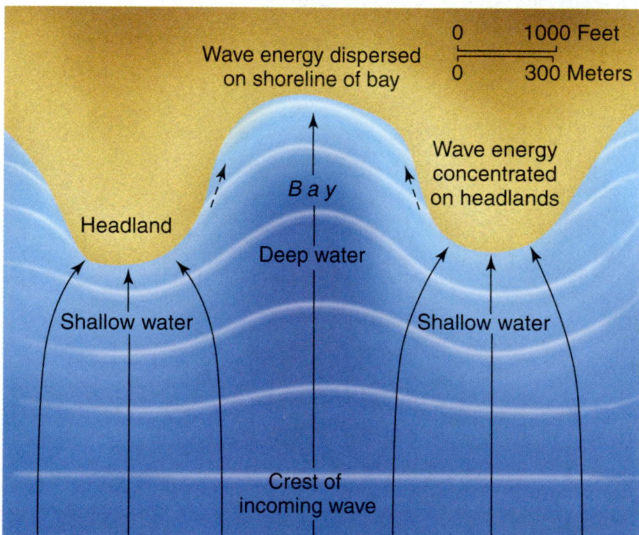

Figure 4.49

Schematic map showing the refraction of waves approaching an irregular shoreline. Wave energy is concentrated on the headlands and dispersed in the bays. The arrows represent lines of equal wave energy. These are equally spaced where the water depth is below the wave base but curved toward the headlands when one part of the wave crest strikes shallow water before the rest of the wave. Dashed arrows show the direction of longshore drift.

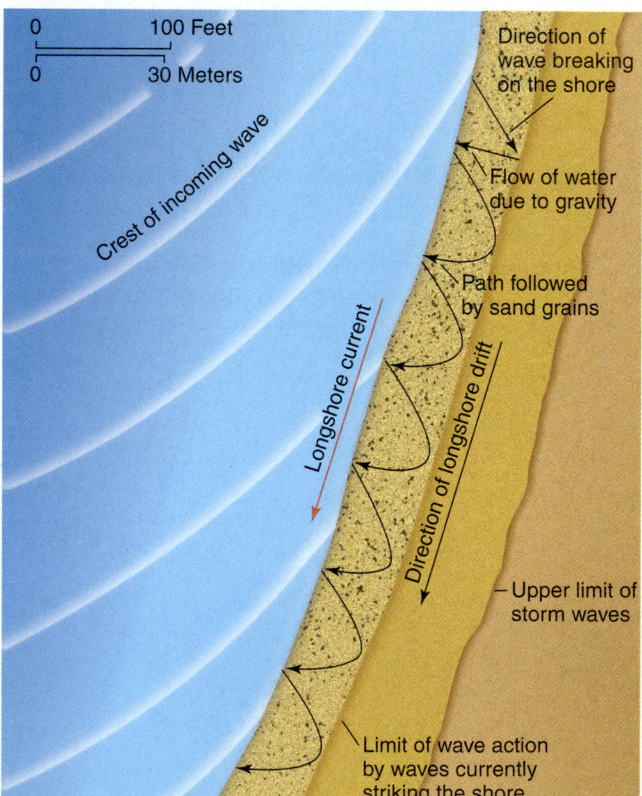

Figure 4.50

Schematic map showing the relationship of refracted waves breaking on the shore to the longshore current and longshore drift.

cliffs. A result of both erosion and deposition is an island (erosional remnant) tied to the shore by a **tombolo** (depositional), a gravel or sandbar connecting the island to the mainland or another island (fig. 4.51).

As a wave-cut cliff retreats under constant wave erosion at its base, a wave-cut platform is formed. A **wave-cut platform** is a gently sloping rock surface lying below sea level and extending seaward from the base of a wave-cut cliff.

Evolution of a Shoreline

Sea level was about 400 feet lower than at present during parts of the Pleistocene when continental glaciers covered about 30% of the earth's land surface. We also know that during the last interglacial, sea level stood about 20 feet higher than it does today. As the last Pleistocene ice sheets melted, sea level rose to its present level. A **eustatic** rise in sea level results from an increase in the volume of water in the oceans, a **eustatic** fall from a decrease in volume.

The shorelines around many coasts at the end of the Pleistocene were irregular and consisted of long embayments formed by the encroachment of the sea into the mouths of rivers. These **drowned river mouths** or **estuaries** were separated by headlands, which were the sites of intense wave erosion. So long as sea level remained unchanged, the headlands became wave-cut cliffs, wave-cut platforms were formed, and drowned river mouths were cut off from the sea by the growth of spits and baymouth bars. Through the process of wave erosion and deposition, the original post-Pleistocene shoreline changed to a straighter form.

Sea level does not remain constant over time, however. A local uplifting of the land along a coast has the same effect as a drop in sea level. Because it cannot always be determined whether sea level has risen or fallen eustatically, whether the land has risen or subsided through tectonic activity, or whether a combination of these events has occurred, it is common practice to refer to **relative** changes in sea level.

Generally, a relative fall in sea level produces an **emergent shoreline,** and a relative rise in sea level produces a **submergent shoreline.** The shorelines of the world's coasts at the end of the Pleistocene were generally submergent, but in coastal regions that were formerly near or within the borders of the continental glaciers, the land began to rise in response to the retreat of the ice caps whose weight had depressed the earth's crust. This uplift did not begin immediately upon retreat of the ice, so there was time for erosional and depositional shore features to be built. These shoreline features now lie above modern sea level, where they are exposed to the normal processes of subaerial weathering and erosion that modify and eventually destroy them.

Deltas

A **delta** is a nearly flat plain of riverborne sediment extending from the river mouth to some distance into a body of water. The deltaic sediments are deposited by **distributaries,** channels that branch out from the main channel of the river. Distributaries are bordered by natural levees, narrow ridges of river sediment built by a river overflowing its banks during flood stages. A delta with many distributaries is called a **bird-foot delta** because of its resemblance in plan view to the outstretched claws of a bird's foot.

New distributaries are formed from time to time, and older distributaries are abandoned. An entire delta may be abandoned when the course of the river feeding it shifts to a new course many miles upstream. The abandoned delta then becomes an **inactive delta.** The inactive delta no longer receives sediment and is modified by wave action to the extent that its seaward margin becomes smooth and it loses its characteristic bird-foot form.

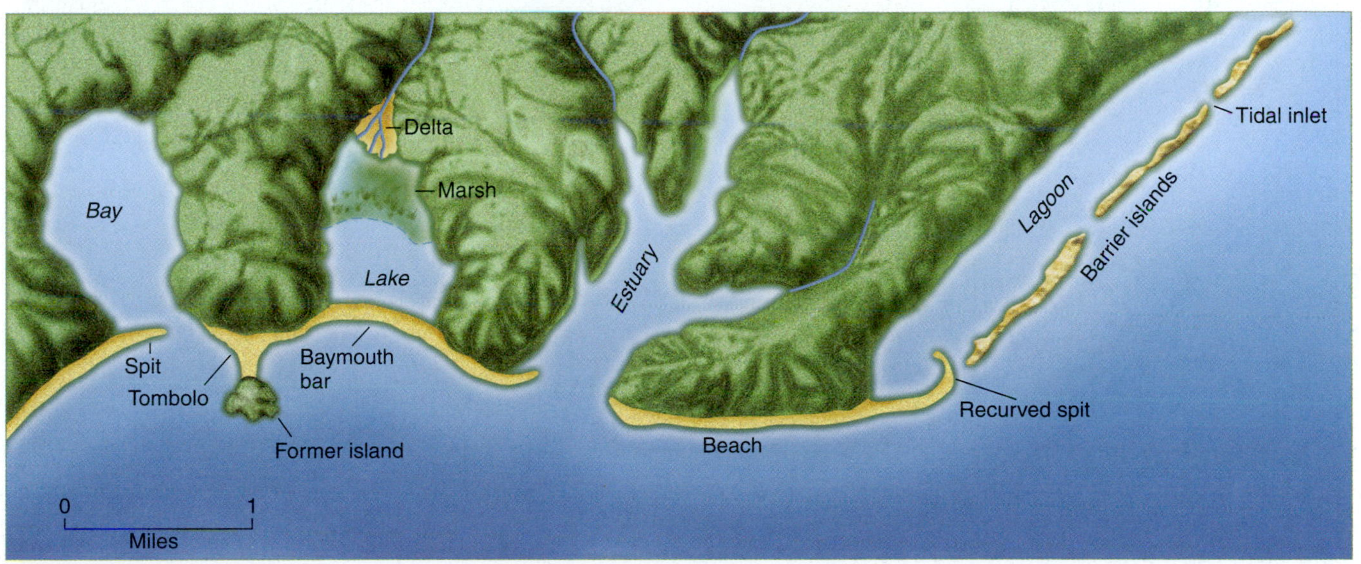

Figure 4.51
Schematic map showing landforms associated with wave action and longshore drift of sand (from left to right in the figure).

An active delta expands seaward if the supply of sediment is too large to be dissipated by wave action and longshore currents. A shoreline that moves seaward by deposition and accumulation is called a **prograding shoreline.** Progradation ceases when a delta becomes inactive, and its seaward margin may in fact retreat due to wave action and subsidence of the deltaic deposits. Eventually, waves and currents redistribute sand along the old delta front to create lagoons behind barrier islands, tidal inlets, and saltwater marshes. An inactive delta can be reactivated if the river course shifts upstream and discharges once again into the site of the former active delta.

The Mississippi Delta

The Mississippi River drainage system carries an enormous load of sediment that eventually is dumped into the Gulf of Mexico along the coast of Louisiana. The delta currently being built is the Balize-Plaquemines delta complex, commonly referred to as the "Birdfoot" Delta. This delta complex has been functional during the last 800–1,000 years. Over the past 8,000 years, four previous delta systems, now inactive, have been built by ancestral courses of the Mississippi River. The most recent of these inactive delta systems, the Lafourche, lies to the west of the now active Birdfoot delta complex and was active between 2,500 and 800 years ago. Other delta complexes that have been identified include the St. Bernard, the Teche and the Maringouin.

In the mid-twentieth century, the Mississippi River was discharging about 25% of its flow into the Atchafalaya River near the city of Angola, some 100 miles north of Atchafalaya Bay on the Gulf (fig. 4.52). This natural diversion was forestalled in 1963 when the U.S. Army Corps of Engineers completed a control dam at the diversion site. Without this dam the flow would have increased until the entire volume of the Mississippi River was discharged into the Gulf of Mexico through the Atchafalaya River. Had this natural event occurred, the modern course of the river south of Angola and the Birdfoot delta complex would have been abandoned. In spite of this human intervention, the sediment deposited by the Atchafalaya River into Atchafalaya Bay has started what might become the next delta system of the Mississippi River.

As a footnote to this discussion it should be noted that Hurricane Katrina had a major impact on the coastal lowlands adjacent to the channel and distributaries of the Birdfoot Delta. Aerial photos show that severe flooding and erosion occurred, and the width of the land areas adjacent to the channels was reduced so that in many places all that remains are the levees themselves (see fig. 4.56B).

Figure 4.52

Map of the Mississippi River Valley and deltas associated with it.

BOX 4.1

New Orleans and Katrina

New Orleans has been described as the lowest, flattest, and geologically youngest major city in the United States. It has been estimated that over 50% of the city is at or below mean Gulf level, some of it more than 7 feet (box fig. 1). Initially the city was built on the natural levees along the Mississippi River. These natural levees vary in width from 1–3 miles on either side of the Mississippi. A second natural levee system, known as the Metairie-Gentilly Ridge, formed by a distributary course on the older delta of the Mississippi, lies between the present course of the Mississippi and Lake Pontchartrain. Other older distributary channels are also present, and these "high ground" areas have played a major role in the location of highways and in foundation design for large buildings. The city was protected from Mississippi River floods by artificial dikes and levees built on top of the natural levees adjacent to the Mississippi River. Major diversion structures were constructed upstream from New Orleans to reduce potential flood risk.

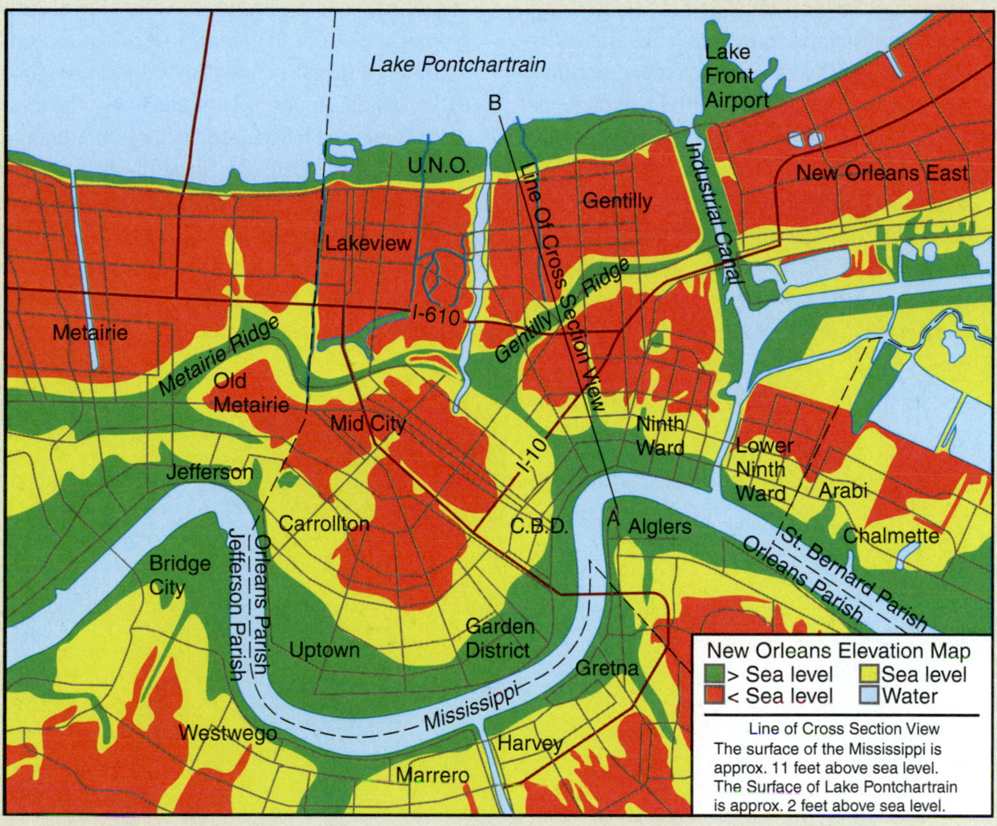

Figure 1 New Orleans Elevation Map.

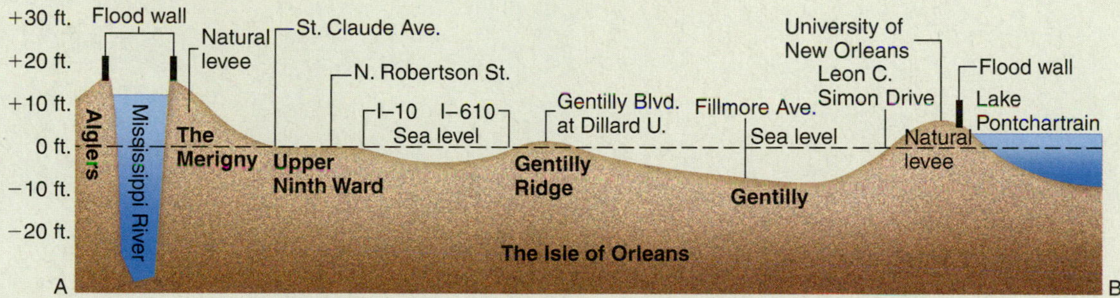

Figure 2

Cross section of the city of New Orleans looking toward the West from the line of crossing indicated on the New Orleans Elevation Map (Box Figure 1).

As the city grew the urban areas spread out away from the river and off the natural levee systems into the wetlands adjacent to the present course of the Mississippi. In these areas the soils are characterized by high moisture content, are compressible, and those with high organic content will upon dewatering shrink to less than 50% of their original thickness. Thus, differential settling in these areas has resulted in cracked foundations, uneven streets and driveways, and in some cases broken utility lines. As urbanization in these areas expanded, the demand for pumping of the groundwater to keep the new areas dry became necessary. Increased runoff resulted from the decrease in surface infiltration areas as roads, driveways, and buildings covered more and more of the surface available to absorb the runoff. As the development of these low-lying areas continued, larger and larger pumps and canals to carry the water away into Lake Pontchartrain were constructed to keep these areas dry. The result was further settling, and the thickness of the organic layer was reduced by as much as 75%. Studies have shown that the settling in some areas has been as much as 8–10 feet in the last 75 years.

There are about 200 miles of drainage canals and 22 pumping stations that operate in the city to keep the low-lying areas relatively dry. The pumps are capable of lifting 35 billion gallons per day. Along the major drainage and navigation canals that run through the city to Lake Pontchartrain, artificial dikes were also emplaced. Box Figure 1 shows the relationship between these canals and the areas of the city lying below the level of the Gulf sea level. It should be noted here that the average level of the Mississippi River at New Orleans is 10–11 feet above sea level (the average high water level is 14 feet above sea level) and Lake Pontchartrain is 1–2 feet above sea level (box fig. 2).

The flooding that inundated large areas of New Orleans did not come from the Mississippi River but rather resulted from the storm surge caused by Hurricane Katrina on August 29, 2005. The first impact was from storm surge waters pushed up the Intercoastal waterway into southeastern New Orleans where they smashed into the Industrial canal. The earthen levees were overtopped and then breached. As the storm moved northward, the storm surge from Lake Pontchartrain forced water up the drainage and navigational canals toward the city. The levees along these canals apparently were not over-topped but rather were breached by the storm surge water, and the resulting damage was both from flooding and from the velocity of the water as it poured through the breached levees. Over 75% of the city was flooded, and those areas not flooded were behind dikes that were not breached (box fig. 3).

What the future holds for New Orleans is still a matter of much discussion.

Figure 3

Aerial photo from New Orleans shows the effect of flooding after two levees were breached. A break is seen in the levee of the 17th Street Canal. The other side of the canal was dry.

Smiley Pool (DMN).

Ancient Shorelines Around the Great Lakes

During the retreat of the continental glaciers from the Great Lakes region, precursors of the modern Great Lakes formed around the lobate front of the ice margin (fig. 4.53). The shorelines of these ancestral Great Lakes are preserved in beach ridges and wave-cut cliffs that now lie above the elevations of the present Great Lakes. Several different ancestral lakes were formed at various times during the retreat of the ice, and each of these lake stages has been assigned a name to distinguish one from the other. Figure 4.53 shows the extent of glacial lakes Maumee and Chicago in the basins now occupied by Lake Erie and Lake Michigan. The levels of these and other stages can be determined by using the elevations of their shorelines to infer the elevations of the water planes of the lakes. For example, the base of an abandoned wave-cut cliff would mark the water plane of the lake that produced it.

Where two or more ancestral shorelines are related to the basin of a modern lake, we will assume that the highest shoreline is the oldest and the lowest is the youngest. While this relationship is not universally true for all the former lake stages, it will suit our purposes for this discussion. Note that a higher lake may have a different outlet; for instance, Lake Maumee and Lake Chicago and the other drainage systems presented on figure 4.53 had outlets to the Mississippi River at this time because the present outlet of the lakes through the St. Lawrence River was blocked by the Ontario lobe.

The Modern Great Lakes

The Great Lakes straddle the U.S.–Canadian border and have a combined surface area of almost 100,000 square miles (fig. 4.54). The lakes came into being during the retreat of the Pleistocene ice sheet that once covered the entire area more than 15,000 years ago. The volume of water in the modern lakes is mainly a function of surface runoff from the surrounding drainage basin, evaporation from the lakes themselves, and the outflow of water to the Atlantic Ocean via the St. Lawrence River. The volume of water stored in the Great Lakes is reflected in their water levels. These levels are

Figure 4.53

Map showing the extent of the last, or Wisconsin, ice sheet in the Great Lakes region about 14,000 years ago. The Glacial Lake Chicago and the Glacial Lake Maumee drained through outlets to the Gulf of Mexico via the Mississippi River.

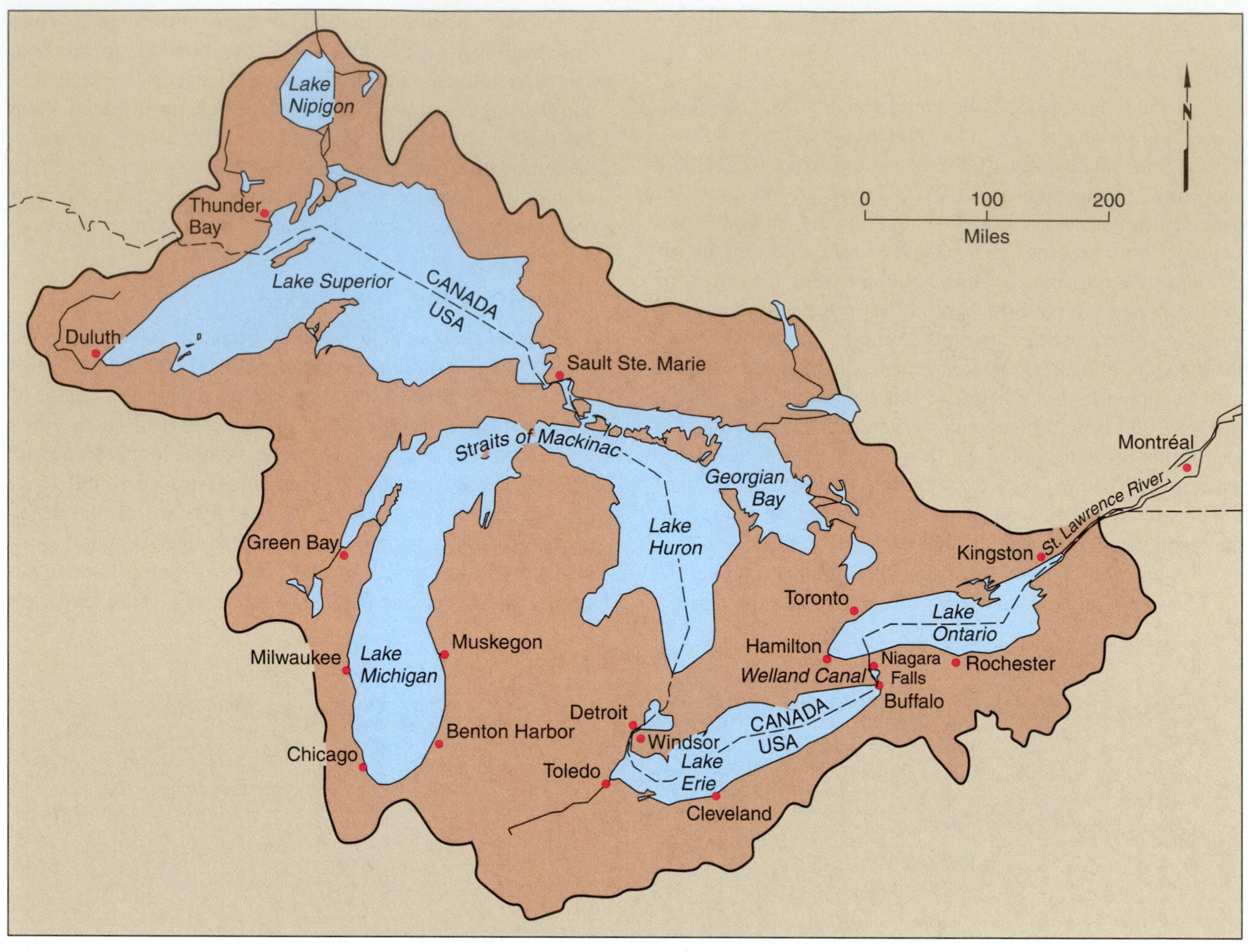

Figure 4.54

Map of the Great Lakes and the land area supplying runoff to them. The five lakes have a combined surface area of almost 100,000 square miles, and the drainage basin containing them is roughly twice that size.

published monthly for each lake by the U.S. Army Corps of Engineers (see fig. 4.60) and are a matter of public record.

The shorelines of the Great Lakes are sites of thousands of vacation homes and permanent dwellings. Those with lake frontage are desirable from an aesthetic point of view, but they are in considerable jeopardy during periods of high water levels when storm waves destroy sandy beaches formed during low water stages and cause severe recession of the bluffs and cliffs by wave erosion at their bases.

Much of the eastern shoreline of Lake Michigan, for example, is characterized by steep bluffs that are formed in old sand dunes and other unindurated Pleistocene sediments. Some of the bluffs rise to more than 100 feet above the lake. Private dwellings on the tops of these bluffs were built during periods of low water when the shores between the bases of the cliffs and the water's edge were characterized by wide, sandy beaches. Those who purchased or constructed homes during low water stages believed that the wide beaches fronting their properties were a permanent part of the landscape and provided adequate protection from any future wave erosion. These were false assumptions, as many who owned property on the shores of Lake Michigan learned at great cost during the period 1950 through 2006. Attempts to stabilize beaches were made by constructing offshore barriers (see fig. 4.62). These efforts generally were unsuccessful.

Four times during this period, 1952–1953, 1972–1976, 1985–1986, and 1996–1997, the water levels of Lake Michigan stood at extraordinarily high levels, and three times during the same period, 1958–1959, 1964–1965, and 2000–2007, the levels were low (see fig. 4.60). The water level of 581.6 feet in 1986 was the highest on record since 1900, and the water level of 575.4 feet in 1964 was the lowest on record for the same period. Thus, in the 22-year period between 1964 and 1986, the level of the lake varied by about 6 feet, mainly by natural causes.

The damage to property during high water stages is all too apparent in the photograph of figure 4.61. The house in figure 4.61 was abandoned by the time it was photographed in 1986.

One might ask why these houses and hundreds of others like them were built in the first place. The answer lies in the lack of understanding of the relatively short time it takes for the lake level to change drastically and in the inability of anyone to predict future levels over a time period of a decade or so. No one would think of building a house next to the one in figure 4.61 today. The consequences of such folly are all too apparent. But when the houses along Lake Michigan and other Great Lake shores were built during low water stages, they seemed secure from the damage and destruction to which they were subjected in later years.

The lesson to be learned from this is that those who contemplate purchasing or building homes should be aware of geologic hazards and should try to acquire as much information as possible about building sites from the public record before proceeding.

Name

EXERCISE 19A

Section Date

Deltas of the Mississippi River

Figure 4.55 shows the Birdfoot Delta of the Mississippi and the plume of sediment being discharged to the Gulf of Mexico. The major distributaries of the Birdfoot Delta are called *passes,* and their names and the names of associated interdistributary bays are shown in figure 4.56. The inactive Lafourche Delta lies on the western flank of the Birdfoot Delta (see also fig. 4.52).

1. What are the narrow ridges visible on either side of Southwest Pass in figure 4.55?

____

____

2. The light blue color around the Birdfoot Delta in figure 4.55 is a plume of suspended sediment. What is the immediate resting place of this sediment?

____

____

3. If the Birdfoot Delta should be abandoned in favor of the delta forming at the mouth of the Atchafalaya River, describe the changes that would occur on the seaward edge of the Birdfoot Delta.

____

____

____

____

____

4. The Mississippi River is a major artery of commerce connecting the Gulf of Mexico with New Orleans and Baton Rouge. Describe the impact on the river between the delta and the cities along its course if human intervention had not been imposed on the site of natural diversion of the Mississippi River near Angola.

____

____

____

____

____

5. What features of the Lafourche Delta indicate it was once an active delta?

____

____

____

____

____

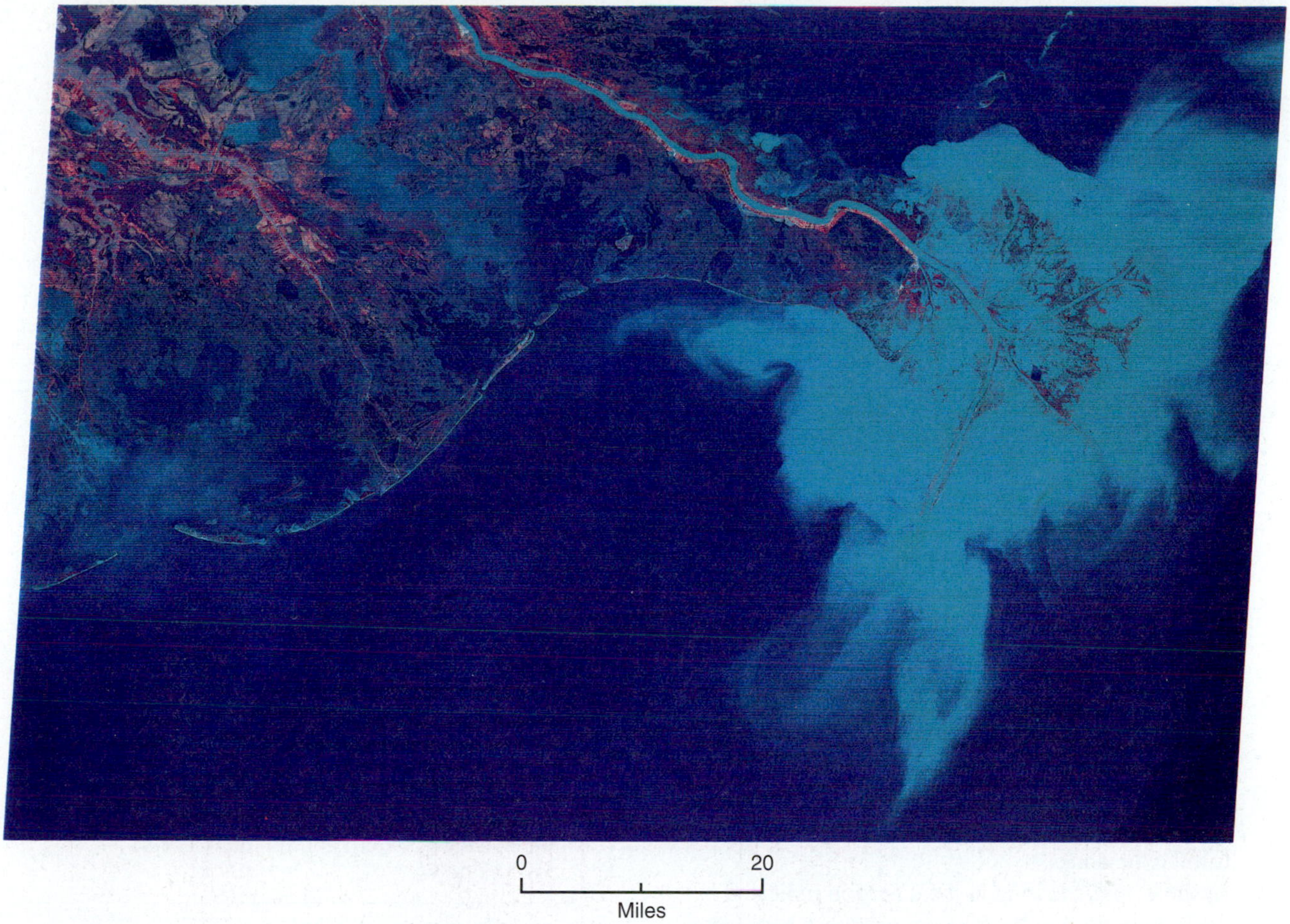

0 20
Miles

Figure 4.55

False color image of the Mississippi Delta made from Landsat on April 3, 1976.

Courtesy of NASA and the U.S.G.S. EROS Data Center, Sioux Falls, South Dakota 57198.

Figure 4.56A

Map of the Birdfoot Delta of the Mississippi River, pre-1989.

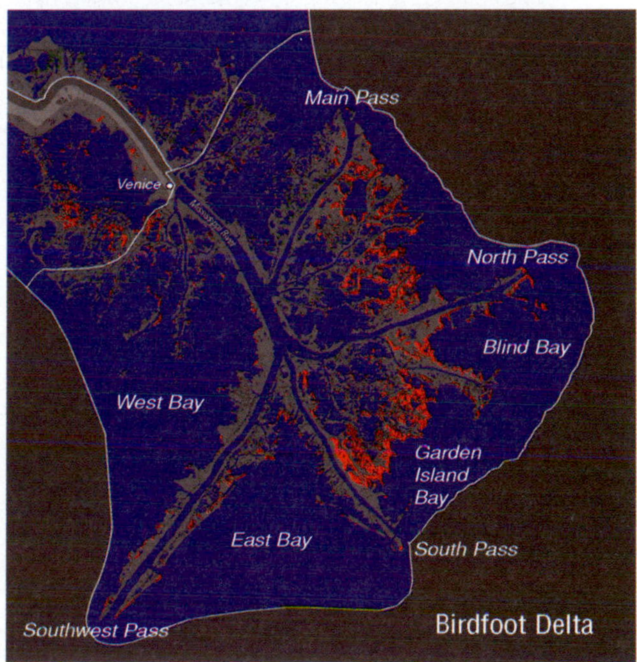

Figure 4.56B

Composite image of the Birdfoot Delta of the Mississippi River showing erosion in red from August 2005 to October 2006.

Courtesy of U.S.G.S. EROS Data Center, Sioux Falls, South Dakota 57198.

EXERCISE 19B

Name

Section Date

Coastal Processes and Shoreline Evolution

The aerial photograph of figure 4.46 shows three waves approaching the shore.

1. Draw a red line along the crest of each of the three waves and show by a red arrow on the water surface the direction of the longshore current. Show by a black arrow on the beach the direction of beach drift.
2. Were the waves in figure 4.46 produced by a wind blowing in the same direction as the prevailing winds that formed the sand dunes?

____

____

3. The Point Reyes map of figure 4.57 shows part of the California coast near San Francisco. North is toward the bound margin of the map. The configuration of the contour lines along Point Reyes Beach reflects the elongate orientation of coastal sand dunes. Draw several short red arrows along the coast that show the prevailing wind direction that formed the dunes.
4. Assume wind blowing in the direction indicated by the red arrows you have drawn produces waves that strike the shoreline of Point Reyes Beach. Based on the wave refraction pattern of these hypothetical waves, draw a red arrow along the beach showing the direction of beach drift.
5. Draw a red arrow along Limantour Spit to show the direction of beach drift and a dashed red arrow to show the direction of the longshore current just seaward of the spit.
6. What is the origin of the horseshoe-shaped lake east of the D Ranch?

____

____

____

____

7. Explain the irregular shape of the embayment formed by Drakes Estero and its many tributaries.

____

____

____

8. Use black dashed lines to extend seaward the 80-foot land contours along the west shore of Drakes Bay between the entrance of Drakes Estero and Point Reyes to show how the shoreline might have appeared before wave action produced the present smoothly curved shore. What is the name applied to the features outlined by the extension of the 80-foot contours?

____

____

____

9. What features in Drakes Estero indicate the ultimate fate of Drakes Estero and its many arms?

____

____

____

10. Point Reyes is marked by a steep cliff on its seaward side. What is the origin of this cliff and the many small islands lying just off shore at its base?

____

____

____

____

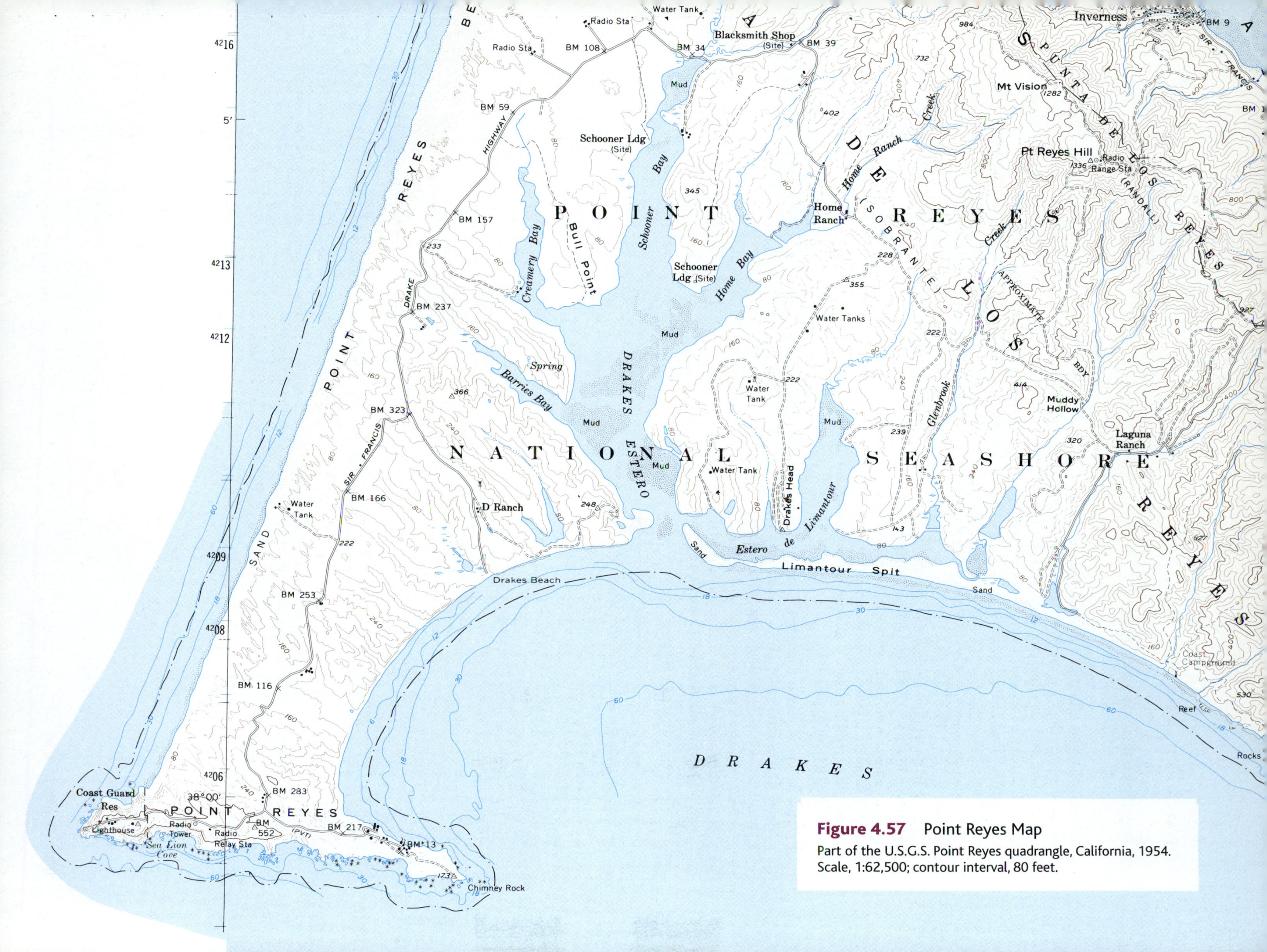

Figure 4.57 Point Reyes Map

Part of the U.S.G.S. Point Reyes quadrangle, California, 1954. Scale, 1:62,500; contour interval, 80 feet.

EXERCISE 19C

Name ______________________

Section ______________ Date ______________

Ancestral Lakes of Lake Erie

The North Olmstead map (fig. 4.59) covers an area just west of Cleveland, Ohio. Detroit Road and Center Ridge Road lie along two shorelines ancestral to Lake Erie. These will be referred to as the Detroit shoreline and the Center Ridge shoreline. The Center Ridge shoreline was formed about 13,000 years ago during the waning phases of the Pleistocene, approximately 1,000 years after glacial lake Maumee.

Figure 4.58 is a north–south profile showing the location of Center Ridge Road and Detroit Road. The profile is aligned along Columbia Road, which is just west of Forest View School in the upper right-hand corner of the map. The generalized profile extends north into Lake Erie and south to the 740-foot contour line.

1. Draw the line of profile with a black pencil on figure 4.59.
2. With reference to the profile, compare the Detroit and Center Ridge shorelines with the shoreline of Lake Erie. Which of these three shorelines are erosional and which are depositional?

____

____

____

3. Use a sharp pencil and a straightedge to draw the water surfaces of the lakes that produced the two ancestral shorelines. Write the elevations of each above the two lines you have drawn.
4. Draw a solid black line on the profile between Detroit Road and Wolf Road to show what the original profile might have looked like before the Detroit shoreline was formed.
5. Based on the elevations of the two shorelines, which is older?

____

____

____

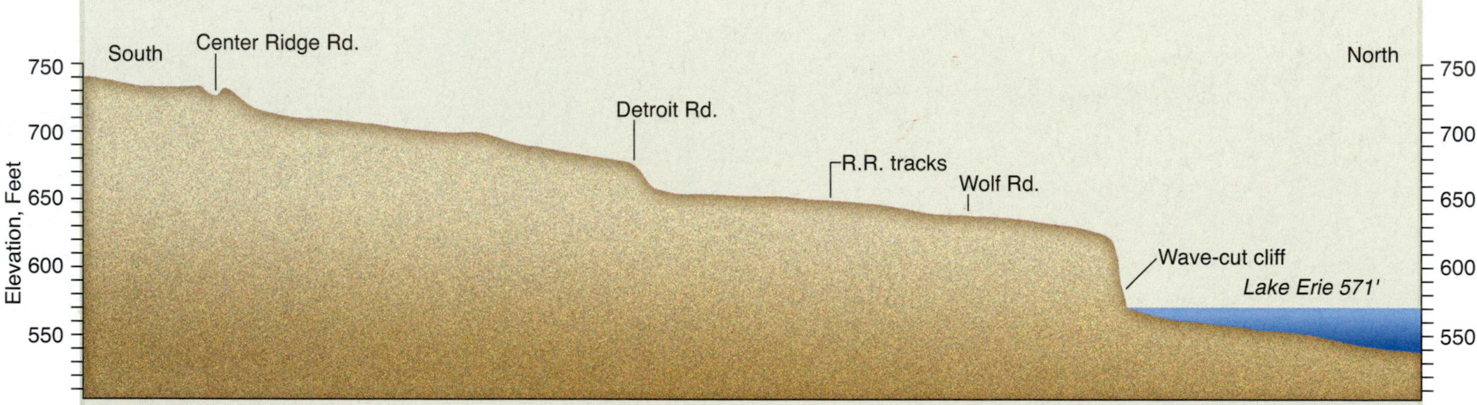

Figure 4.58

A generalized north–south topographic profile based on the North Olmstead map (fig. 4.59) showing some ancient shorelines related to the ancestral stages of Lake Erie, Ohio. (See Exercise 19C for exact location of the profile.)

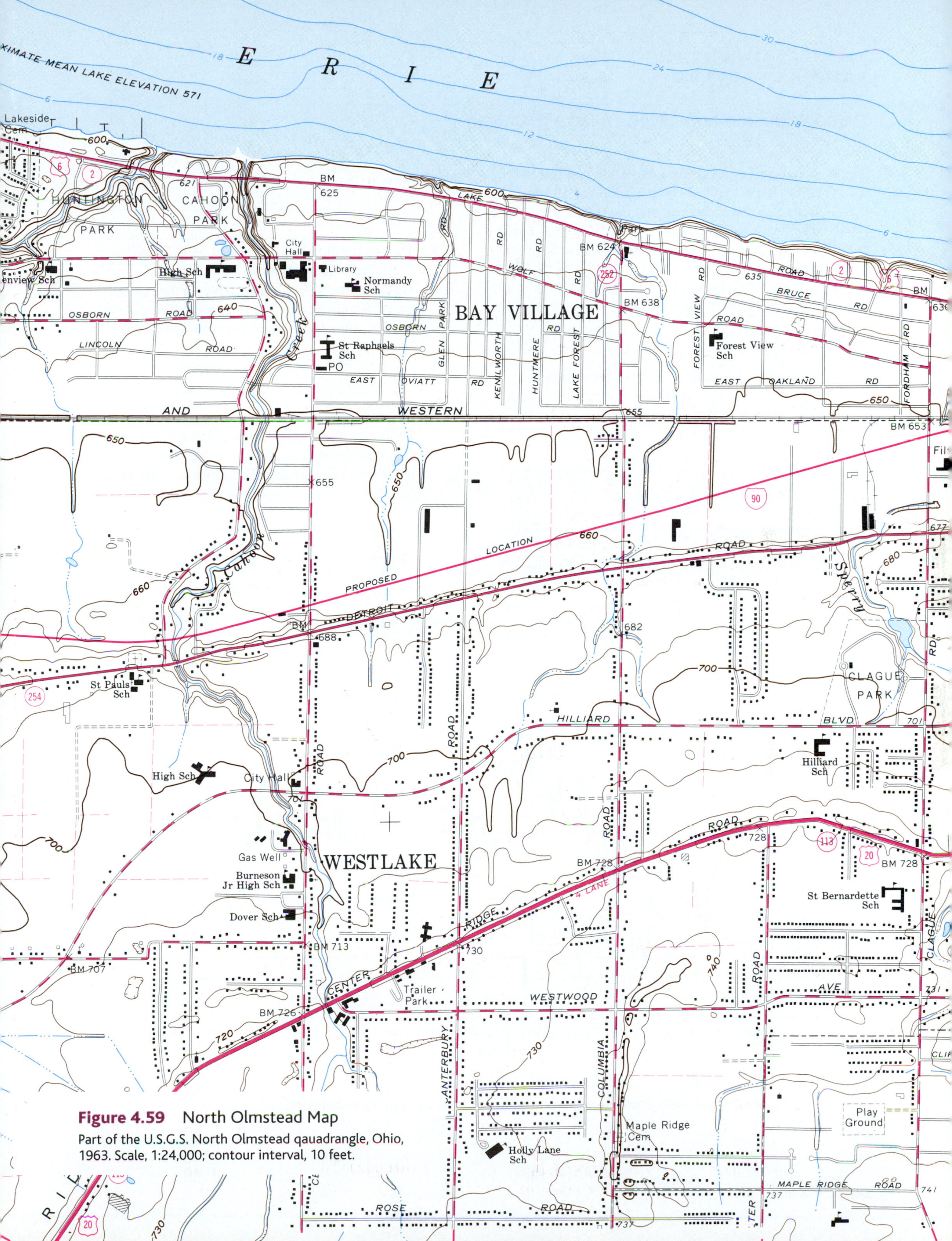

Figure 4.59 North Olmstead Map

Part of the U.S.G.S. North Olmstead qauadrangle, Ohio, 1963. Scale, 1:24,000; contour interval, 10 feet.

EXERCISE 19D

Name ____________________

Section ____________________ Date ____________________

Shore Erosion and Levels of Lake Michigan

1. The record of levels for Lake Michigan (fig. 4.60) shows highs and lows over a period of 59 years. Does this record suggest that the fluctuations of the lake levels follow a regular periodicity that would permit the forecasting of future lake levels? Explain your answer.

____

____

____

2. What physical characteristics of the sediment exposed in the wave-cut cliff of figure 4.61 made this cliff particularly susceptible to the wave erosion during the high lake levels of 1987?

____

____

____

3. The chart of figure 4.60 shows a pronounced drop in the level of Lake Michigan between 1987 and 1990. Describe the impact of this drop on the sand bluff in figure 4.61.

____

____

____

4. The face of the bluff in figure 4.61 is littered with debris from the abandoned house perched on top. What additional signs are visible on the face of the bluff to indicate that it was undergoing severe erosion at the time the photograph was taken?

____

____

____

5. Variations of the water level in Lake Superior and the other Great Lakes have considerable economic impact. Other than shoreline erosion, what are some of the effects of higher or lower levels of the lakes?

____

____

____

(continued)

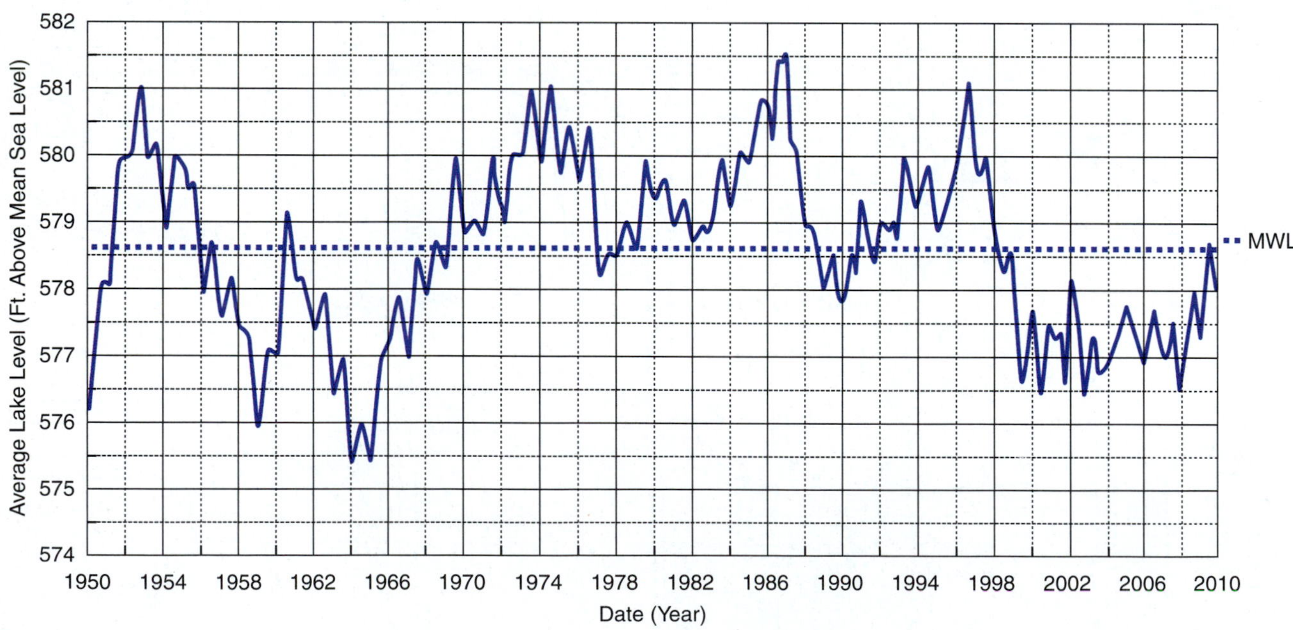

Figure 4.60

Water levels of Lake Michigan, 1950–2009; mean water level (MWL), 1860 through the end of 2009.

Figure 4.61
An ancient sand dune exposed to wave erosion on the shore of Lake Michigan near Muskegon, Michigan, November 1986.
Photo by Marge Beaver, Muskegon, Michigan.

Name

EXERCISE 19D

Section Date

Shore Erosion and Levels of Lake Michigan *(Continued)*

6. What evidence is there in figure 4.62 that an attempt was made to impede, if not stop, the erosive action of storm waves?

7. The cliff in figure 4.62 has receded tens of feet between the time the house was built in the 1930s and the date of the photograph. What happened to the material eroded from the cliff during this period?

8. What must happen to the level of Lake Michigan before the stability of cliffs and bluffs along its shore can be restored?

Figure 4.62

Photograph of a part of the Lake Michigan shoreline near Benton Harbor, Michigan, taken April 16, 1974.

Hann Photo Service, Hartford, Michigan.

Landforms Produced by Volcanic Activity

Background

A volcano is a vent in the earth's crust through which magma, ash, and gases erupt. Over 1,300 **active volcanoes** (an active volcano is one that has erupted in historic times) occur on the earth today, and it is estimated that over a million extinct volcanoes are spread over the land surface and on the ocean floor. **Dormant volcano** is a term applied to an active volcano during periods of quiescence. An **extinct volcano** is one in which all volcanic activity has ceased permanently.

In this section, we will deal with two of the major types of volcanoes: the shield volcano, as exemplified by those on the island of Hawaii, and the composite or stratovolcano, as represented by Mount St. Helens in the state of Washington. A **shield volcano** is a gently sloping dome built of thousands of highly fluid (low viscosity) lava flows of basaltic composition. A **composite volcano,** or **stratovolcano,** is a conical mountain with steep sides composed of interbedded layers of viscous lava and pyroclastic material. **Pyroclastic** refers to all kinds of clastic particles ejected from a volcano, the most ubiquitous of which is commonly called **volcanic ash.** The lavas in a composite volcano are rhyolite, dacite, or andesite (see table 1.7). (The composition of dacite, an aphanitic igneous rock, lies between that of rhyolite and andesite.)

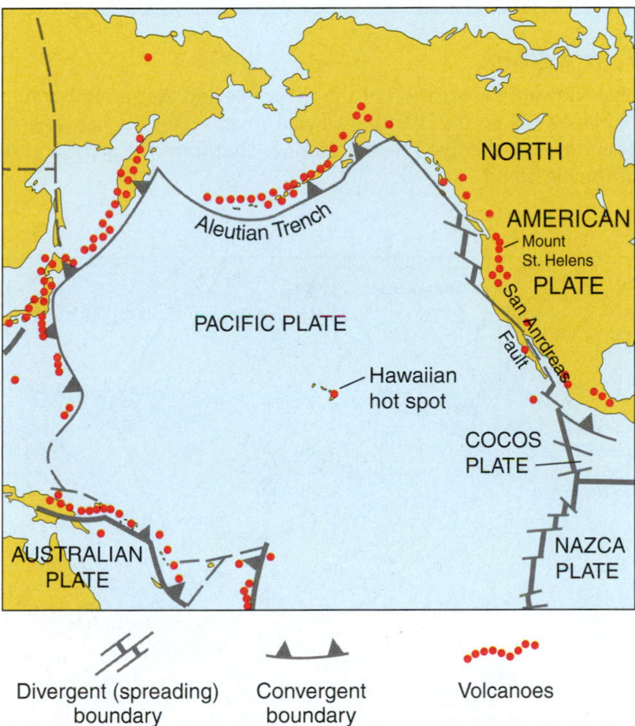

Figure 4.63

Map showing locations of the Hawaiian hot spot and Mount St. Helens with respect to part of the Pacific Plate and its margins.

Distribution of Volcanoes

Active volcanoes are widespread around the world, but their specific locations are controlled by conditions existing in the earth's crust and upper mantle. Composite volcanoes are concentrated along the margins of tectonic plates that abut each other, and shield volcanoes generally occur over **hot spots,** deep-seated zones of intense heat whose geographic coordinates remain fixed for several millions of years. We will become more familiar with hot spots and tectonic plates in Part VI of this manual. For purposes of the discussion here, it is sufficient to know that the zone formed where two plate margins converge is characterized by intense volcanic and earthquake activity. The most active convergent plate boundaries occur along the margins of the Pacific Plate "Pacific Rim of Fire," part of which is shown in figure 4.63. Mount St. Helens is one of the many composite volcanoes lying on the North American segment of the Pacific Rim. The hot spot over which the island of Hawaii lies is not associated with a plate margin but occurs beneath the middle of the Pacific Plate (fig. 4.63).

The shape and size of a shield volcano and a composite volcano are shown for comparison in the topographic profile of figure 4.64.

The Shield Volcanoes of Hawaii

The island of Hawaii, also called the Big Island, is part of the Hawaiian Ridge, a partially submerged chain of volcanic mountains that rise from the deep ocean floor and extend from the Big Island to the northwest for a distance of more than 2,000 miles. All of the islands along the Hawaiian Ridge

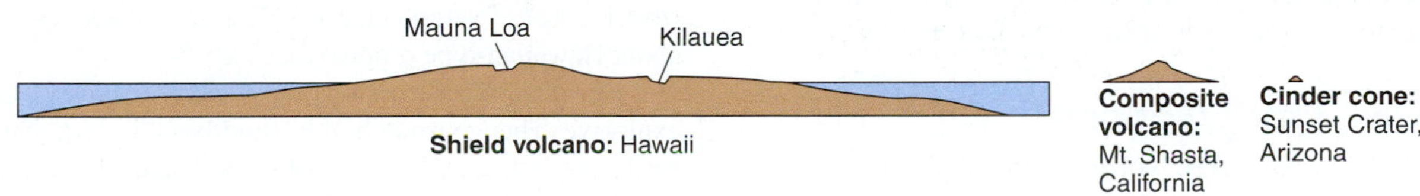

Figure 4.64

Topographic profiles showing the comparison in size and shape of Mauna Loa, Hawaii, (shield volcano), Mount Shasta, California, (composite volcano), and Sunset Crater, Arizona (cinder cone). [Profile drawn to same scale.]

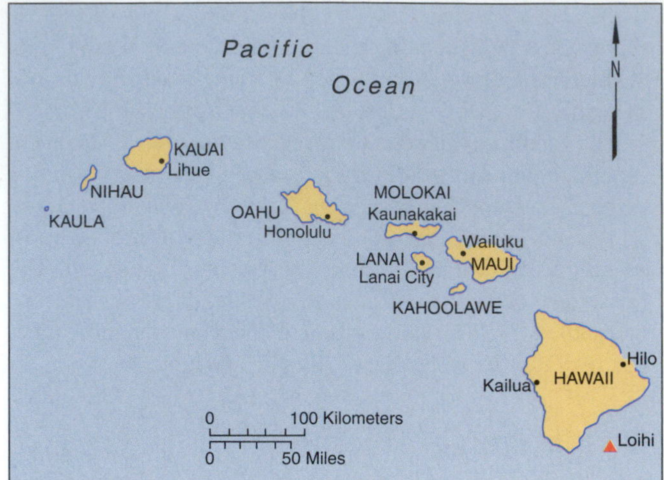

Figure 4.65
Map of the Hawaiian Islands in the Pacific Ocean.

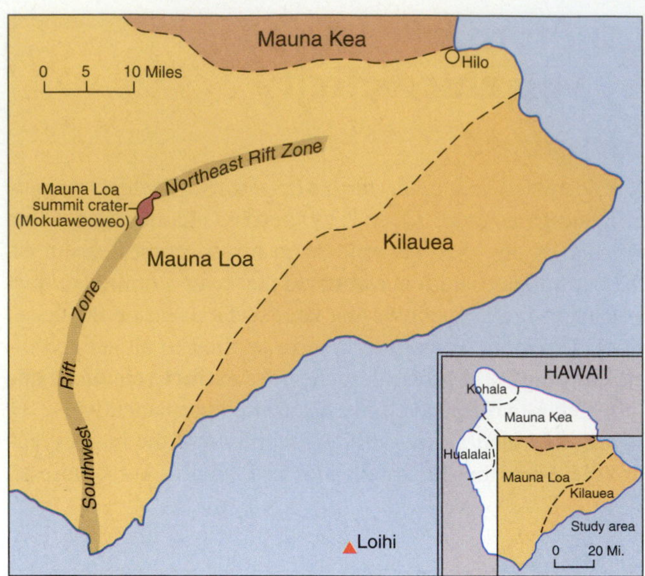

Figure 4.66A
Map showing the location of the rift zones and the summit crater of Mauna Loa on the island of Hawaii. Inset shows the names and boundaries of the five shield volcanoes that form the architecture of the Big Island.

are shield volcanoes that were formed over the Hawaiian hot spot as the ocean floor moved progressively to the northwest over the last 40 million years.

The Big Island of Hawaii is the largest of the Hawaiian Islands (fig. 4.65) and consists of five shield volcanoes (see inset of fig. 4.66A), of which only one, Kilauea, is currently active (fig. 4.66B). Together, these five volcanoes rise 30,000 feet above the floor of the Pacific Ocean to an elevation more than 13,000 feet above sea level. The oldest of the five is Kohala, which became inactive about 60,000 years ago; Mauna Kea, the next oldest, ceased its eruptive activity about 3,000 years ago. Hualalai and Mauna Loa are dormant, Hualalai last erupted in 1801, Mauna Loa in 1984. Loihi, an active submarine volcano off the southeast coast of the Big Island, rises about 10,000 feet above the ocean floor to its summit 3,000 feet below the ocean surface. Loihi is the youngest volcano on the Hawaiian Ridge and may become the next island in the Hawaiian Islands.

Recent studies have shown that Kilauea Volcano is separate from Mauna Loa and has its own magma system. Kilauea Volcano has had 34 eruptions since 1952 and has been in continuous eruption since 1983, mainly from vents along the East Rift Zone of Kilauea Volcano. The lavas erupted from the main cinder-and-spatter cone at Pu'u O'o, east of Kilauea caldera, flow through a tube system about 6.8 miles to the ocean. This continuous eruption has changed both the physical and cultural landscape of the island of Hawaii. A town, Kalapana, was buried by the lavas, roads have been covered, and shorelines are changing on almost a daily basis (fig. 4.66B).

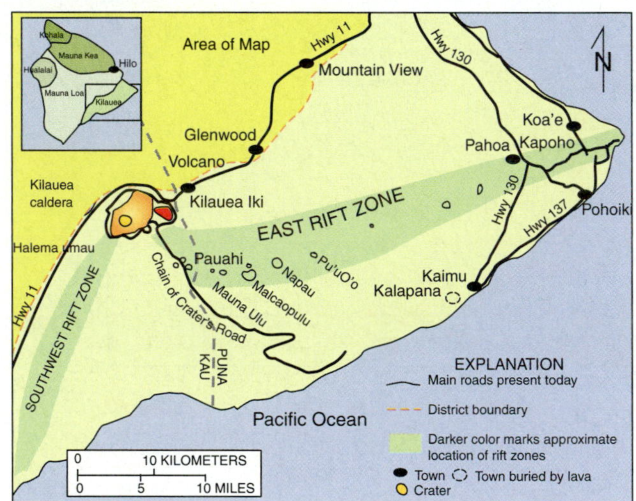

Figure 4.66B
Map showing the location of the rift zones, summit caldera, Pu'u O'o, roads, and several communities of Kilauea on the island of Hawaii.
Courtesy of Jenda Johnson.

Hawaiian-Type Eruptions

Almost all Hawaiian-type volcanic eruptions on the Big Island during historic times (roughly the period from 1843 to the present) have occurred on Mauna Loa and Kilauea (figs. 4.66A and 4.66B). These two volcanoes have been intensely studied by volcanologists in recent years, and it is from their observations that we are able to know a great deal about Hawaiian-type eruptive activity.

Hawaiian-type eruptions are weakly explosive or nonexplosive. They extrude highly fluid basaltic lava that flows easily down the gentle slopes of the volcano's flanks. Indeed, it is the low viscosity of the lavas that accounts for the gentle slopes to begin with. Lava is erupted not only from the crest of the volcano but also from rift zones along its flanks. The rift zones on Mauna Loa are called the Southwest Rift Zone and the Northeast Rift Zone (fig. 4.66A), and the rift zones on Kilauea are called the Southwest Rift Zone and the East Rift Zone (fig. 4.66B). The lava extruded from these rifts comes from magma reservoirs a few miles beneath the summits.

In the early stages of a Hawaiian-type eruption, lava spouts from fissures in the rift zones as **lava fountains.** These spectacular "curtains of fire" rise to heights of 100 feet or more; the largest lava fountain ever recorded rose to 1,900 feet on Kilauea in 1959. Lava derived from lava fountains or oozing from fissures flows downhill as incandescent rivers. These lava flows may reach speeds of 35 miles per hour before they congeal into solid rock. Two types of surface forms result, **pahoehoe** (pronounced "pa-hoy-hoy") and **aa** (pronounced "ah-ah"). Pahoehoe is commonly called "ropy lava" because it looks like a profusion of snarled ropes. Aa is a jagged, blocky mixture of irregular shapes and sizes. Eruptions produce both pahoehoe and aa lavas in widely varying proportions. Pahoehoe may turn into aa lava downstream from the point of extrusion due to the loss of gases, but aa flows cannot change into pahoehoe. Aa flows are cooler and usually move more slowly than pahoehoe flows.

Some lava flows reach the sea, where they plunge into the water and cool rapidly to form **pillow lavas.** Pillow lavas resemble a pile of pillows formed when chunks of molten lava, ranging in diameter from several inches to more than a foot, develop an outer solid skin and become draped over each other like a pile of water-filled balloons as they settle on the sea floor. Pillow lavas also form where lava is extruded directly into the sea through submarine vents. Most of the mass of oceanic shield volcanoes is believed to consist of pillow lavas.

During the infrequent explosive phases of Hawaiian-type volcanoes, lava and its contained gases are ejected violently into the atmosphere, where the pieces solidify and fall back to the ground as **tephra,** a collective term assigned to all sizes of airborne volcanic ejecta. The constituent particles in tephra are called **pyroclasts,** a term that refers to any clastic particle derived from a volcanic eruption—preexisting rocks fragmented by a volcanic explosion or particles of lava that solidified while aloft. The smallest pyroclasts are collectively known as volcanic ash. Larger pyroclastic particles are referred to as cinders, scoria, pumice, and volcanic bombs. Cinders, scoria, and pumice are formed from the rapid cooling of frothy lava from which gases were escaping during a fountain eruption. Pumice contains so many gas-bubble cavities that it is light enough to float on water. Pyroclastic materials account for only 1% of the mass of a Hawaiian-type volcano above sea level.

Composite Volcanoes

Composite volcanoes are steep-sided, conical mountains consisting of flows formed from viscous lava interbedded with pyroclastic deposits. Composite volcanoes erupt with great violence because the gases contained in the viscous lava build to a very high pressure before the stiff lava finally breaks through to the surface. The initial explosion blasts pulverized rock fragments several miles into the atmosphere.

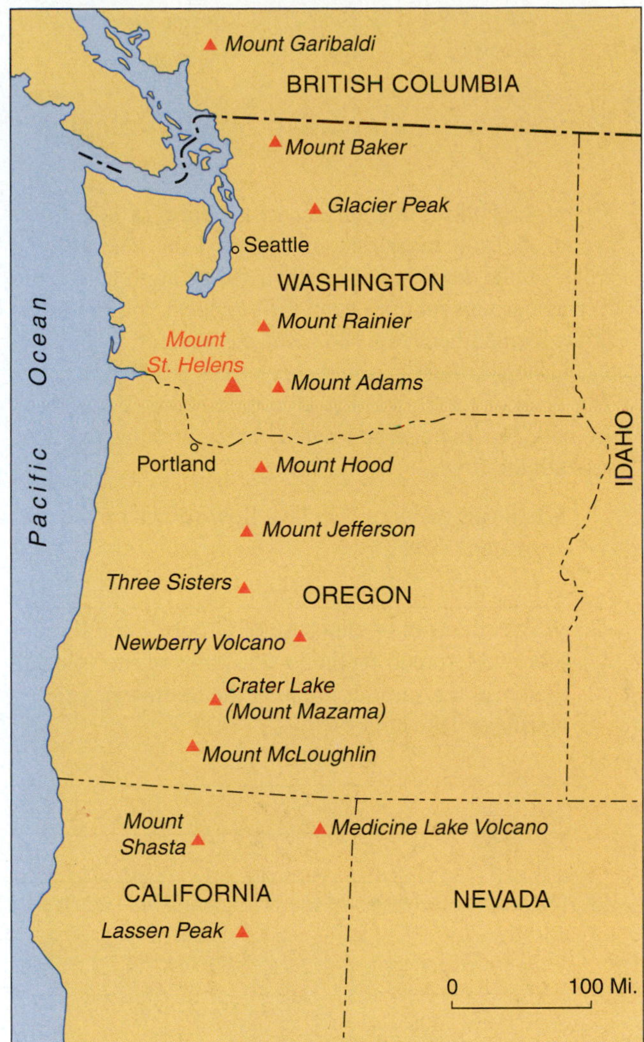

Figure 4.67

Map showing the location of Mount St. Helens and other composite volcanoes in the Cascade Range from northern California to southern British Columbia. The Cascade Range lies on a segment of the Pacific Rim where the Juan de Fuca Plate is subducting under the North America Plate (see fig. 6.1).

The smallest particles form volcanic ash that is carried by prevailing winds and deposited eventually in measurable thicknesses over hundreds of square miles downwind from the eruptive site. Volcanic ash produced by large explosive eruptions may be carried around the world.

A composite volcano may lie dormant for a hundred years or more before it is reactivated by an explosive eruption. Violent eruptions of composite volcanoes cause widespread destruction of flora and fauna and extensive damage to or total loss of property. They have been responsible for the loss of at least 250,000 lives in recorded history.

The violent eruption of Mount St. Helens in the Cascade Range of the northwestern United States (fig. 4.67) in 1980 gave volcanologists an unprecedented opportunity to study the eruptive stages of a typical composite volcano, and much of what follows on page 192 and in Exercise 20B (page 195) is based on their published accounts.

EXERCISE 20A

Name ______________________________

Section ______________ Date ______________

Mauna Loa, A Hawaiian Shield Volcano

Figure 4.68 shows the surface distribution of lava flows produced from eruptions of Mauna Loa. The colored areas of the map represent lava flows aggregated into five age groups ranging from the youngest, the Historical flows formed between 1843 and the present, to the oldest, Group I, those formed more than 4,000 years ago. The ages of the flows that are contained in each group are based on radiocarbon-dated charcoal recovered from beneath the flows.

1. When was the youngest lava flow shown on the map deposited?

   ____

2. Why is the area of exposure of Groups I and II more widespread on the lower flanks of the volcano than near the summit or along the Southwest or Northeast Rift Zones?

   ____

   ____

   ____

3. Did any of the lavas in the five groups *not* reach the coastline?

   ____

   ____

   ____

4. The aa lavas are shown on the map by a stippled pattern. Aa lavas occur at any elevation on the volcano's flanks, but they are most common on the lower flanks. What is the reason for this?

   ____

   ____

   ____

5. The colored bands on the map that represent Historical and Group IV lavas generally are narrower than the colored bands representing lavas in Groups I, II, and III. Account for this difference.

   ____

   ____

   ____

   ____

6. Some of the lavas that reached the coast were pahoehoe (nonstippled areas), and some that reached the coast were aa. Why would a diver be more likely to encounter pillow lavas off the coast of pahoehoe lavas than off the coast of aa lavas?

   ____

   ____

   ____

   ____

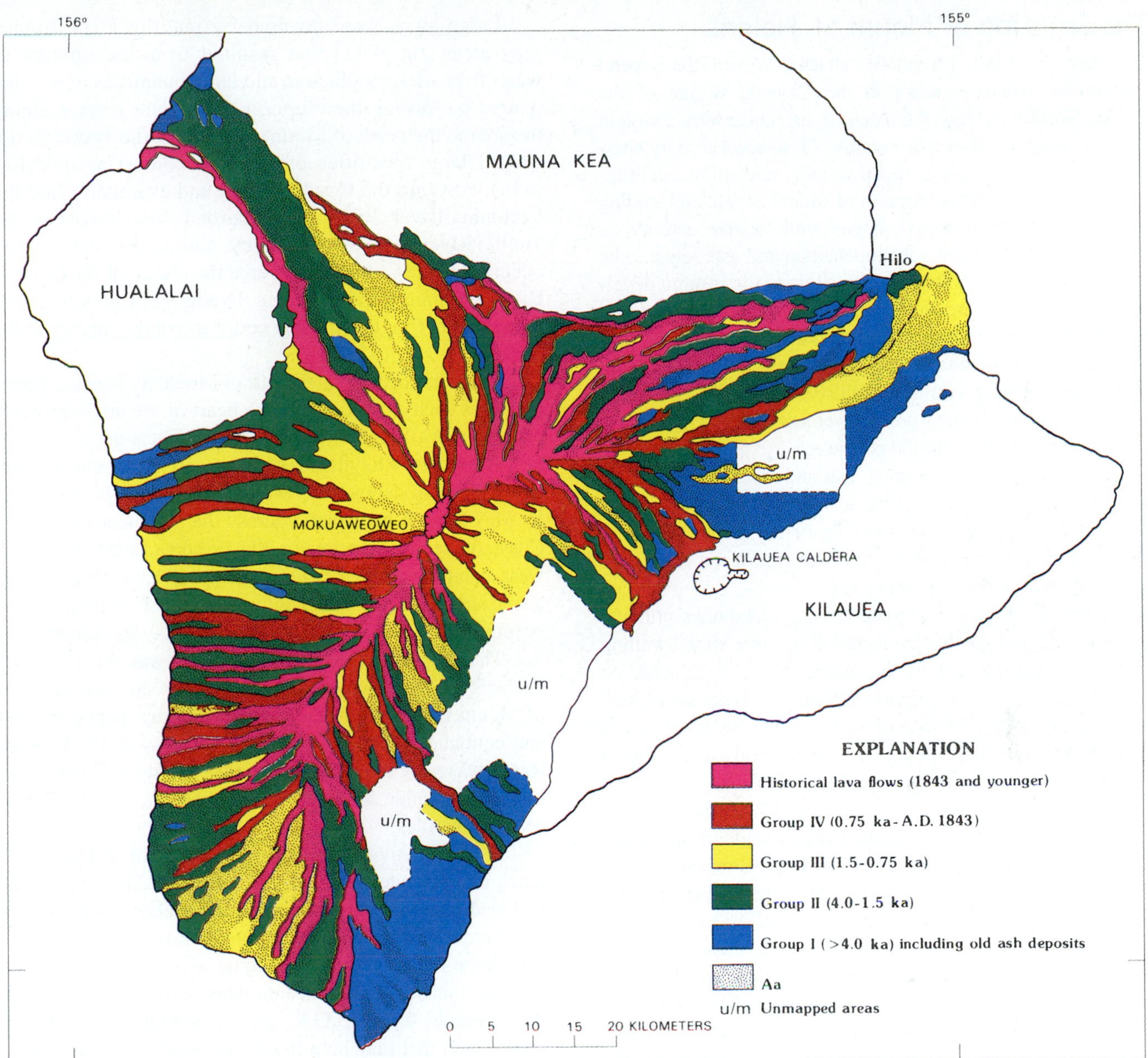

Figure 4.68

Map of Mauna Loa showing the surface distribution of lava flows in five different age categories. The notation "ka" stands for thousands of years before the year 1950. Thus, 0.75 ka = 750 years, 1.5 ka = 1,500 years, and 4.0 ka = 4,000 years.

The Eruption of Mount St. Helens

On May 18, 1980, Mount St. Helens, one of the several spectacular volcanic peaks in the Cascade Range of the Pacific Northwest (figs. 4.69A and 4.70), underwent a violent volcanic eruption after two months of low-level activity characterized by earthquakes, steam venting, and small ash eruptions (fig. 4.69B). This eruption of Mount St. Helens, ending a 123-year dormant period, began with seismic activity on March 20, 1980. Steam vents opened up and ash began to be erupted by March 27. Small craters close to the summit developed, ash plumes were erupted, and ash avalanches took place in early April. Seismic activity continued, and by mid-April, a significant crater had formed. The summit area of Mount St. Helens began to swell, enough so that Goat Rocks on the northern flank had a measured movement of 20 feet vertically and 9 feet horizontally to the northwest.

As this activity continued, ash and steam continued to be produced, but in late April, seismic activity was greatly reduced. In early May, as swelling of the peak continued, the U.S.G.S. reported that the northern rim of the crater was rising at a rate of 2 to 4 feet per day.

Through the use of remote sensing techniques utilizing infrared film, hot spots were recorded in early May. Swelling continued, seismic activity increased, and at 8:32 A.M. on May 18, 1980, a violent eruption of Mount St. Helens occurred, an event that claimed 57 lives, devastated over 200 square miles of timberland and recreational areas, and spread ash downwind and eventually around the world (fig. 4.69B).

The eruption itself was marked by an earthquake that triggered a major landslide down the north side of the mountain, followed quickly by a violent explosion. The initial landslide and lateral blast removed 0.64 cubic miles of material from the north side of the mountain and 0.03 cubic miles of ice and snow (location 7, fig. 4.71). The material moved down the north side of the mountain as a mixture of rock, ash, steam, and glacial ice (location 6, fig. 4.71). Additional fluidization occurred when this avalanche mass hit the water of Spirit Lake and Toutle River. The velocity of the avalanche has been estimated at over 150 mph.

A portion of this enormous avalanche flowed down the valley of the Toutle River for 13 miles, depositing materials in a swath up to 1.2 miles wide and with a thickness up to 450 feet (fig. 4.71). Another part of the avalanche continued to the north, rose over a ridge that was 1,000 feet high, depositing over 100 feet of debris on top of the ridge before pouring over into the valley of South Coldwater Creek on the north side of the ridge (location 4, fig. 4.71). New lakes were formed as stream valleys were dammed by debris (location 3, fig. 4.71), and new islands were formed in Spirit Lake (location 5, fig. 4.71). The heavy black line on figure 4.71 marks the southern edge of the "Eruption Impact Area" as defined by the U.S. Geological Survey.

Large areas were covered by mudflows (unstippled gray areas, fig. 4.71) that resulted from the mixture of water from melting glaciers and large quantities of ash that poured out during the eruption. The major river draining the area to the north of Mount St. Helens, the Toutle River, carried large quantities of sediment almost as mudflow to the west into the Cowlitz River and eventually into the Columbia River. Previously recorded flood stages on the Toutle River were exceeded by almost 30 feet. Silting occurred in the Columbia River at the mouth of the Cowlitz River, trapping ships upstream. Dredging of a channel was necessary before these ships could move downriver to the ocean.

Ash falls occurred to the east of Mount St. Helens, affecting cities such as Yakima, in the heart of the apple-growing district of Washington, and the major wheat-growing areas farther east. At Ritzville, 205 miles east of Mount St. Helens, a fine ash deposit of 70 millimeters was recorded, and by May 21, ash had spread across the continent to the East Coast. A later ash eruption on May 25 spread ash to the northwest, mantling an area lying roughly between the Columbia River and Olympia, Washington. Small ash eruptions have occurred since, but none as large as the May 25 activity.

The eruption of Mount St. Helens was the first such volcanic event in the contiguous 48 states since the eruption of Mount Lassen in northern California that began in 1914 and continued until 1921. Volcanic activity in the Cascade Range was recorded during the 1800s, and several peaks such as Mount Rainier, Mount Baker, and Mount Hood still have active fumaroles.

Volcanic activity has continued on Mount St. Helens at a much reduced rate since the 1980 eruption. This activity has included gas and ash emissions, earthquakes, rockfalls, and extrusion of a lava dome in the center of the crater (fig. 4.69C). The dome-building phase is adding about 35 million cubic feet per month to the volcano. This activity is continually monitored by the U.S.G.S. and is providing considerable information that may help in the development of earthquake and volcanic prediction models.

References

Tilling, R.I. 1987. *Eruptions of Mount St. Helens: Past, Present, and Future,* U.S. Geological Survey.

Tilling, R.I., Heliker, C., and Wright, T.L. 1987. *Eruptions of Hawaiian Volcanoes: Past, Present, and Future,* U.S. Geological Survey.

The material presented in this section of this laboratory manual was drawn freely from these two outstanding publications. They are excellent sources of information on the eruptions of the Hawaiian-type and composite volcanoes.

Figure 4.69A

Mount St. Helens one day before eruption. Elevation of Peak, 9,677 feet. View to southwest.

U.S.G.S. photograph taken by Harry Glicken on May 17, 1980.

Figure 4.69B

Oblique aerial view of Mount St. Helens on day of eruption. View to northeast.

U.S.G.S. photograph taken by Austin Post on May 18, 1980.

Figure 4.69C

Mount St. Helens four years after eruption. Elevation of peak, 8,363 feet. View to south. Note fireweed, one of the first plants to appear in the ecological succession.

U.S.G.S. photograph taken by Lyn Topinka in August 1984. Note spine forming in center of crate.

Name

EXERCISE 20B

Section Date

The Impact of the Eruption of Mount St. Helens on Surrounding Topography

The eruption of Mount St. Helens resulted in several changes in the topography of the Eruption Impact Area, the southern boundary of which is shown as a heavy black line in figure 4.71. Figure 4.70 is a pre-eruption map published in 1980, and figure 4.71 is a map of the same area made after the eruption and was published in 1981. The pre-eruption map will be referred to as the "Old Map," and the post-eruption map will be called the "New Map." Notice that the contour interval of both maps is 50 meters.

1. The elevation of Spirit Lake was 974.7 meters on the Old Map. Determine the elevation of the contour line nearest the shoreline of Spirit Lake on the New Map and answer the following questions:
   (a) Was the elevation of Spirit Lake increased or decreased because of the eruption?

   ______

   (b) Was the area of Spirit Lake increased or decreased because of the eruption?

   ______

   (c) Estimate the *maximum* amount of increase or decrease in the elevation of Spirit Lake in feet.

   ______

2. The North Fork of Toutle River was the outlet of Spirit Lake on the Old Map. Why is there no outlet of Spirit Lake shown on the New Map?

   ______

   ______

   ______

3. What impact did the eruption have on the valley of the North Fork of the Toutle River?

   ______

   ______

   ______

4. Draw in blue pencil on the Old Map the outlines of the lakes shown on the New Map that occur along the courses of Coldwater Creek, South Coldwater Creek, and Castle Creek. How did these lakes come into existence?

   ______

   ______

   ______

5. The flanks of Mount St. Helens are draped with alpine glaciers on the Old Map. How have the area and length of these glaciers been changed by the eruption?

   ______

   ______

   ______

   ______

6. The shape of Mount St. Helens is a nearly symmetric cone on the Old Map with a summit elevation of 2,950 meters. The elevation of the summit of Mount St. Helens on the New Map is 2,550 meters, and it lies at a different position than it did before the eruption. Trace the 2,250-meter contour line around the top of the mountain on both maps to show how the mountaintop changed because of the eruption. Explain why the summit elevation is lower on the New Map and why it has changed positions from the Old Map summit.

   ______

   ______

   ______

   ______

7. The landslide-debris flow shown on the New Map altered the terrain over which it moved. Would you describe the *new* landscape as a constructional or destructional surface? Explain why you called it one or the other.

   ______

   ______

   ______

   ______

8. Assuming the 1980s average growth of the central dome of 17 million cubic yards per year, how long will it take to rebuild the 2,200-year-old volcanic cone to its pre-1980 height and volume?

   ______

   ______

   ______

   ______

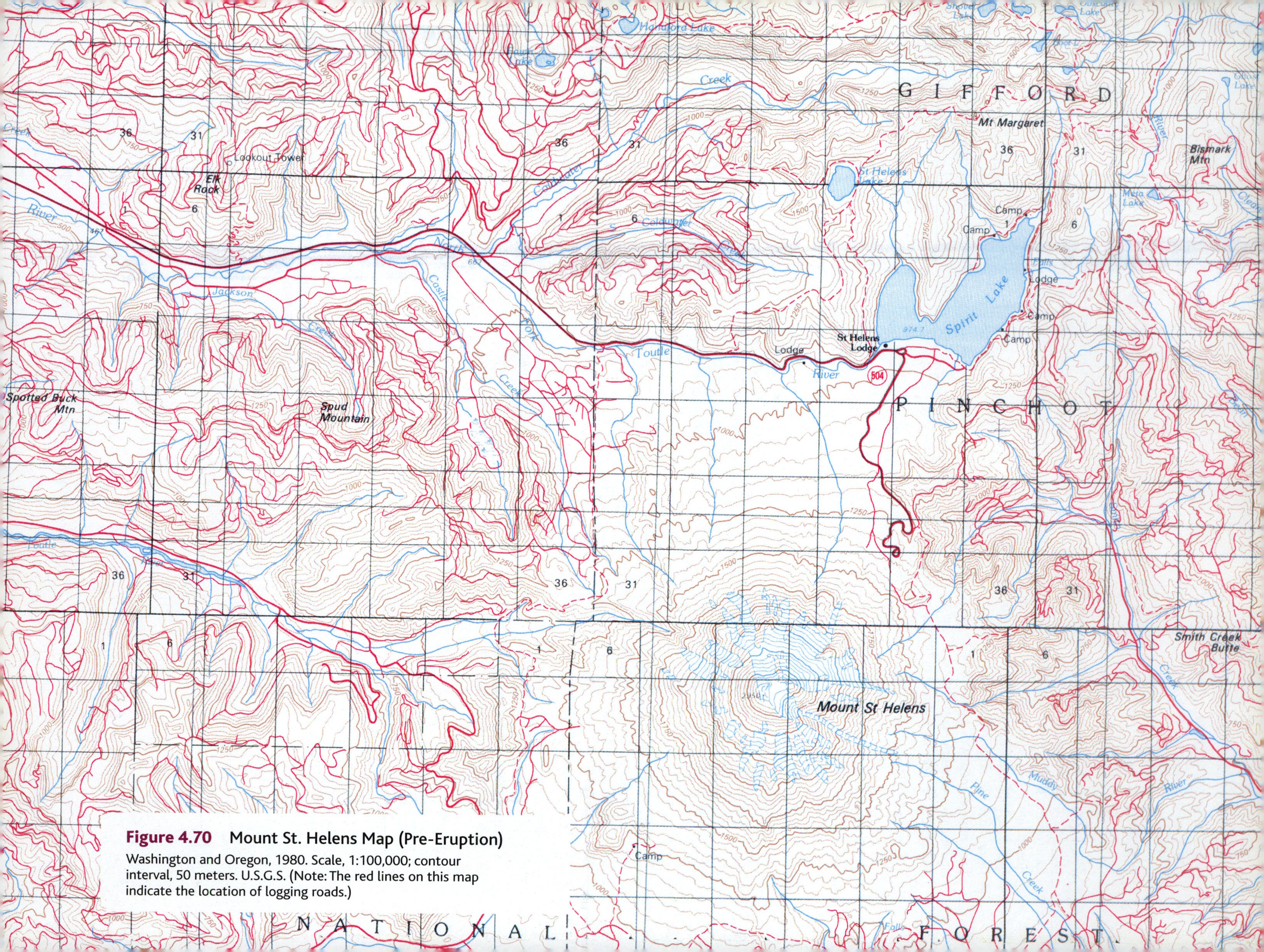

Figure 4.70 Mount St. Helens Map (Pre-Eruption)

Washington and Oregon, 1980. Scale, 1:100,000; contour interval, 50 meters. U.S.G.S. (Note: The red lines on this map indicate the location of logging roads.)

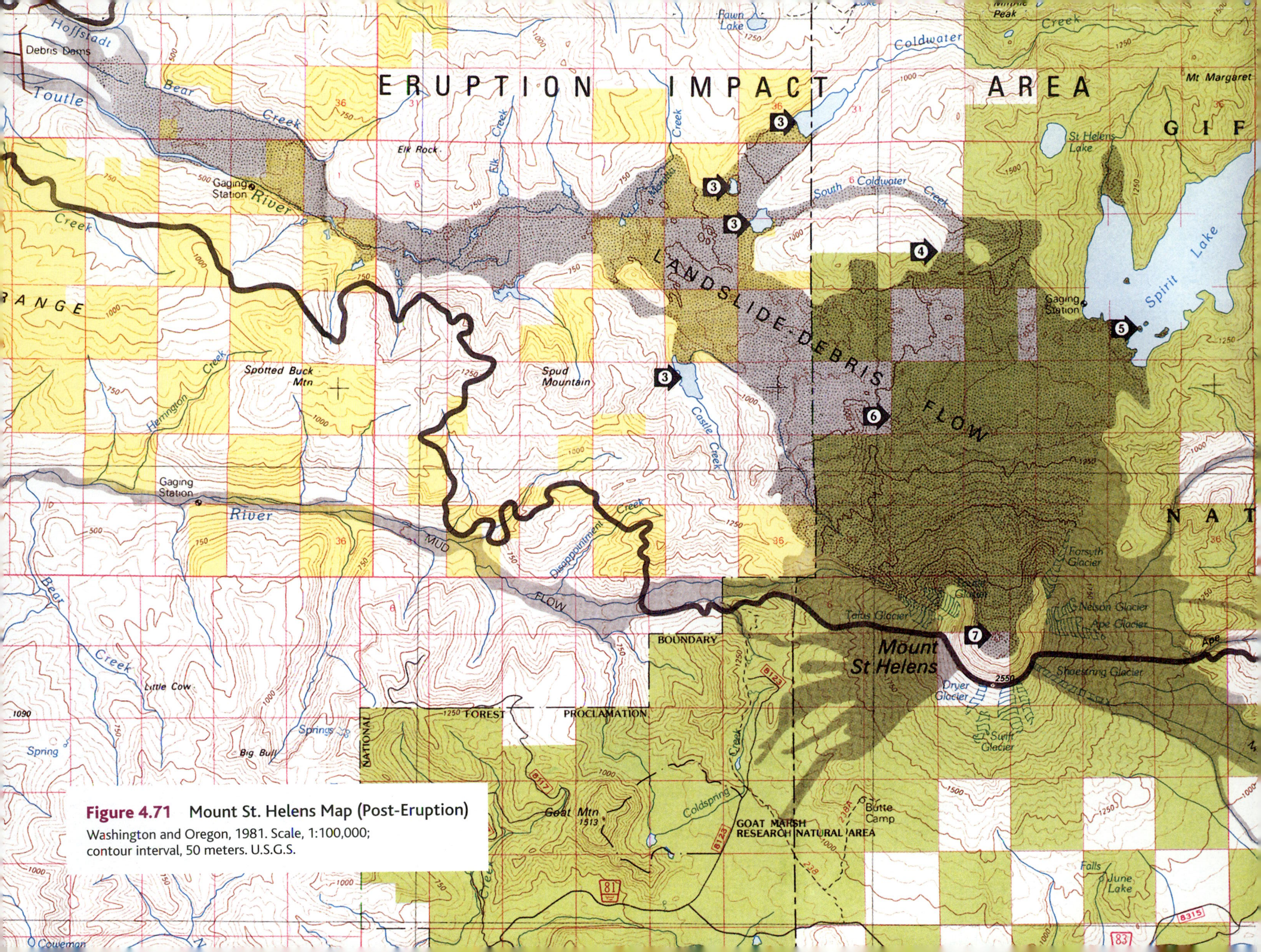

Figure 4.71 Mount St. Helens Map (Post-Eruption)

Washington and Oregon, 1981. Scale, 1:100,000; contour interval, 50 meters. U.S.G.S.

PART V

Structural Geology

Background

Structural geology deals with the architectural patterns of rock masses found in nature. In Parts I and II you were introduced to some basic concepts and principles about the occurrence of rocks. In Part V we will build on those concepts and related terminology with the introduction of geologic maps and block diagrams to show the relationships between outcrop patterns and topography.

Structural geology involves all three rock types—igneous, sedimentary, and metamorphic. In this part of the manual, we concentrate on structures in which sedimentary rock layers are dominant. A sedimentary rock unit that is characterized by a distinct lithologic composition is called a **formation.** A formation is the basic stratigraphic unit depicted on a geologic map. The boundary between two contiguous formations is called a **contact.**

For the sake of simplicity, the formations illustrated herein will be homogeneous in their lithology and will be referred to by some name such as "limestone formation," "sandstone formation," or some other name. In reality, however, a formation may consist of several thinner layers or beds. The contacts between these beds are called **bedding planes** and are more or less parallel to the contacts of the formation itself.

It is common geologic practice to assign a name to a formation. In some cases, the name of the formation will include a lithologic descriptor such as the Madison Limestone or the Pierre shale. In other cases, the name will consist only of a proper name such as the Nelson Formation or the Sunflower Formation. Formational names such as these will be found only on published geologic maps. On simple diagrammatic maps that are used herein, it will suffice to use only a lithologic definition such as "shale formation" or "sandstone formation."

Sedimentary strata occur in a variety of three-dimensional geometric forms and are easy for geologists to work with because they usually contain primary sedimentary structures (i.e., cross beds, ripple marks, etc.) that allow for the interpretation of a stratigraphic younging direction. The pattern of contacts between sedimentary formations portrayed on a map shows the distribution of the formations in only two dimensions. The visualization of a three-dimensional geometric form from a two-dimensional geologic map is one of the main objectives of this part of the manual. By learning certain basic principles and following standard procedures described in the pages that follow, the ability to formulate a mental three-dimensional picture of the geometric configuration of strata beneath the earth's surface can be mastered.

Geologic structures are produced when strata are deformed by forces that contort the strata from their original position of horizontality into geometric forms called **folds.** Folded rock layers retain their continuity as layers; that is, the contact between two formations on a geologic map can be followed as an unbroken line. When exposed by erosion, these folds are revealed in a two-dimensional **outcrop pattern** on a geologic map that is diagnostic of the three-dimensional forms of the structures.

In many cases, deformation causes the strata to break or **fault** along **fault planes** so that the outcrop pattern shows a discontinuity of the formations on either side of the fault. Movement along a fault plane produces an **earthquake.**

Folds, faults, and earthquakes therefore are the subject of Part V and will be treated in that order in the pages that follow.

Structural Features of Sedimentary Rocks

Deformation of Sedimentary Strata

During the course of geologic history, sedimentary strata have been subjected to vertical and horizontal forces that may alter the original horizontal position of the rock layers. Some strata may be uplifted in a vertical direction only, so their original horizontality remains more or less intact. In other cases, the forces of deformation produce architectural patterns ranging from simple to extremely complex structures.

In order to decipher these structures, geologists measure certain features of a given formation where it crops out at the surface of the earth. These measurements define the position of the formation with respect to a horizontal plane of reference. The precise orientation of a contact, bedding plane, or any planar feature, including a fault surface, associated with a rock mass is called the **attitude.** When attitudes from many outcrops are plotted on a base map, such as a topographic map or aerial photograph, and combined with the contacts between formations, the overall geometric pattern or structural configuration of the strata can be determined.

Components of Attitude

The **attitude** of a structural surface such as a bedding plane or a fault plane consists of two parts, **strike** and **dip,** that collectively define the position of the surface at a given location with respect to a horizontal plane and compass direction (fig. 5.1).

1. **Strike** is defined as the line of intersection between a horizontal plane and a structural surface expressed as a compass direction.
2. **Dip** is measured at right angles to the strike and is defined as the maximum angle of slope of a surface measured in a vertical plane downward from the horizontal and is measured in degrees. **Dip direction** is the compass direction of the dip measured at right angles to the strike.

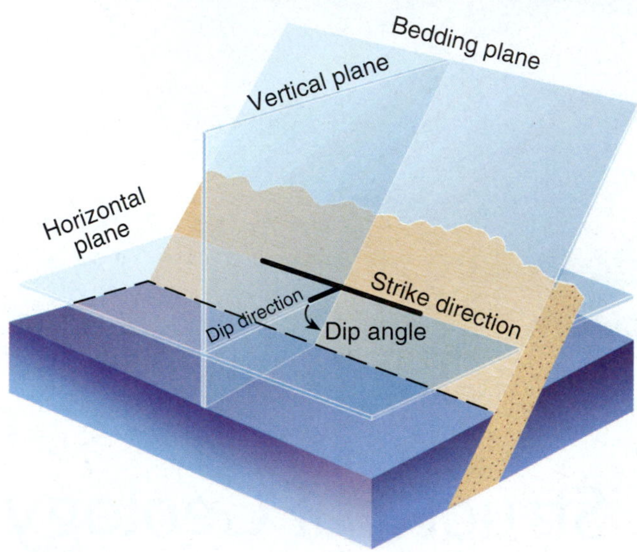

Figure 5.1

Three-dimensional view of an outcropping of sandstone in which the **attitude** of a bedding plane is measured with respect to horizontal and vertical planes. The shaded slanting plane represents the bedding plane of the layered sandstone. The intersection of the bedding plane and a horizontal plane results in a line called the **strike** of the formation. This line is expressed as a compass direction. The angle formed by the horizontal plane and the bedding plane is the **dip** of the formation. The dip, measured in degrees, is always measured in a vertical plane that is perpendicular (at right angles) to the direction of strike. **Dip direction** is the compass direction of that vertical plane and is measured at right angles to the strike.

The number that appears next to a strike and dip symbol on a geologic map refers to the angle of dip of the rock layers as measured by a field geologist at a specific rock outcrop or exposure. When these numbers occur near a line on the map along which a geologic cross section is to be constructed, they should be considered approximations of the dip angles rather than absolute values. Dip angles within a few tens of feet of each other can vary as much as 5 to 10 degrees.

As an example of a verbal description of the attitude of a formation at a particular site, the following notation would be used: On the south side of the Arbuckle Anticline, the Kindblade Formation strikes N50°W or 310° azimuthal and

Figure 5.2
"Tombstone topography" formed on the Kindblade Formation, a limestone interbedded with easily erodible shale on the south flank of the Arbuckle Anticline, Oklahoma. The view is to the southeast along the strike of the beds. The dip is to the south (right) at 50 degrees.
Photo by Robert Rutford.

dips to the south at an angle of 50 degrees (fig. 5.2). On a geologic map, however, the attitude of a formation would be shown by a **strike and dip symbol.** Various forms of this symbol are given in figure 5.3. In illustrations used in parts of this manual, the strike and dip symbols may appear without the notation of the angle of dip.

Methods of Geologic Illustration

Geologic information, gathered by the study of outcrops at the surface and through the use of subsurface information obtained from wells, is displayed in a number of ways in order to depict the overall structural features and relative age relations of the strata involved. The three main types of geologic illustrations or diagrams and their relationships are shown in figure 5.4 and are described below.

1. **Geologic map:** A map that shows the distribution of geologic formations (fig. 5.4A). Contacts between formations appear as lines, and the formations themselves are differentiated by various colors and symbols (refresh your memory by looking again at fig. 2.3). The map may also show topography by standard contour lines.
2. **Geologic cross section:** A diagram in which the geologic formations and other pertinent geologic information are shown in a vertical section (fig. 5.4B). It may also show a topographic profile, or it may be schematic and show a flat ground surface.
3. **Block diagram:** A perspective drawing in which the information on a geologic map and geologic cross section are combined (fig. 5.4C). This mode of geologic illustration is used to show the three-dimensional aspects of a geologic structure.

Sedimentary Rock Structures

Sedimentary strata that have been subjected to forces of deformation may form one of three fold structures as shown in the block diagrams of figure 5.5.

1. **Monocline:** A one-limb flexure (fold) in which the strata have a uniform direction of strike but a variable angle of dip.
2. **Anticline:** A fold, generally convex upward, whose core contains the stratigraphically older rocks.
3. **Syncline:** A fold, generally concave upward, whose core contains the stratigraphically younger rocks.

Notice that in figure 5.5B the arch of the anticline is not reflected in a corresponding topographic arch and that the synclinal trough in figure 5.5C is a geologic trough, not a topographic one. The surface topography of the parallel ridges in figures 5.5B and 5.5C is controlled by a formation that is more resistant to erosion than the other formations in the structure.

Geometry of Folds

The geometry of a fold is more precisely defined by the attitude of the **axial plane** of the fold, an imaginary plane that separates the **limbs** of the folds into two parts (as symmetrically

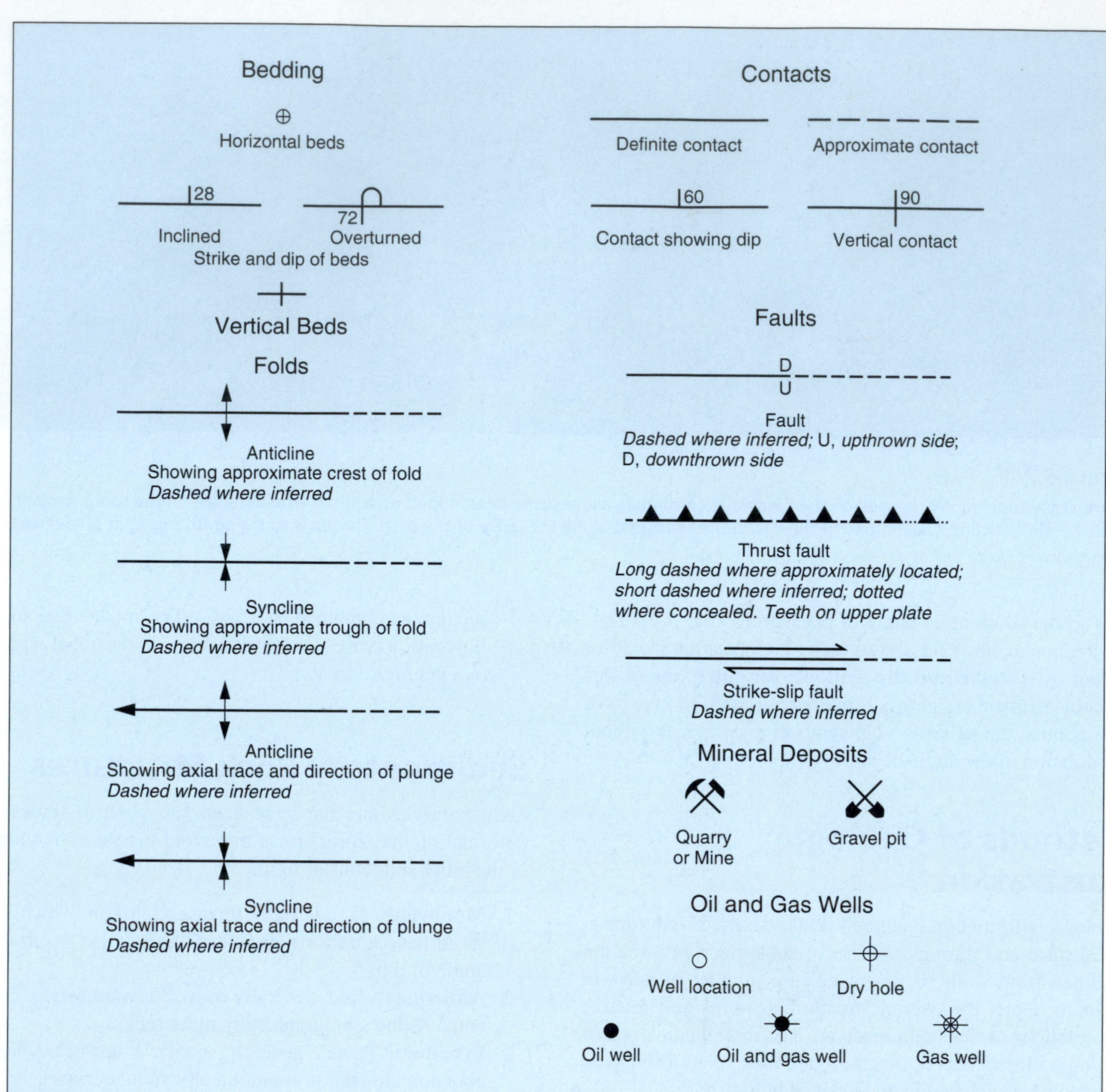

Figure 5.3

Standard symbols used on geologic maps.

as possible) as shown in figure 5.6. The **axial trace** of the fold appears as a line on a geologic map. The **hinge line** marks the axis along which the curvature of the fold is greatest.

If the axial surface is essentially vertical, the fold is said to be **symmetric** (fig. 5.7A); if the axial surface is inclined so that the limbs dip in opposite directions but one limb is steeper than the other, the fold is **asymmetric** (fig. 5.7B); and if the axial surface is inclined to the extent that one limb of the fold has tilted beyond the perpendicular, the fold is **overturned** (fig. 5.7C). A **recumbent fold** is an overturned fold in which the axial surface is nearly horizontal. The symbols used on geologic maps to show the traces of axial surfaces are shown in figure 5.3.

The folds shown in figures 5.4, 5.5, 5.6, and 5.7 are **nonplunging folds** because the strikes of the limbs are parallel. Another way of describing a nonplunging fold is to say that the strikes of the folded formations are all parallel, as shown in figure 5.4.

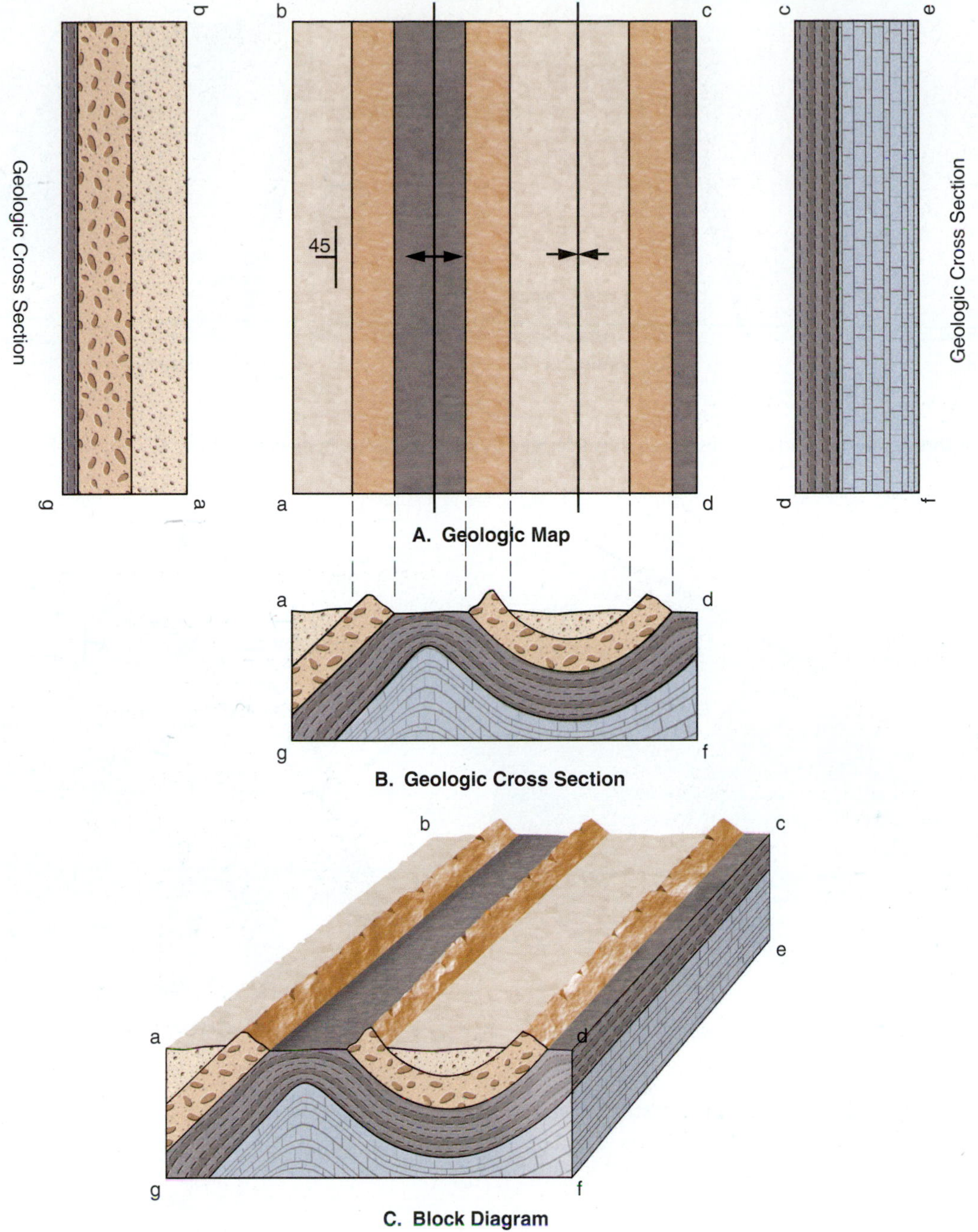

Figure 5.4

(A) A **geologic map** shows the areal extent of geologic formations at the earth's surface and exhibits certain symbols that further define the geometry of the rock masses as they extend beneath the surface. (B) A **geologic cross section** is a view of the geologic formations in a vertical plane. (C) A **block diagram** is a three-dimensional drawing in which the geometric configuration is depicted. (Note that there is no surface expression of the limestone unit. This subsurface information came from local wells.)

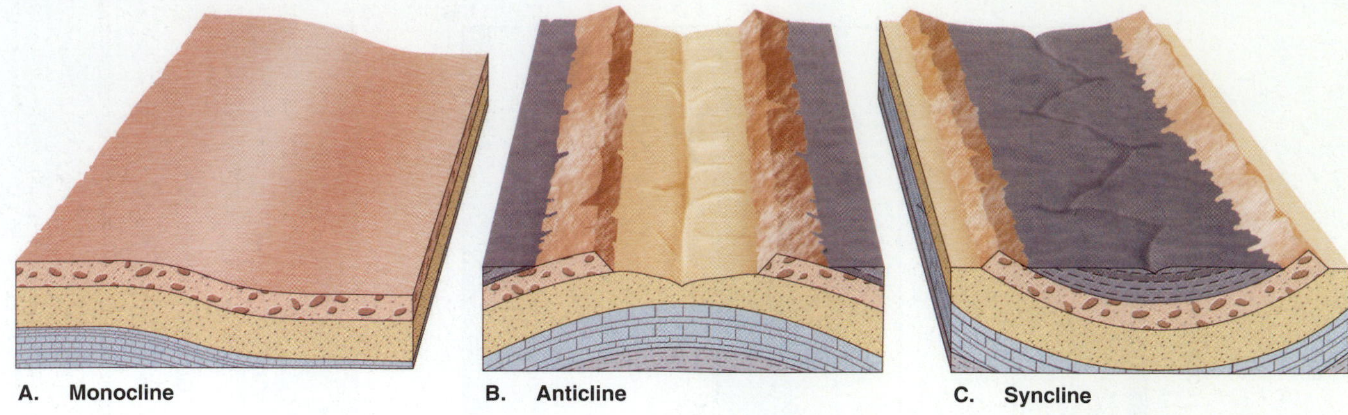

Figure 5.5

Block diagrams of three common folds. The dissected ridges formed by resistant layers in diagrams B and C are called hogback ridges.

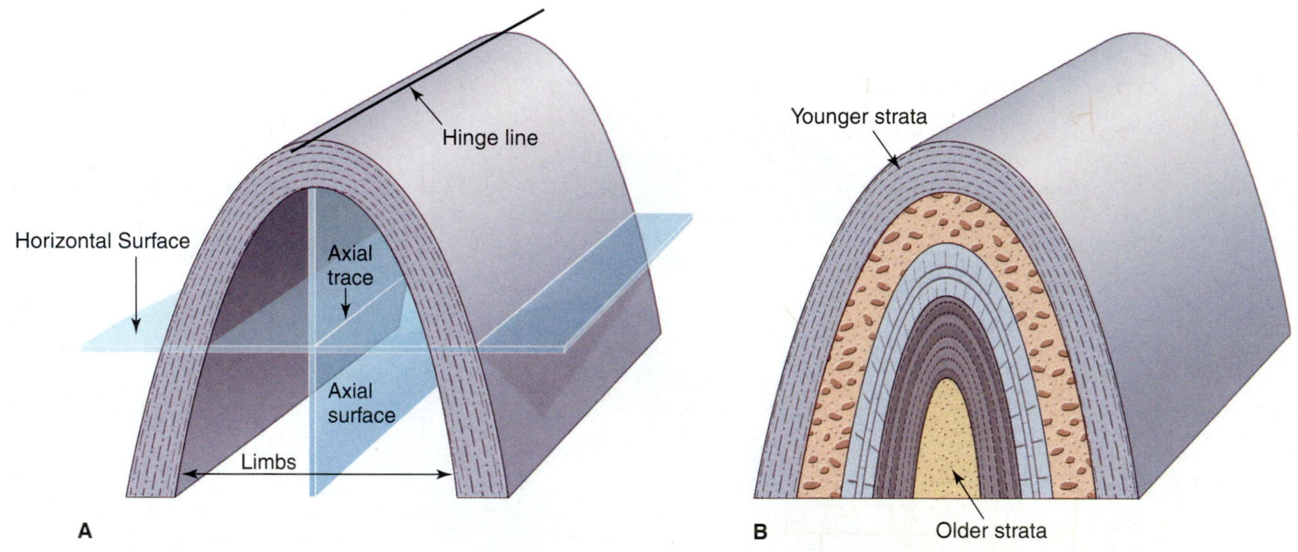

Figure 5.6

(A) Nomenclature of a fold. (B) Age relationships of strata in an anticline.

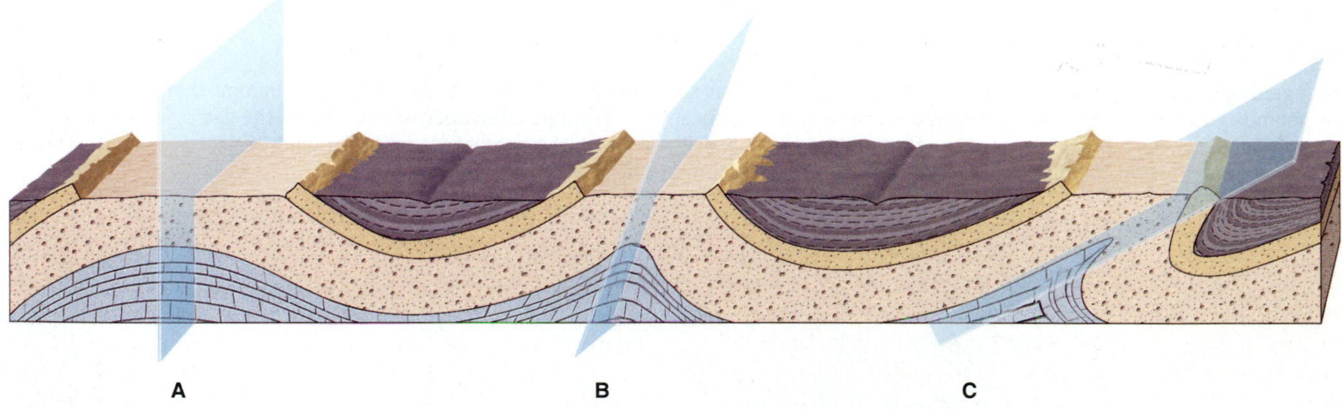

Figure 5.7

Block diagram in which three variations of a fold are shown: (A) symmetric anticline; (B) asymmetric anticline; (C) overturned anticline. Note the different attitudes of the three axial planes.

to be found in the ordinary layer cake. When a layer cake is viewed from above, all that can be seen is the frosting; the "structure" of the cake is obscured. However, if the cake is cut vertically and the two halves are separated, the component layers of the cake constitute a cross section of the cake so that its structure will be revealed.

Geologic formations do occur in "layer-cake" structures, but they commonly occur in much more complex structures, and it is through the construction of a geologic cross section that these complexities are unraveled. Following are some general rules and guidelines for use in constructing a geologic cross section from a geologic map.

1. A geologic cross section is constructed on a vertical plane. The cross section is shown on the corresponding geologic map by a line that is equivalent to the line along which the cake was cut in the layer-cake analogy. Information on or near the line of the cross section on the map is transferred to the cross section as the first step in its construction. Such notations as directions and angles of dip, formational contacts, traces of axial surfaces, supplemented by subsurface information from cuttings, and well logs provide the basic elements used to make a geologic cross section from a geologic map.
2. Sedimentary formations to be drawn on cross sections in the exercises in this manual are assumed to have a constant thickness. That is to say, they do not thicken or thin with depth or along the strike.
3. Dip angles from strike and dip symbols on the map can be used as a basis for estimating the inclination of strata on a cross section. If dip angles are not shown, keep the dip angles as small as possible but consistent with the thickness of the strata and structural relationships. (Note no vertical exaggeration for dip measurements.)
4. The relative ages of sedimentary strata in some of the maps and cross sections used herein are designated by arabic numerals. For example, if four formations are shown on a map or block diagram, the oldest formation is assigned the number "1," and the youngest, a number "4" (fig. 5.11A).
5. If you are required to draw a geologic cross section from a geologic map on which no strike and dip symbols are present, the direction of dip can be determined in the following manner.
   (a) Where a formation contact crosses a stream on the map, it forms a V, the apex of which points in the direction of dip as shown in the geologic map of figures 5.11B and 5.11C. (This rule is not to be confused with the "law of V's" as applied to contour lines when they cross a stream.)
   (b) The shape of a V formed by a contact that crosses a stream may be used to estimate the angle of dip of the contact. A broad V is indicative of a steep dip angle, and a narrow V is indicative of a shallow dip angle. Where no V is formed, the formation contact is vertical, as shown in figure 5.11D. The foregoing method for determining the direction of dip takes precedence over the method described next.
   (c) In a sequence of formations, none of which has been overturned, the oldest beds dip toward the youngest, as shown in the geologic map of figures 5.11B and 5.11C.

Width of Outcrop

Strata exposed to erosion at the earth's surface appear as bands on the geologic map. The width of a single band is called the **width of outcrop,** although the full thickness of the formation may not be exposed in a single outcrop. The width of outcrop is controlled by three factors: the thickness of the formation, the angle of dip of the formation, and the slope of the land surface where the outcrop is exposed.

To illustrate these controlling factors in the simplest case, consider the three horizontal formations of equal thickness in figure 5.12. The geologic cross section in figure 5.12A shows how the thickness of each formation varies with the slope of the land surface. A gentle slope results in a width of outcrop that is greater than the thickness of the formation, as in the case of the shale formation; and a steeper slope produces a width of outcrop that is less than the thickness of the formation, as in the case of the sandstone and limestone formations.

Two other cases of the relationship of thickness to the width of outcrop are shown in figures 5.12B and 5.12C. In figure 5.12B, where the beds are dipping 30 degrees, the thickness of each formation is shown on the cross section, and the corresponding width of outcrop is shown on the geologic map. In figure 5.12C, the formations are vertical; that is, they dip 90 degrees. In this case, the true thickness of a formation is the same as the width of outcrop. The general rule, however, is that **the width of outcrop on a geologic map is not necessarily the same as the true thickness of the formation as seen in a geologic cross section.** This rule must be kept in mind when drawing cross sections from a geologic map or a block diagram in the exercise that follows.

Figure 5.11

Block diagram and maps showing the relationship of topography to outcrop patterns. In all cases, the stream flows from north to south. (A) Horizontal strata dissected by a drainage system. Numbers refer to relative ages of the formations. The formation labeled 1 is the oldest. The apex of the V formed will point upstream and will be parallel to the contours. (B) Tilted rock strata dipping downstream at an angle steeper than the stream channel. The oldest beds (i.e., 1, 2, and 3) dip toward the youngest beds (5 and 6). The apex of the V formed points downstream and in the direction of dip. (C) Tilted strata dipping upstream. The apex of the V formed points upstream and in the direction of dip, but the contact crosses contours. (D) Vertical sedimentary beds, one of which is more resistant to erosion than the other two. In this case, the law of V's cannot be used, because no V's are formed. Thus, the age relationship cannot be determined from the information either on the block diagram or on the map.

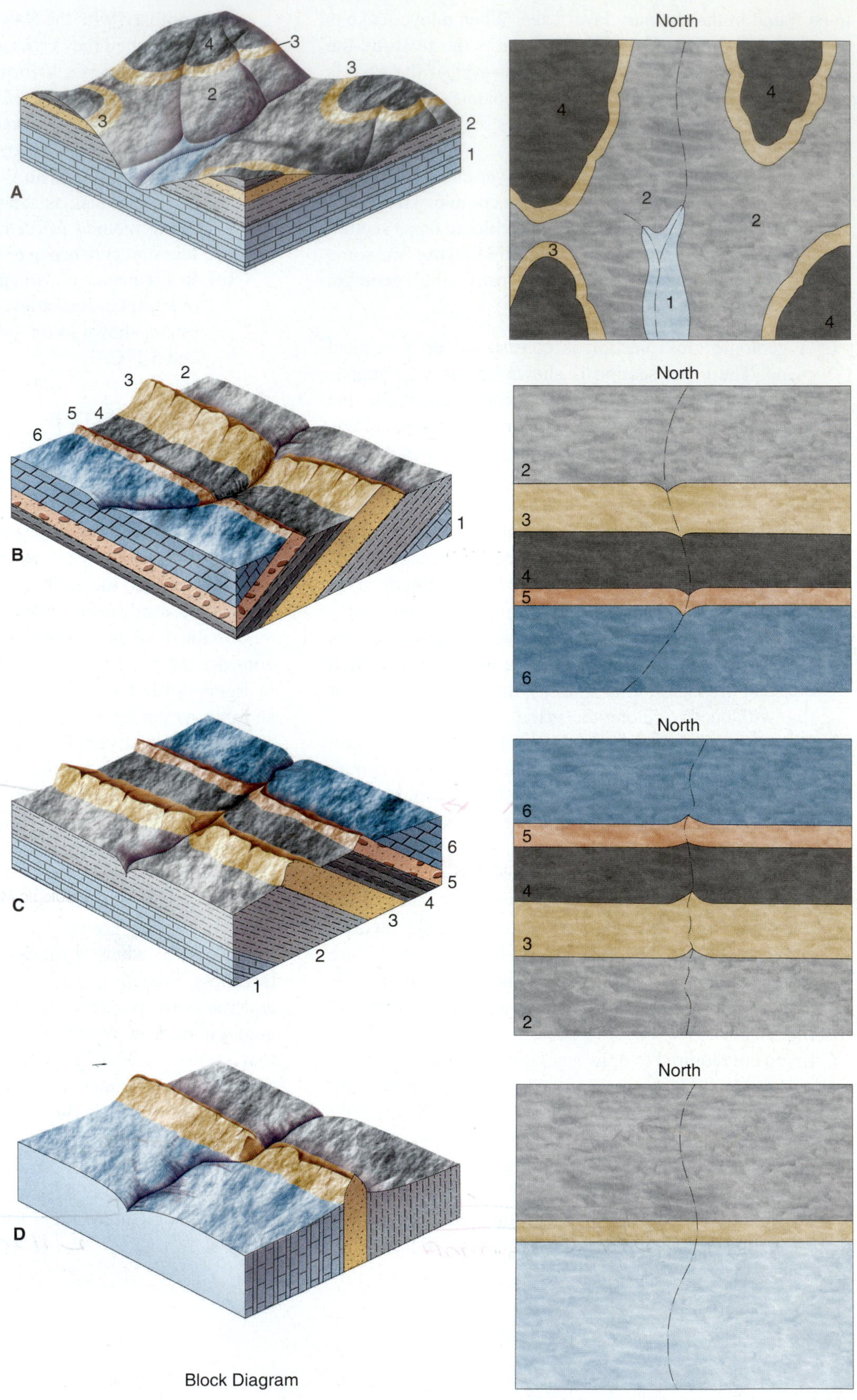

NOTES

EXERCISE 22A

Name

Section Date

Geologic Mapping on Aerial Photographs

Chase County, Kansas

Figure 5.14 shows the outcrop pattern of sedimentary strata. The light and dark gray tones correspond to different lithologic characteristics of the various formations. At A, near the center of the photograph, the contacts of a light gray formation are shown by two black lines.

1. Extend these lines along the contacts as far as possible on the photo. Do the same for formation B shown in the upper left-hand corner of the photo.
2. What is the general attitude of these two formations? (Compare the outcrop pattern of fig. 5.14 with fig. 5.11A.)

____

____

____

3. If it is assumed that the thickness of these two formations is constant, why do their widths of outcrop change from place to place?

____

____

____

4. Applying the law of superposition in this case, which of the two formations is relatively older than the other?

____

____

____

Figure 5.14

Aerial photograph, Chase County, Kansas. Scale, 1:20,000.

Photo by U.S. Department of Agriculture.

EXERCISE 22B

Name ______

Section ______ Date ______

Geologic Mapping on Aerial Photographs

Fremont County, Wyoming

This stereopair (fig. 5.15) shows sedimentary strata cropping out in the area. Study the stereopair with a stereoscope, and using figure 5.11B for guidance, answer the following questions.

1. Draw several strike and dip symbols on the right-hand photo of the stereopair, and write a verbal description of the attitude of the formations.

______

______

______

______

2. Are the oldest beds in the northern or southern part of the area? What rule is applied here that allows you to answer this question?

______

______

______

______

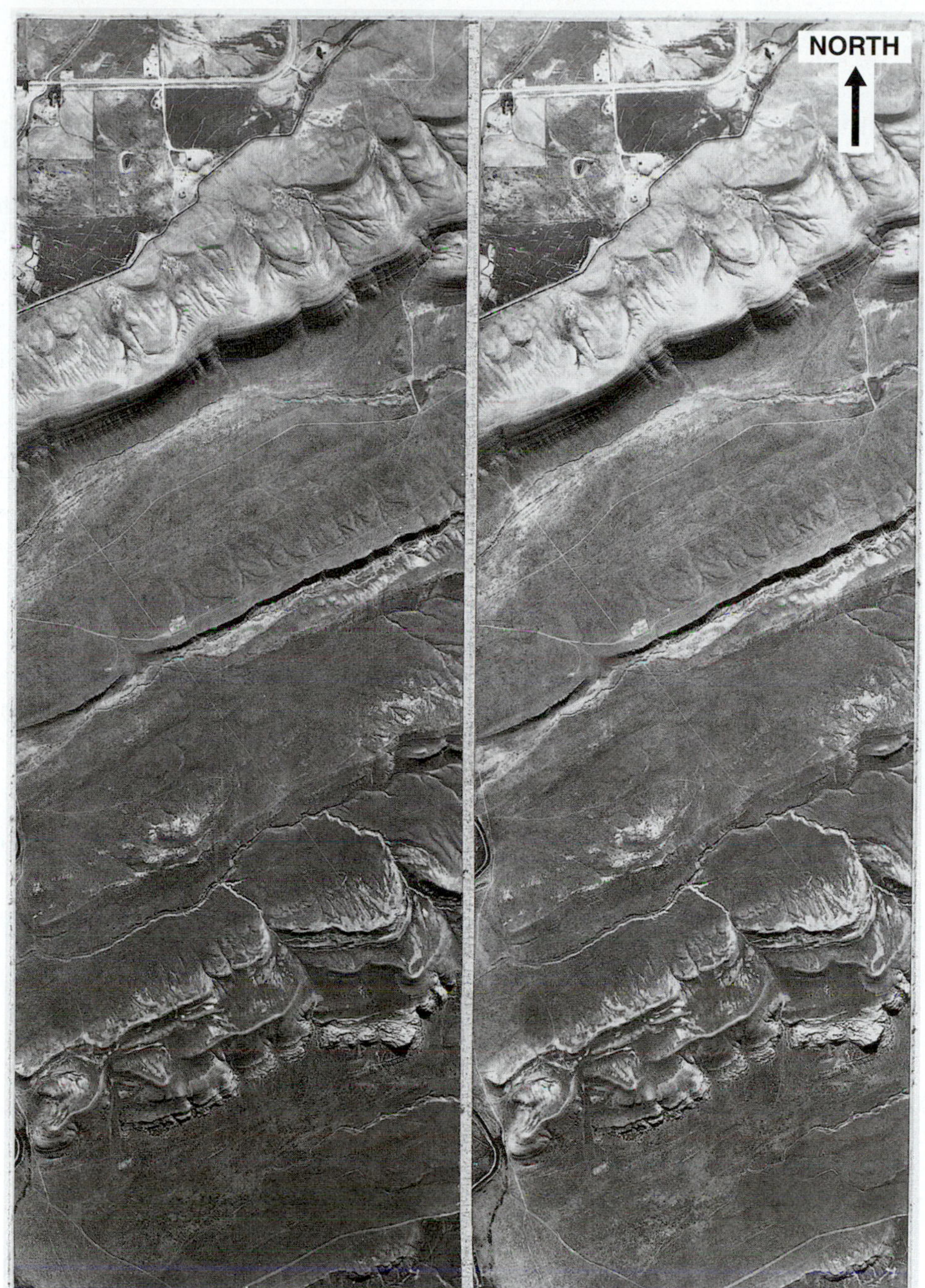

Figure 5.15

Stereopair of part of Fremont County, Wyoming. Scale, 1:21,500; July 13, 1960.

U.S. Geological Survey.

EXERCISE 22C

Name

Section Date

Geologic Mapping on Aerial Photographs

Aerial Photograph, Arkansas

The pattern of curved ridges in this photograph (fig. 5.16) is the result of differential erosion of sedimentary strata. The ridges are composed of rocks that are more resistant to erosion than the rocks that form the intervening valleys. The structure displayed in the photograph is the nose of a steeply plunging anticline.

1. Draw lines on the photo to show contacts between formations of different resistance to erosion.
2. Draw the trace of the axial plane of the fold on the photograph with a red pencil, and add other appropriate symbols on the axial trace and elsewhere on the photograph to indicate all relevant structural information. (Refer to fig. 5.3 as a reminder of the appropriate symbols to use for the fold axis.)

Figure 5.16

Aerial photograph showing the nose of a plunging anticline in Arkansas. Scale, 1:24,000; November 9, 1957.

U.S. Geological Survey.

EXERCISE 22D

Name ______________________________

Section ____________________ Date ____________________

Geologic Mapping on Aerial Photographs

Little Dome, Wyoming

The structure shown here (fig. 5.17) is an elongate dome or a doubly plunging anticline. Use a stereoscope to study the stereopair while formulating the answers to the following questions.

1. Draw strike and dip symbols on the right-hand photograph of the stereopair. Use red pencil.
2. Draw the trace of the axial plane and other symbols that are appropriate for this structure on the right-hand photograph. Use red pencil.
3. What is the evidence that the angle of dip changes as one follows the ridges northward along the eastern flank of the structure?

____

____

____

____

4. If a hole were drilled on the axis of the fold at the center of the structure, would the drill encounter any of the formations that crop out on the surface in the area covered by the photographs? Explain your reasoning.

____

____

____

____

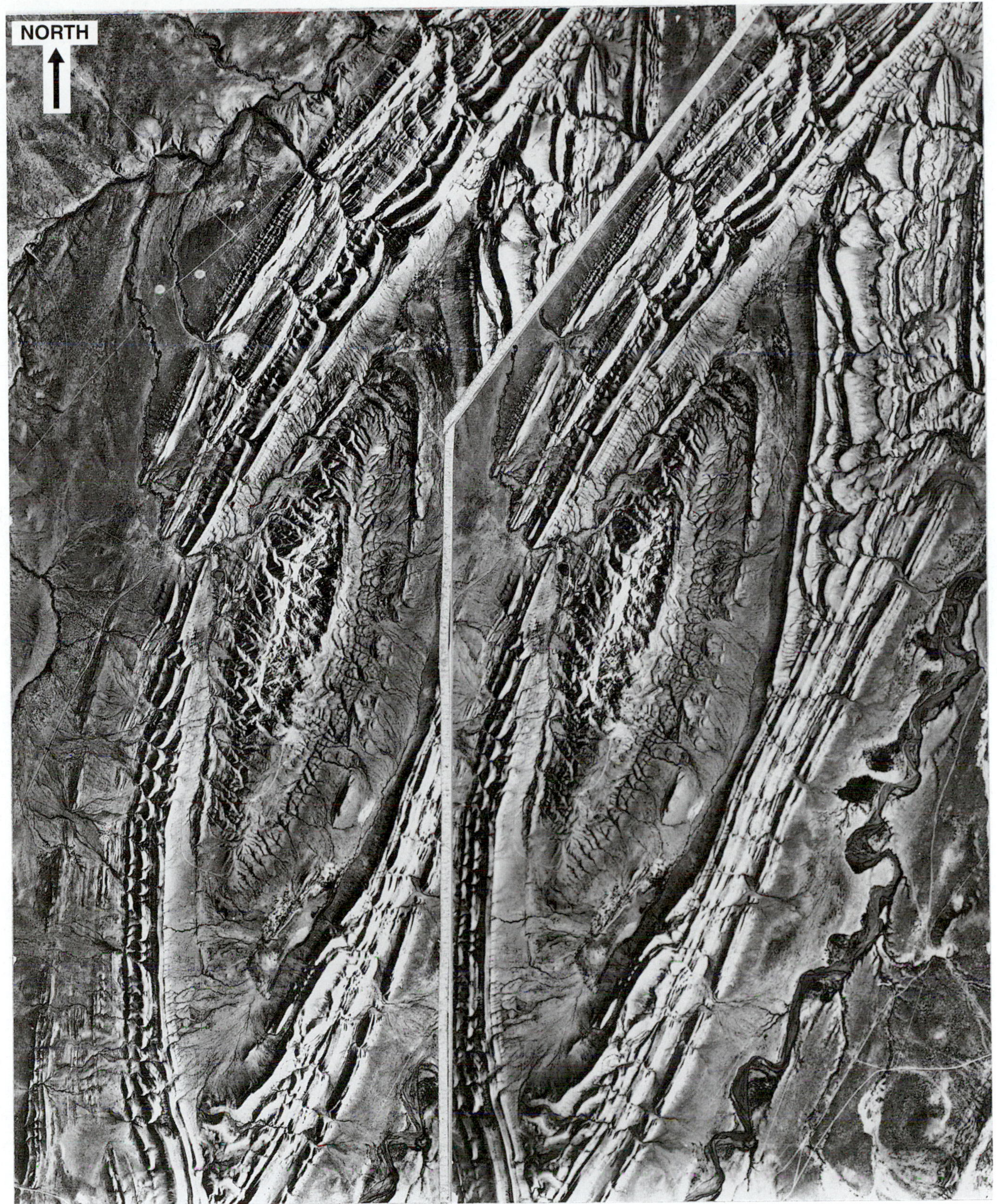

Figure 5.17

Stereopair of Little Dome, Wyoming. Scale, 1:23,600; October 20, 1948.

U.S. Geological Survey.

EXERCISE 22E

Name ______________________________

Section ______________________ Date ______________________

Geologic Mapping on Aerial Photographs

Harrisburg, Pennsylvania

Figure 5.18 looks like an aerial photograph but, in fact, is a side-looking airborne radar (SLAR) image that enhances surface features. A SLAR image is particularly useful in deciphering the structural geology of an area.

The patterns of ridges and valleys shown here are the result of differential erosion of a series of plunging folds in the Appalachian Mountains of Pennsylvania. The river flowing through the area is the Susquehanna River, and the city of Harrisburg lies at the upper margin of the figure near the river.

Orient the map, figure 5.18, with the legend box in the upper left-hand corner. North direction is toward the bottom of the figure. The reason for this is that the "shadows" produced by the imaging process must be toward the viewer in order for the ridges on the ground to appear as ridges on the image. By rotating the figure 180 degrees, you may see "reverse topography"—that is, the ridges appear as valleys and the valleys appear as ridges. Examine the image to get a feel for the structural pattern it reveals.

1. Locate the zig-zag ridge that is cut by the Susquehanna River at four places on the image. Using an easily erasable pencil, draw the trend of this formation on the figure. Or, to put it another way, draw a continuous strike line along the crest of the ridge throughout its length. For the purpose of identification, this ridge-forming formation will be called formation A. It forms the flanks of plunging synclines where it is intersected by the Susquehanna River.
2. When you are satisfied that you have identified formation A by the method just described, draw over your pencil line with a yellow felt-tipped pen or yellow pencil to distinguish the ridge from other ridges in the figure.
3. Locate the next youngest ridge-forming formation and trace its strike as you did for formation A. This younger formation will be called formation B, and it should be identified with a color on the image that contrasts with the one used for formation A.
4. Using a pencil, draw the axial traces of the folds that occur on the figure showing the direction of plunge and the symbol for an anticline or syncline. Reinforce your pencil line with a red pencil when you are certain of your interpretation.
5. Are the rocks west of the Susquehanna River (i.e., between the river and the right-hand margin of the figure) generally older or younger than those near the left-hand margin of the figure? Explain your reasoning.

____

____

____

____

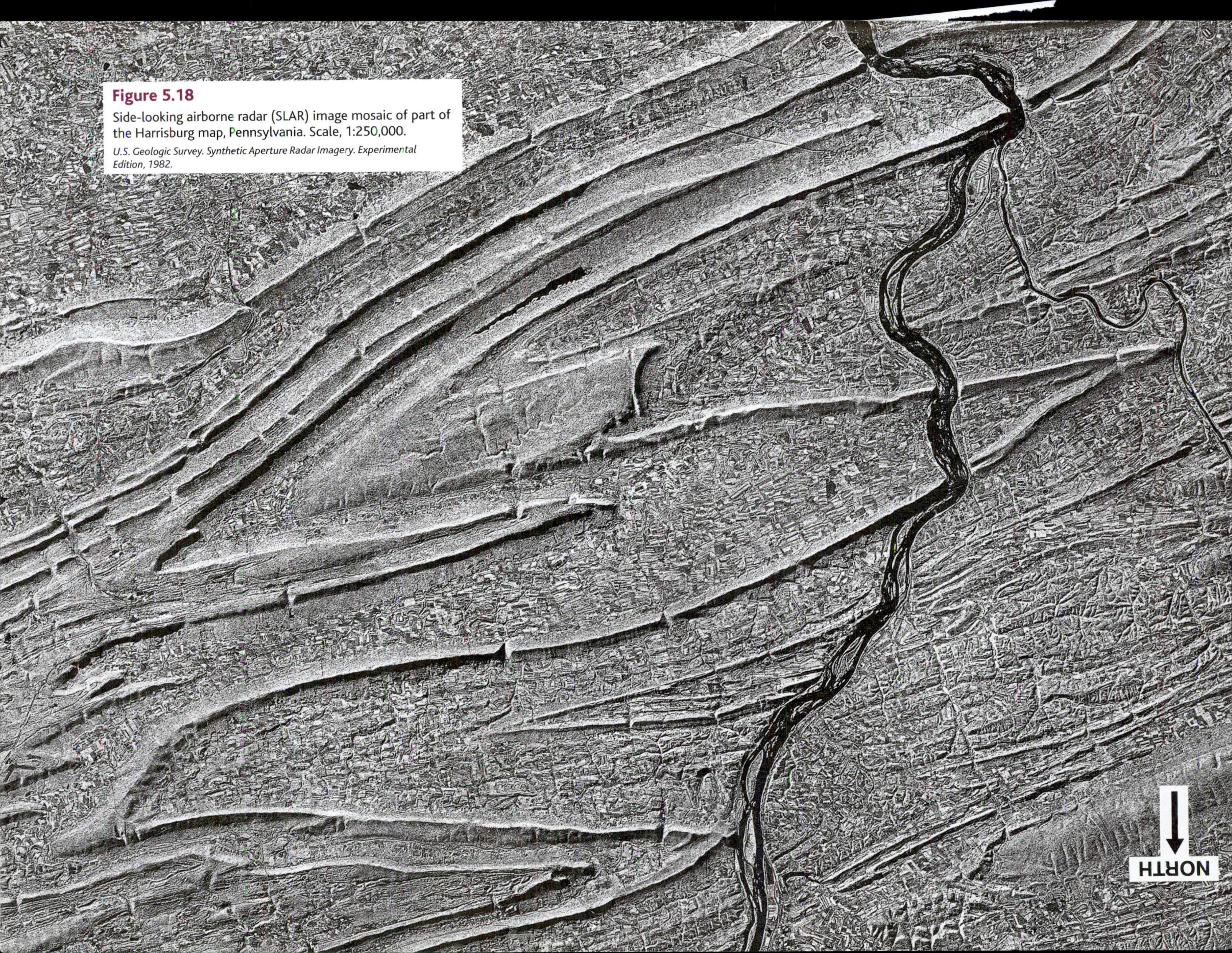

Figure 5.18
Side-looking airborne radar (SLAR) image mosaic of part of the Harrisburg map, Pennsylvania. Scale, 1:250,000.

U.S. Geologic Survey. Synthetic Aperture Radar Imagery. Experimental Edition, 1982.

EXERCISE 23A

Name

Section Date

Interpretation of Geologic Maps

Lancaster Geologic Map, Wisconsin

Six formations are shown on this map (fig. 5.19), each of which is identified by an abbreviation. The abbreviations and the names they represent are, in alphabetical order: Od, Decorah Formation; Ogl, Galena Dolomite; Op, Platteville Formation; Opc, Prairie du Chien Group; Osp, St. Peter Sandstone; and Qal, Alluvium. A **group** consists of two or more formations with significant features in common.

The contacts of these formations are more or less parallel to the topographic contours, thereby indicating that the formations are more or less horizontal. Another set of contours, shown in red, defines the top of the Platteville Formation (Op). The red numbers associated with these red contour lines indicate the elevation of the contour line above sea level.

1. Determine the oldest and youngest formations on the map and those of intermediate age. (Label oldest with number 1.) Complete the geologic column in figure 5.19 by printing the **abbreviation** of a formation in the appropriate box and printing the **name** of the formation on the line immediately below the box (refer to p. 58).
2. Using the topographic contour lines, estimate the thickness of the Decorah Formation (Od), the Platteville Formation (Op), and the St. Peter Sandstone (Osp). Print the estimated thickness in feet of each of these formations to the right of the appropriate box in the geologic column. The thickness of a formation is determined by subtracting the elevation of the bottom of the formation from the elevation of the top of the formation. These elevations can be estimated from contour lines on either side of a contact.
3. Why is it impossible to determine the thickness of the Galena Dolomite and the Prairie du Chien Group?

____

____

____

____

4. Locate Cement School in the northeast part of the map area. A road intersection near the school has an elevation of 1,076 feet and is so marked on the map. Using a nearby contour line showing the top of the Platteville Formation, determine how deep a well must be drilled at the road intersection to reach the top of the Platteville Formation.

____

____

____

____

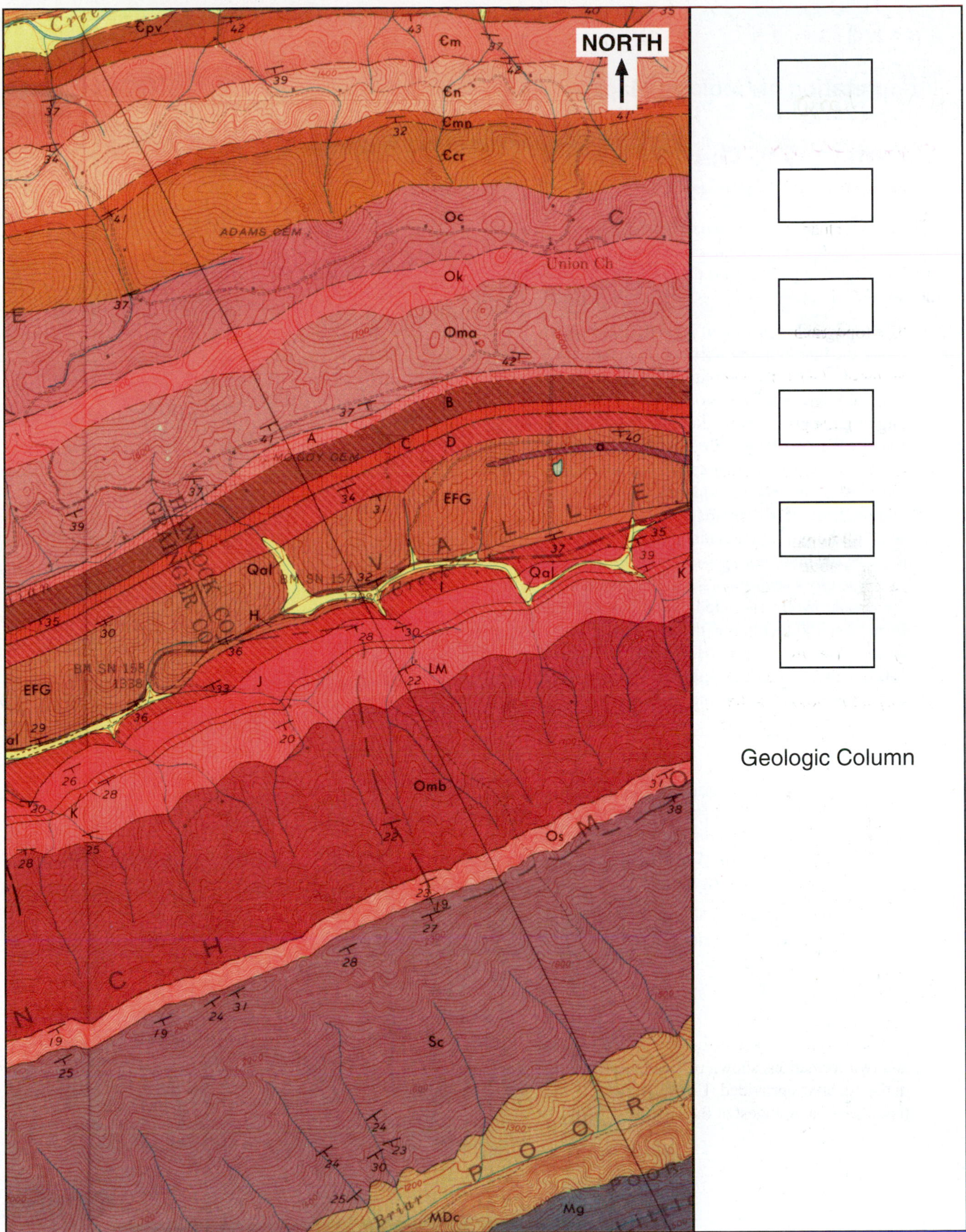

Figure 5.21 Swan Island Map

Geologic map of part of the Swan Island quadrangle, Tennessee, 1971. U.S. Geological Survey. Scale, 1:24,000; contour interval, 20 feet. The black line is the line of topographic profile of figure 5.20.

Courtesy of the U.S.G.S.

EXERCISE 23C

Name

Section Date

Interpretation of Geologic Maps

Coleman Gap Geologic Map, Tennessee–Virginia

This area (fig. 5.23) is underlain by sedimentary rock formations of different thicknesses. Note the many strike and dip symbols that occur on the map. (North is toward the top of the page.)

1. The topographic profile of figure 5.22 is drawn along the line trending northwest to southeast across the map area from margin to margin, passing through the location of Brooks Well. Draw a geologic cross section along this line. Align the Brooks Well on the profile with the position of the Brooks Well on the map to achieve the proper correlation between the topography of the profile and the contours on the map. Note that the vertical scale of the profile is identical with the horizontal scale of the map. For assistance in drawing the geologic cross section, it should be noted that the Brooks Well penetrated the base of the Ꞓc formation 300 feet below the ground surface. Color formations Ꞓc and Ꞓcr on the cross section, and label all formations with their correct symbols.
2. What is the thickness of formation Ꞓcr?

____

____

3. Complete the geologic column to the right of the map. Put the symbol of the formation in the appropriate box and the name of the formation on the line below the box. The symbols and formational names are as follows:

   Ocl, Lower Chepultepec Dolomite
   Ꞓmn, Maynardville Limestone
   Ꞓcr, Copper Ridge Dolomite
   Ꞓc, Conasauga Shale

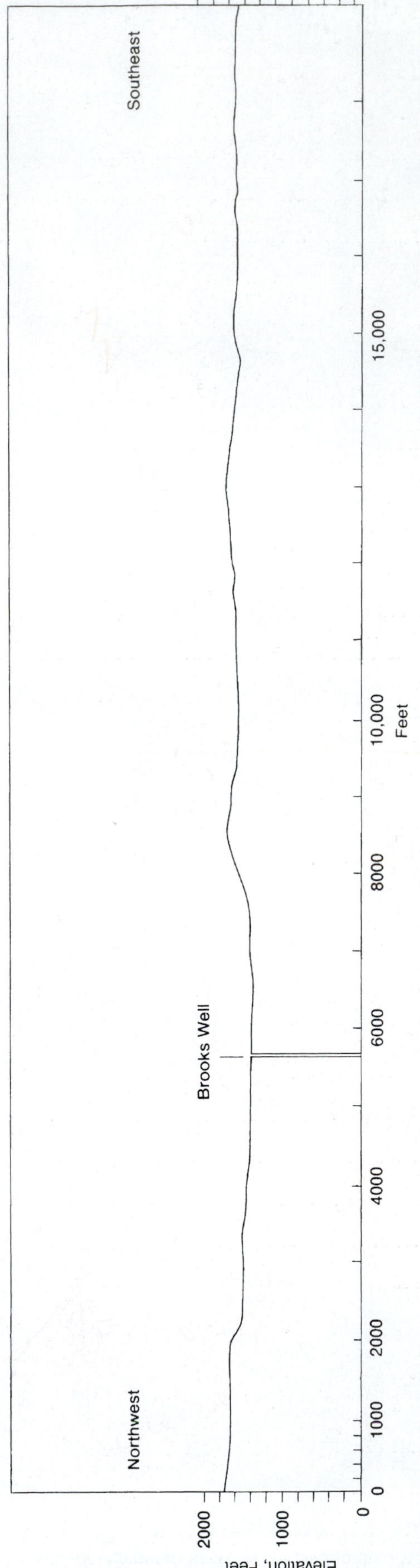

Figure 5.22 Topographic profile from northwest to southeast across the Coleman Gap map. (See fig. 5.23 for location.)

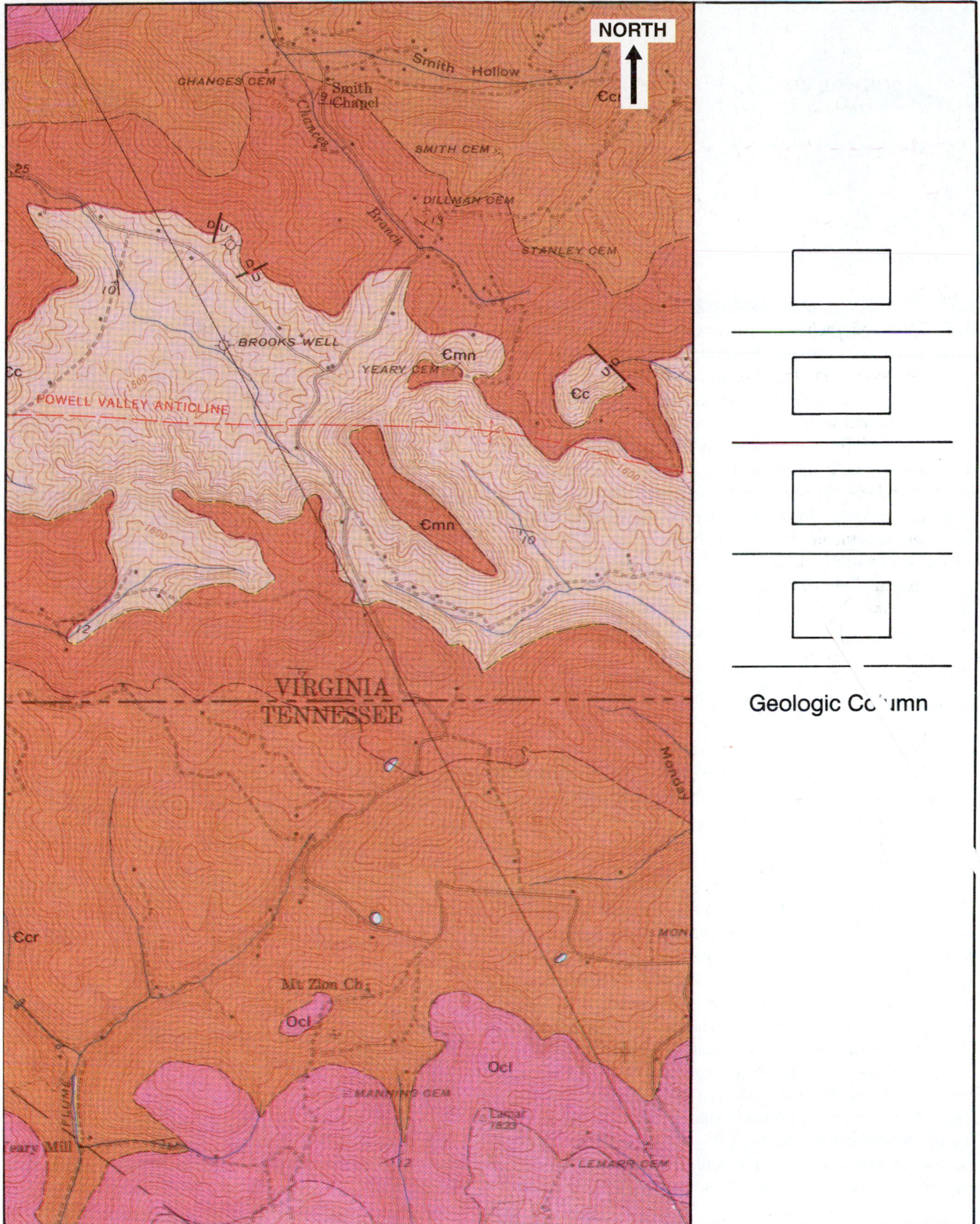

Figure 5.23 Coleman Gap Map

Geologic map of part of the Coleman Gap quadrangle, Tennessee–Virginia, 1962. U.S. Geologic Survey. Scale, 1:24,000; contour interval, 20 feet. The black line passing through the Brooks Well profile is the line of topographic profile of fig. 5.22.

Faults and Earthquakes

Earth (tectonic) stresses that produce folds also produce faults. A **fault** is a fracture or break in the earth's crust along which differential movement of the rock masses has occurred. Movement along a fault causes dislocation of the rock masses on each side of the fault so that the contacts between formations are terminated abruptly.

Faults may be active or inactive. **Active faults** are those along which movement has occurred sporadically during historical time. Earthquakes are caused by movement along active faults. **Inactive faults** are those along which no movement has occurred during historical time. They are treated as part of the structural fabric of the earth's crust.

In this section, we will deal first with inactive faults as part of structural geology and, second, with active faults and their relationship to earthquakes.

Inactive Faults

A fault is a planar feature, and therefore its attitude can be described in the same way that any geologic planar feature can be described. Figure 5.3 shows the various symbols used on a geologic map to define faults.

Nomenclature of Faults

Figure 5.24 is a block diagram of a hypothetical faulted segment of the earth's crust. The **fault plane** is defined as a-b-c-d. The fault plane strikes north–south and dips steeply to the east. A single horizontal sedimentary bed acts as a reference marker and shows that the displacement along the fault plane is equal to the distance between *x* and *y*. This is called the **net slip.** The arrows show the direction of relative movement along the fault plane. Block A has moved up with respect to block B, and conversely, block B has moved down with respect to block A. Block A is called the **upthrown side** of the fault, and block B is the **downthrown side.** Block B is also known as the **hanging wall** and block A as the **footwall.** Both terms are derived from miners who drove tunnels along fault planes to mine ore that had been emplaced there.

Faults generally disrupt the continuity or sequence of sedimentary strata, and they cause the dislocation of other rock units from their prefaulted positions. On geologic maps, the intersection of the fault plane with the ground surface is called a **fault trace.** Fault traces are depicted on geologic maps by the use of standard symbols (fig. 5.3).

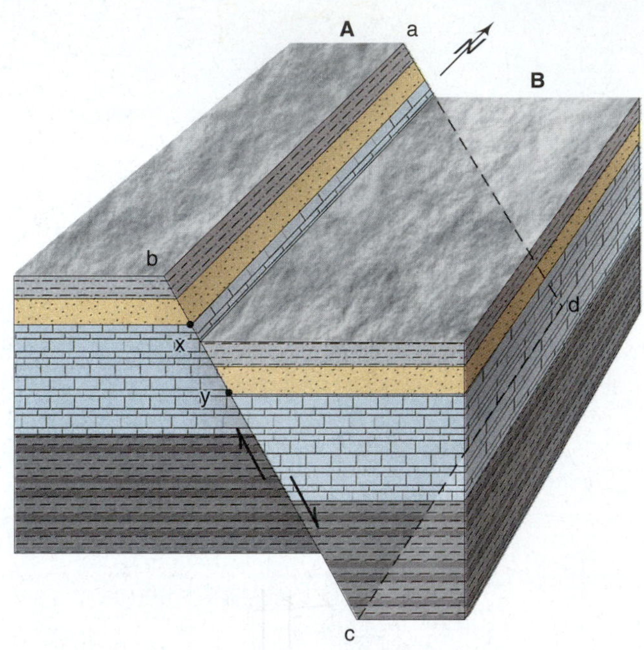

Figure 5.24

Block diagram of a fault. Arrows show the relative movement of block (A) with respect to block (B). The horizontal beds have been dislocated the distance between *x* and *y*. The fault plane is that surface defined by a-b-c-d.

After faulting occurs, erosion usually destroys the surface evidence of the fault plane, so with the passage of time, the **fault scarp** (the exposed surface of the fault plane in figure 5.24) is destroyed. Only the fault trace remains.

Types of Faults

Faults are divided into three major categories, each of which is defined by the relative displacement along the fault plane. In the first type, **dip-slip faults,** displacement has been vertical, more or less parallel to the dip of the fault plane. A **normal fault** is one in which the hanging wall has moved down relative to the footwall (figs. 5.24, 5.25, and 5.26A). A **reverse fault** is one in which the hanging wall has moved upward relative to the footwall (fig. 5.26B). A

Figure 5.25

Normal fault in the Oil Creek sandstone, U.S. Silica Quarry, Mill Creek, Oklahoma.

Photo by Robert Rutford.

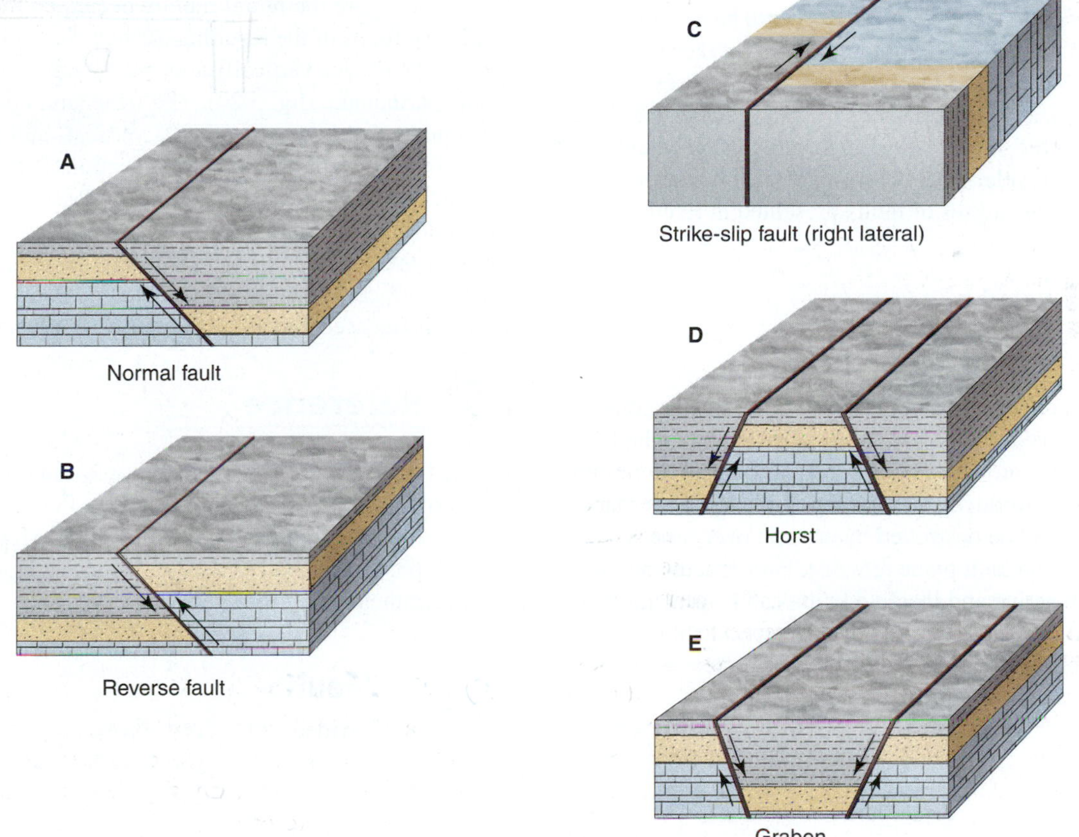

Figure 5.26

Block diagrams illustrating major fault types. Arrows indicate relative movement along the fault plane. Note that these diagrams illustrate the map and cross section views *after* erosion has removed the fault scarp, so on the surface, only the fault trace remains.

low angle reverse fault (dip of less than 45 degrees) is called a **thrust fault.** Figure 5.26D shows a **horst,** an upthrown block bounded on its sides by normal faults. Figure 5.26E shows a **graben,** a downthrown block bounded on its sides by normal faults.

A second category of fault is defined by relative horizontal displacement parallel to the fault plane and are known as **strike-slip** faults (fig. 5.26C).

A special type of strike-slip fault, the **transform fault,** associated with plate boundaries (see figure 6.1B).

The third category, and possibly the most common are those faults with oblique motion along the fault plane. These **oblique-slip faults** have both vertical (dip-slip) and horizontal (strike-slip) displacement.

A fault shown on a geologic map can be analyzed to determine what kind of fault is involved. The analysis of normal and reverse faults will reveal the hanging and footwalls that lead to an understanding of the relative movement along the fault plane. In a strike-slip fault, off-setting of a marker bed as in figure 5.26C is the most direct evidence of the direction of movement. In this case, the displacement is to the right as one looks across the fault, and this would be called a **right-lateral fault.** If movement was to the left as you look across the fault, it would be a **left-lateral fault.**

A normal or reverse fault that cuts across the strike of inclined or folded sedimentary beds presents one of the most common situations for the analysis of movement along the fault plane. In such cases, there will be an apparent migration of the beds in the direction of dip of these beds on the upthrown side of the fault as erosion progresses. Stated another way, if an observer were to stand astride the fault trace, the observer's foot resting on the older rock would rest on the upthrown side. This is a simple mental test that can be applied to the analyses of faults presented in Exercise 24.

Active Faults

Fault Scarps and Fault Traces

The nomenclature of active faults is identical with the nomenclature of inactive faults. The attitude of an active fault plane may range from vertical to horizontal. Movement along the fault plane may produce a fault scarp at the earth's surface, but this scarp may be destroyed by erosion over time so that only a **trace** of the fault plane remains. Fault traces and fault scarps of both active and inactive faults can be identified on aerial photographs or images from earth-orbiting satellites by abrupt changes in topography or color patterns.

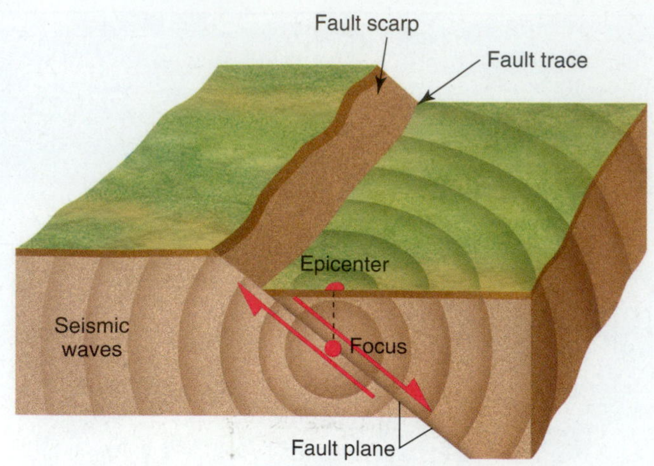

Figure 5.27

Block diagram showing a fault plane and the focus and epicenter of an earthquake generated by movement along the fault. Arrows show direction of relative movement along the fault, and concentric shaded circles show the propagation of seismic waves.

Earthquake Epicenters and Foci

The place where the initial rupture occurs on the fault plane marks the **focus** of the resulting earthquake, and the point on the earth's surface vertically above the focus is the **epicenter** of the earthquake (fig. 5.27). The focus and epicenter of an earthquake that is generated on a vertical fault plane define the fault plane as a line on a cross section. If the earthquake is generated by movement along an inclined fault plane, the fault plane is defined in cross section by a line passing through the focus and the fault scarp or fault trace at the earth's surface. When movement occurs along a fault plane, energy is released and an earthquake is produced.

Reference

Greensfelder, Roger. 1971. Seismologic and crustal movement investigations of the San Fernando earthquake. *California Geology,* April–May 1971: 62–68. California Division of Mines and Geology, Sacramento, California 95814.

814 4509158

EXERCISE 24B

Name

Section Date

Fault Problems

Faulted Sedimentary Strata

In Exercise 24A you determined the map view of eroded and faulted sedimentary structures. In this exercise you are given a geologic map and are asked to determine the structure of the map area.

1. On the Swan Island map (fig. 5.31), formation Ok is cut by a fault trending NW–SE. Label the upthrown and downthrown sides of the fault with the correct geologic symbols.
2. Refer to the fault described under question 1. What would be the direction of dip of the fault plane if this were a normal fault?

   South - North

3. If the fault referred to in question 1 were a reverse fault, what would be the direction of dip of the fault?

   North - South

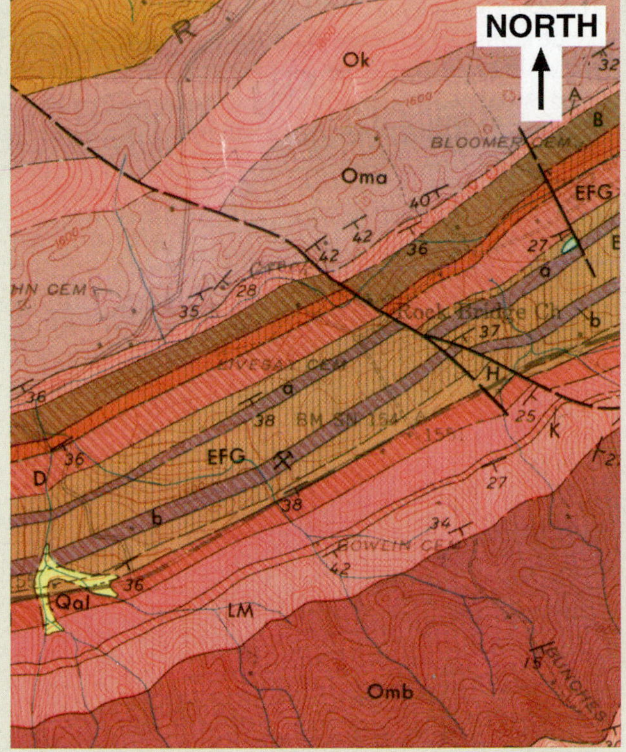

Figure 5.31

Part of the geologic map of Swan Island quadrangle, Tennessee, 1971. U.S. Geological Survey. Scale, 1:24,000; contour interval, 20 feet.

Courtesy of the U.S.G.S.

EXERCISE 25

Name

Section Date

Relationship of Fault Planes to Fault Traces, Epicenters, and Foci

The image of figure 5.32 extends from the Mohave Desert on the north to the Pacific Ocean on the south. Figure 5.33 is a generalized map of the same area showing the traces of *some* of the faults in the area. Those that are easily visible on the image are shown with a solid line, and those that are more difficult to recognize are shown by dashed lines.

1. Transfer the fault traces from figure 5.33 to figure 5.32.

2. Describe the physiographic features associated with the San Andreas Fault and the Garlock Fault.

o San Andreas Fault → The rocks have rose up above the sand making the dessert

o Garlock Fault → The mountains have been pushed up making a half bowl.

(continued)

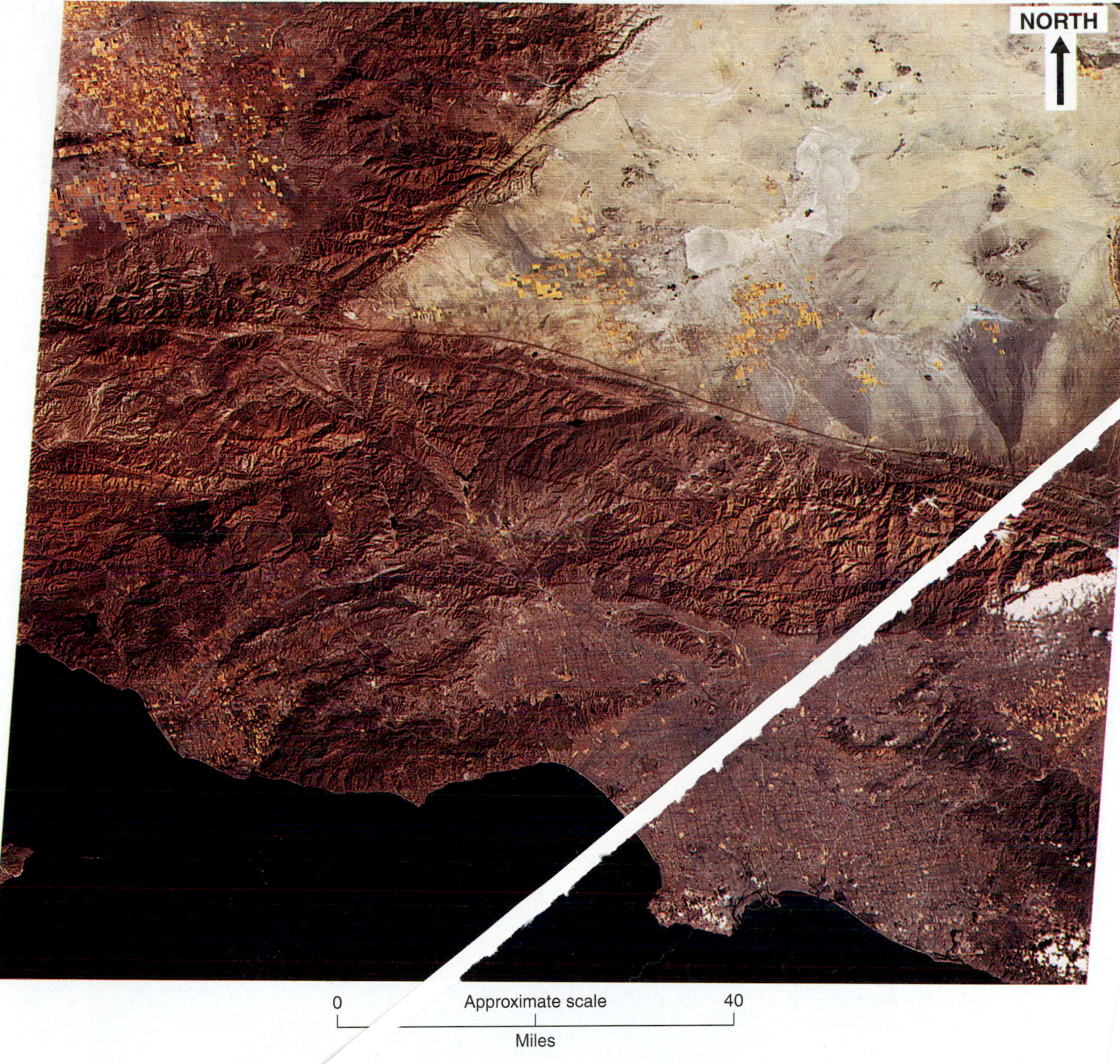

Figure 5.32

False color image of the great[illegible]os Angeles area of southern California, made from Landsat 1, October 21, 1972.
NASA ERTS image E-1090-180 12.

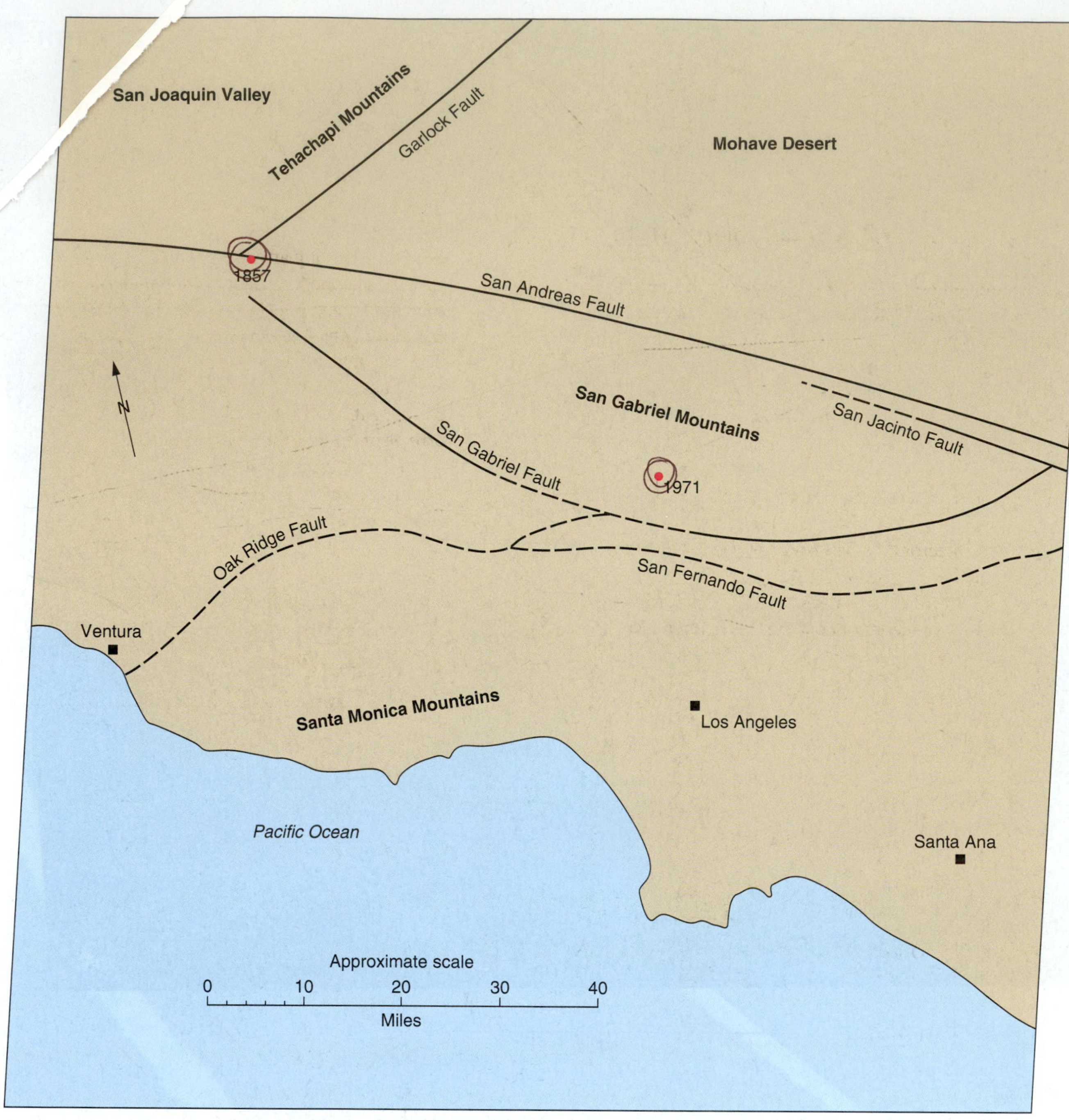

Figure 5.33

Generalized map of the greater Los Angeles area showing traces of some of the faults occurring there; also, the epicenters of the Ft. Tejon earthquake of 1857 and the Sylmar earthquake of 1971 are shown as red dots.

EXERCISE 25

Name ______________________________

Section ______________________ Date ______________________

Relationship of Fault Planes to Fault Traces, Epicenters, and Foci *(Continued)*

3. The San Andreas Fault is a strike-slip fault that extends from the Gulf of California to beyond San Francisco in the Pacific Ocean, a distance of about 600 miles. The Pacific Ocean side of the fault has moved north (west in the area of the image) some 300 to 350 miles in a series of horizontal displacements. The San Francisco earthquake of 1906 was caused by slippage along the San Andreas Fault in the amount of 21 feet.

   Figure 5.33 shows the epicenters of two major earthquakes in the greater Los Angeles area during historical times—the Ft. Tejon earthquake of 1857 and the Sylmar earthquake of 1971.

   (a) Figure 5.34 is a schematic cross section of the earth across the San Andreas Fault. The focus and the epicenter of the Ft. Tejon earthquake are shown. Draw a solid red line on the diagram showing the attitude of the San Andreas Fault. What is the dip of the fault plane?

   ____

   ____

   ____

   (b) Figure 5.35 is a schematic cross section through the focus of the 1971 Sylmar earthquake to the fault scarp caused by the Sylmar earthquake. Precise surveying after the Sylmar earthquake showed that the San Gabriel Mountains increased about 6 feet in elevation as a result of the movement along the San Fernando Fault. This movement was responsible for the Sylmar earthquake. Draw a solid red line on figure 5.35 showing the San Fernando Fault plane. Draw red arrows on each side of the fault indicating the relative movement along the fault plane. Label the hanging wall (H) and the footwall (F). Is the San Fernando Fault a normal, reverse, or strike-slip fault? How is this determined?

   Reverse, ______________________________

   ____

   ____

   ____

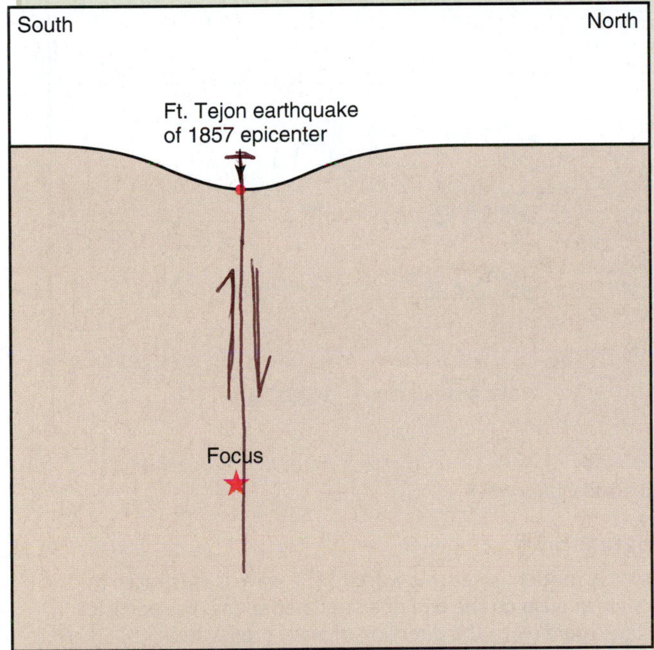

Figure 5.34

Schematic north–south cross section across the trace of the San Andreas Fault showing the epicenter and focus of the Ft. Tejon earthquake of 1857.

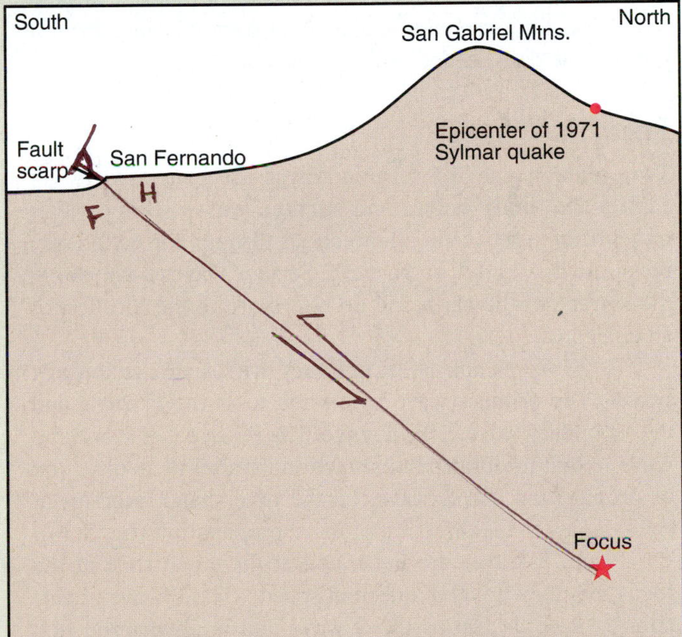

Figure 5.35

Schematic north–south cross section from San Fernando across the San Gabriel Mountains through the epicenter of the Sylmar earthquake of 1971. Vertical scale exaggerated.

The Use of Seismic Waves to Locate the Epicenter of an Earthquake

Seismographs, Seismograms, and Seismic Observatories

The energy released by an earthquake produces vibrations in the form of **seismic waves** that are propagated in all directions from the focus. Seismic waves can be detected by an instrument called a **seismograph,** and the record produced by a seismograph is called a **seismogram.** The geographical location of a seismograph is called a **seismic station** or **seismic observatory,** and it is given a name in the form of a code consisting of three or four capital letters that are an abbreviation of the full name of the station. For example, a seismic observatory on Mount Palomar, in southern California, has the code designation of PLM. A worldwide network of seismic observatories provides records of the times of arrival of the various kinds of seismic waves. Seismograms from at least three different stations located around the focus of a given earthquake but at some distance from it provide the data needed to locate the epicenter.

Seismic Waves

Two general types of seismic waves are generated by an earthquake: **body waves** and **surface waves.** Body waves travel from the focus in all directions through the earth; they penetrate the "body" of the earth. Surface waves travel along the surface of the earth and do not figure in the location of an epicenter.

Body waves consist of **primary waves** and **secondary waves.** The primary wave is referred to as the **P wave,** and the secondary wave is the **S wave.** The P wave is like a sound wave in that it vibrates in a direction parallel to its direction of propagation. An S wave, on the other hand, vibrates at right angles to the direction of wave propagation (fig. 5.36).

P and S waves are generated at the same time at the focus, but they travel at different speeds. The P wave travels almost twice as fast as the S wave and is always the first wave to arrive at the seismic station. The S wave follows some seconds or minutes after the first arrival of the P wave. **The difference in arrival times of the P and S waves is a function of the distance from the seismic station to the epicenter.** The distance from the seismic station to the epicenter is called the **epicentral distance.**

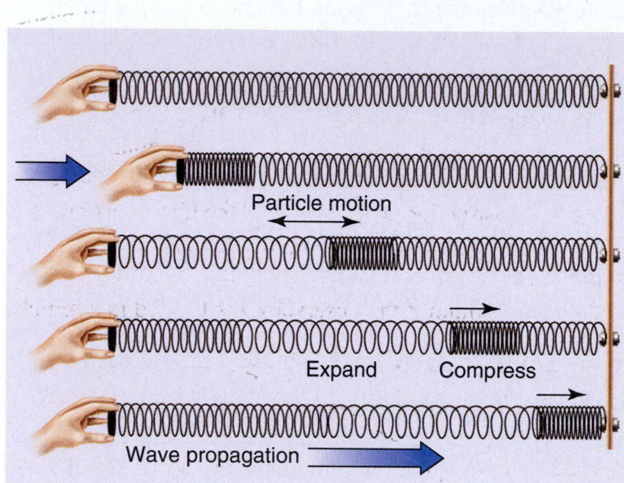

A. Primary wave

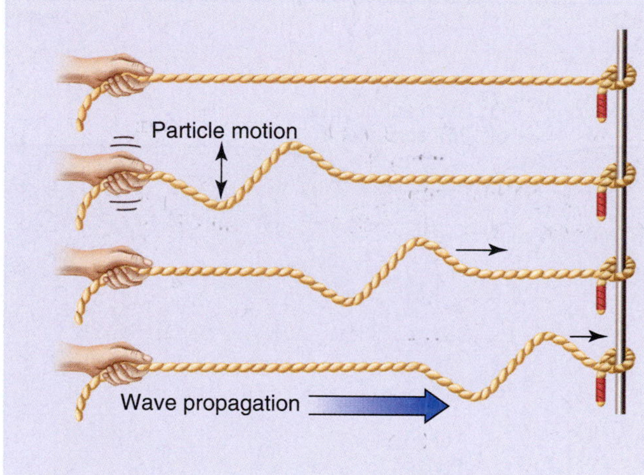

B. Secondary wave

Figure 5.36

Particle motion in seismic waves. (A) P wave is illustrated by a sudden push on the end of a stretched spring. The particles vibrate *parallel* to the direction of wave propagation. (B) S wave is illustrated by shaking a loop along a stretched rope. The particles vibrate *perpendicular* to the direct wave propagation.

Reading a Seismogram

A seismograph records the incoming seismic waves. For years the waves were recorded as "wiggly" lines on a drum rotating at a fixed rate of speed. The records were then digitized so that they could be analyzed using computer software. Today the seismic records are generally recorded by digital systems that are more accurate and allow more rapid analysis of the record. Seismograms contain not only the record of the incoming seismic waves but also marks that indicate each minute of time. Clocks at all seismic stations around the world are set at Coordinated Universal Time (UTC), so no matter what time zones observatories are located in, the seismograms produced at them are all based on a standardized clock.

When no seismic waves are arriving at an observatory, the seismograph draws a more or less straight line (fig. 5.37). Some small irregular wiggles on the seismogram may be **background noise** from vibrations produced by trucks, trains, heavy surf, construction equipment, and the like. Most modern seismographs contain a damping mechanism that reduces background noise to a minimum. In addition, background noise is kept to a minimum if the observatory is located in a remote area where human activities are uncommon.

The time of arrival of the first P wave is noted as T_p. The P wave continues to arrive until the first S wave appears, which is noted as T_s. The S wave has a much larger amplitude than the P wave. (The amplitude is the vertical distance between the peak of the recorded wave and the line on the seismogram recorded when no seismic waves are arriving.)

Figure 5.37 shows a seismogram from the Santa Ynez Peak Observatory on which the arrival times of the P and S waves are shown as T_p and T_s. These were determined by using the time scale on the seismogram to measure the time from the mark labeled 12:40:00 (12 hrs: 40 min: 00 sec) to the times of arrival of the first P and S waves.

On the SYP seismogram of figure 5.37, T_p is 19 seconds after the time mark, or 12:40:19 UTC, and T_s is 44 seconds after the time mark, or 12:40:44 UTC. The difference between the time of arrivals, $T_s - T_p$, is therefore 25 seconds.

Locating an Epicenter on a Map Using Travel-Time Curves

$T_s - T_p$ is measured in units of time, and this time, when converted to a distance, indicates the epicentral distance. Converting this time to distance requires the use of **travel-time curves** for both the P and S waves as shown in figure 5.38. A point on either one of the curves indicates the time required for a P or S wave to travel a certain distance from the epicenter. Time in seconds is shown on the vertical scale, and the corresponding distance in kilometers is shown on the horizontal scale. Following is the procedure for converting $T_s - T_p$ in seconds to an epicentral distance in kilometers:

1. Determine T_p and T_s from a seismogram to the nearest second. Record these values for use in the next step.
2. Subtract T_p from T_s and record as a time in seconds for use in the next step.
3. From the vertical scale of the travel-time graph of figure 5.38, determine the length of a line with the value of $T_s - T_p$. Mark this distance on a sheet of paper.
4. Keep the paper parallel with the vertical axis and slide the paper upward and to the right until one end of the line lies on the S curve. Follow the vertical line down to the horizontal scale and read the epicentral distance.

   As an example of this procedure, let us use the data from the SYP seismogram of figure 5.37. The value for $T_s - T_p$ on this seismogram is 25 seconds. The length of the line corresponding to 25 seconds is marked on the paper from the vertical travel time axis, and the paper is moved to the right until one end is on the P curve and the other end is directly above it on the S curve. Follow this vertical line down and read 192 km, the epicentral distance at station SYP.

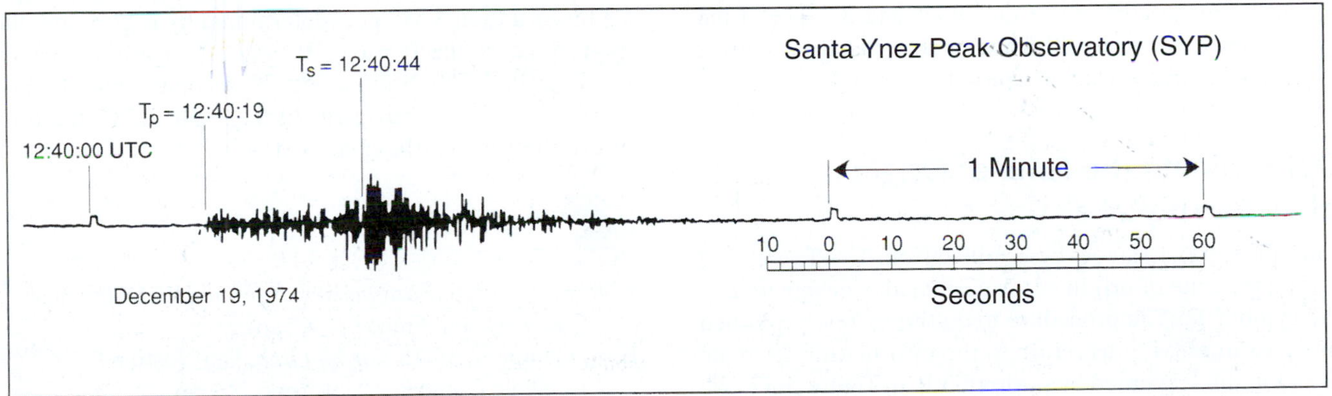

Figure 5.37

A seismogram recorded at the Santa Ynez Peak Observatory (SYP) in California showing an earthquake on December 19, 1974. The time mark automatically recorded on the seismogram is 12:40:00, which is 12:40 P.M. Coordinated Universal Time (UTC). The time of arrival of the first P wave, T_p, is 12:40:19, and the time of arrival of the first S wave, T_s, is 12:40:44.

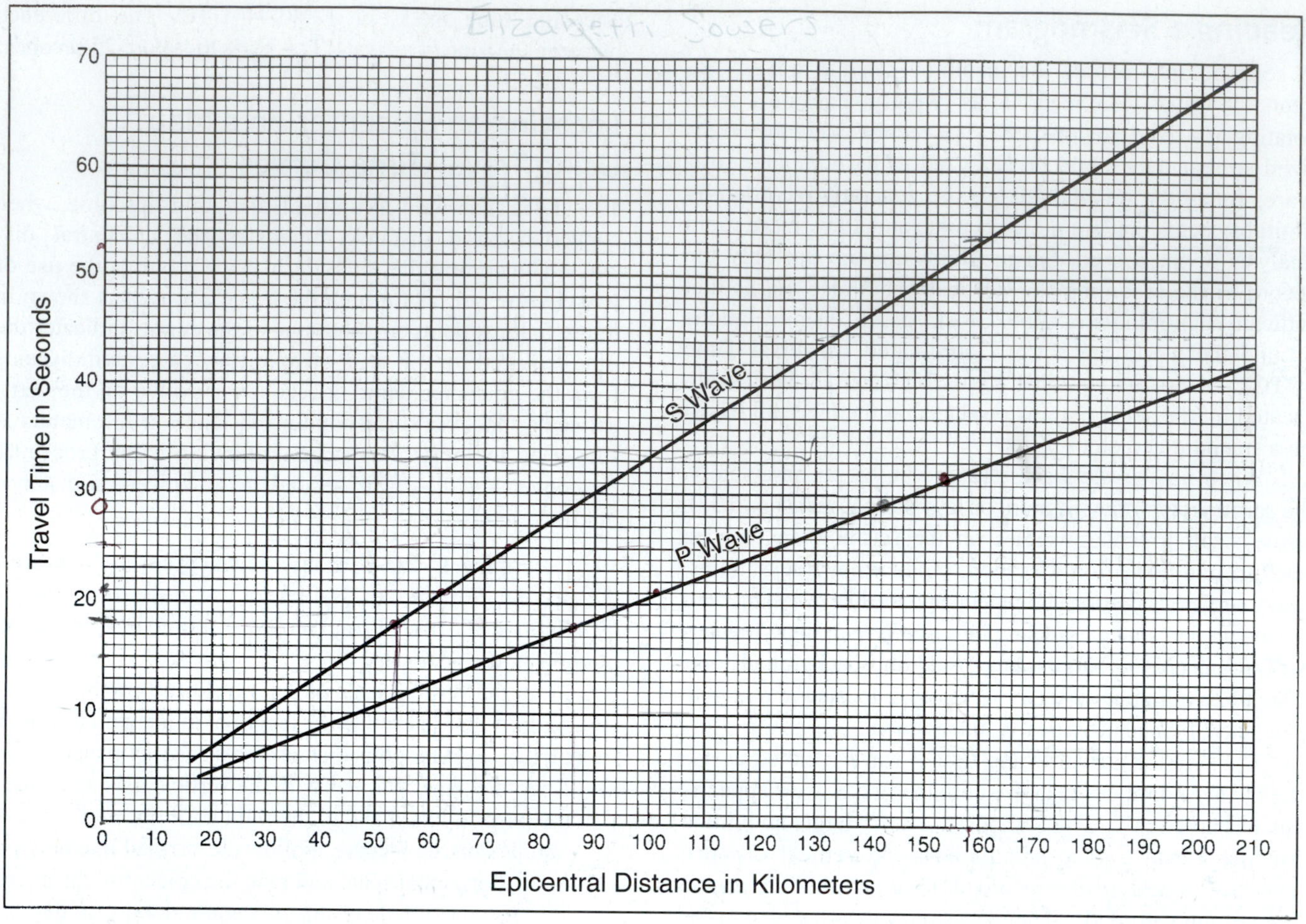

Figure 5.38

Travel-time curves for P and S waves in southern California.

5. On a suitable base map, use the bar scale to set your compass to the epicentral distance determined in step 4. Use this compass setting to draw a circle on the map whose center is at the geographic coordinates of the appropriate seismic station.
6. By following steps 1 through 5 for three different seismograms at appropriate directions and distances from the epicenter, you will draw three circles that intersect or nearly intersect at the epicenter.

Determining the Time of Origin of an Earthquake

The epicentral distances determined from $T_s - T_p$ are used to find the time of origin of the earthquake, designated by the symbol T_o. The procedure to do this is best explained by an example. Let us return to the information from the seismogram, recorded at station SYP, of figure 5.37. We have already determined that the epicentral distance is 192 kilometers. Remember that this is the distance from the seismic station to the epicenter of the earthquake. We want to know the time when this earthquake occurred, T_o. That is also the time when the seismic waves started their journey of 192 km to SYP. Looking at figure 5.38, we see that the point where the P wave curve intersects the 192 km line is 39 seconds. This tells us that it took the P wave 39 seconds to travel from the earthquake epicenter to station SYP. T_o is determined by subtracting the travel time of the P wave, 39 seconds, from T_p, which is 12:40:19 UTC. Subtracting 39 seconds from 12 hrs, 40 minutes, 19 seconds gives us 12:39:40 UTC, the time of origin of the earthquake, or T_o.

References

Bolt, Bruce. 1978. *Earthquakes: A primer.* W. H. Freeman & Company, San Francisco, Chapter 6.

Eiby, George A. 1980. *Earthquakes.* Van Nostrand Reinhold Company, New York. 209 pp.

We are indebted to Charles G. Sammis, Department of Geological Sciences at the University of Southern California, for his assistance in preparing Exercise 26.

P A R T

Plate Tectonics and Related Geologic Phenomena

Background

The theory of plate tectonics is a widely accepted concept that has been guiding geophysical and geological research since the mid-1960s. The term **plate tectonics** refers to the rigid plates that make up the skin of the earth and their movement with respect to one another. Figure 6.1 shows the distribution of the principal plates as they are currently understood. The plates differ greatly in size, although the true size of the plates is distorted in figure 6.1 because the map on which they are displayed is a Mercator projection. This kind of map represents the spherical shape of the earth on a flat surface and shows extreme distortion in the polar regions because the longitude lines on the map are parallel instead of converging (see fig. 3.1). It can be seen from figure 6.1 that some plates include both oceans and continents, as for example the African and south American plates.

The theory of plate tectonics can be used to explain many phenomena on planet earth, both on the continents and in the ocean basins, such as the origin and distribution of volcanoes and earthquakes, the topography of the sea-floor, and a host of other major geologic features.

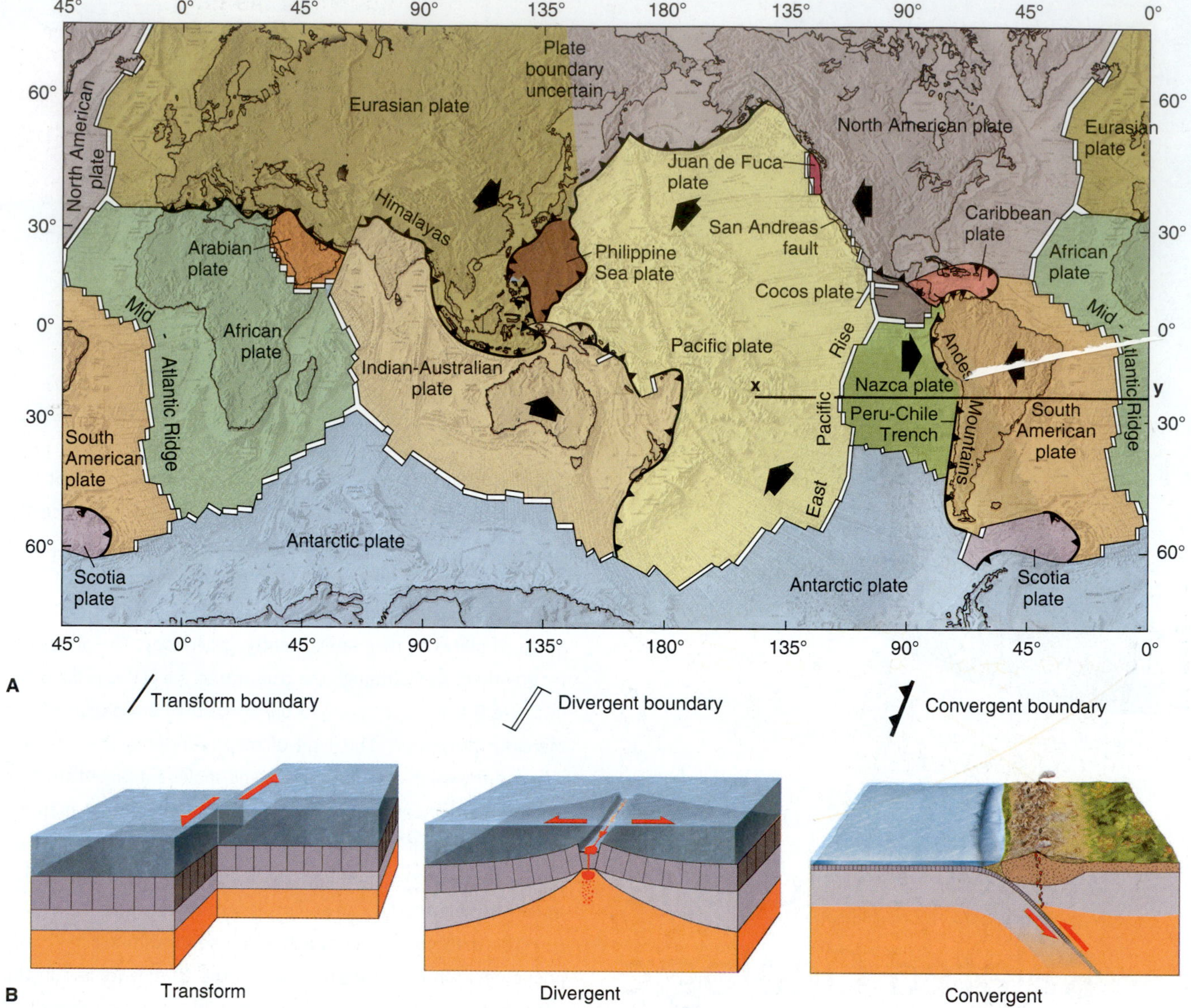

Figure 6.1

(A) Map of the earth showing the name and boundaries of major lithospheric plates. The line labeled x-y is the line of the cross section of figure 6.2. The heavy black arrows show the direction of plate movement. (B) Block diagrams showing the three main types of plate boundaries. Red Arrows show relative movement of plates.

The Major Components of the Earth

Through the use of geophysical techniques, it has been determined that the earth is composed of layers or shells of different composition and mechanical properties. The compositional layers of the earth are the **crust, mantle,** and **core** (fig. 6.2). The crust is subdivided into **oceanic crust,** which is mainly basaltic in composition, and **continental crust,** which is mainly granitic.

The mechanical layers are the **lithosphere,** composed of the crust and uppermost mantle; the **asthenosphere,** a weak, ductile (a ductile substance is one that deforms without fracturing) layer of the earth's upper mantle (beneath the lithosphere) on which the lithospheric plates move; the **transition zone,** which lies between 350–700 km depth (here the mantle minerals are compressed and collapse into denser structures); the **mesosphere,** the transition zone and lower mantle; the liquid metallic **outer core;** and the solid metallic **inner core.**

The relationship of these features is shown in figure 6.2, a scale diagram of a cross section of the earth from the surface to the core. The information shown on figure 6.2 has been gleaned mainly from the interpretation of seismic waves and laboratory experiments on rocks under high temperatures and pressures. None of the features in figure 6.2 has been observed directly in place, except, of course, the outermost crust at the earth's surface. Figure 6.2

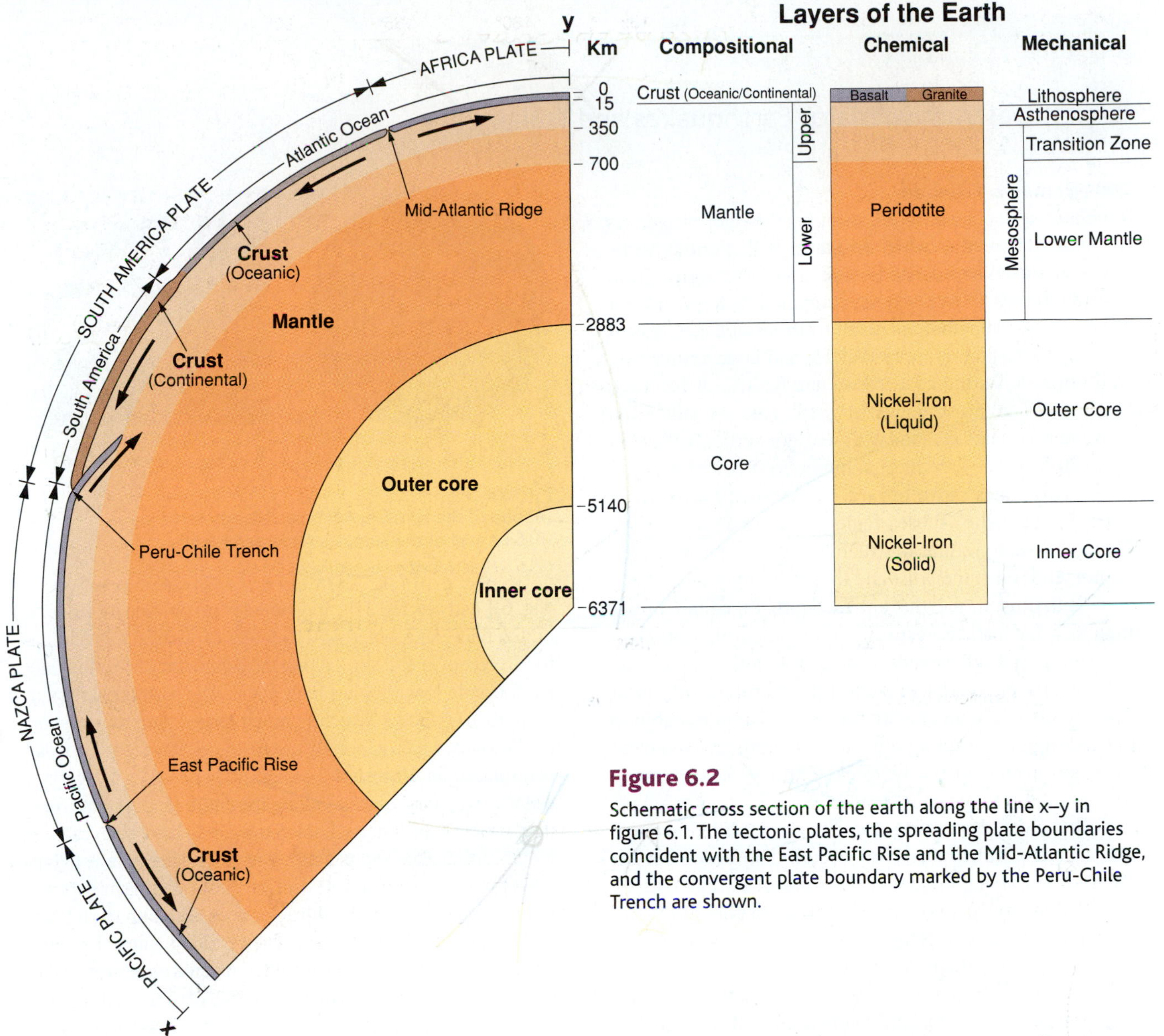

Figure 6.2

Schematic cross section of the earth along the line x–y in figure 6.1. The tectonic plates, the spreading plate boundaries coincident with the East Pacific Rise and the Mid-Atlantic Ridge, and the convergent plate boundary marked by the Peru-Chile Trench are shown.

is therefore an interpretation of the conditions prevailing at depth beneath the surface of the earth, based on the current theory of plate tectonics.

Plate Movement

The arrows in figures 6.1 and 6.2 show the direction of movement of the plates. This movement was initially deduced from the nature of the plate boundaries. Two plates moving away from one another form a **divergent** or **spreading boundary** along an **oceanic ridge** (Mid-Atlantic Ridge) or **rise.** It is at these spreading centers where new oceanic lithosphere is formed by the extrusion of basaltic magma. Continental divergence also occurs; the Red Sea, the Gulf of California, and the Great African Rift System are examples.

Two plates that are moving toward each other collide to form a **convergent boundary** (the Nazca Plate moving against the South American Plate is an example). Three types of convergent boundaries are recognized: (1) oceanic–continental convergence (the example just given), where, by a process known as **subduction,** the denser oceanic material is forced under the lighter continental plate; (2) ocean–ocean convergence, where two oceanic plates converge and form a deep trench and associated volcanic islands (Aleutian Trench is an example); and (3) continental–continental convergence (the collision of the Indian Plate with the Eurasian Plate is an example), resulting in the spectacular mountains we know as the Himalayas. In all three cases, there is a loss of lithosphere either by subduction or by the piling up and greatly increasing the thickness of the continental crust.

At the margin of some plates, the motion is such that the plates are sliding past one another. The San Andreas Fault in California between the Pacific Plate and the North American Plate is an example of such **transform faults** on land. They

BOX 6.1

Recent Destructive Major Earthquakes and Tsunamis

2004 Sumatra Tsunami

Tsunami, a Japanese word meaning "harbor wave," is the most destructive wave on earth, with periods (time between successive crests to pass a point) ranging from tens of minutes to over an hour, compared with seconds for wind waves. Tsunamis can result from earthquakes, volcanic eruptions, submarine landslides, and large extraterrestrial impacts. Earthquakes cause nearly 90% of recorded tsunamis. The most powerful earthquakes result from movement on subsea faults with seafloor vertical offsets of up to 20 feet. They are most common in subduction zones where two plates move toward one another. Large transcurrent (strike-slip) subsea faults where two plates move laterally do not produce destructive tsunamis because they do not tend to lift the crust and thus the overlying water.

Tsunamis move the entire depth of ocean rather than just the surface, resulting in immense energy. They propagate at high speeds (up to 600 miles/hour) and can travel great distances with little overall energy loss. The effects of a tsunami can range from unnoticeable to devastating, pulverizing objects in its path and scouring exposed ground to the bedrock with the loss of many lives.

The most destructive tsunami in recorded history, known as the Sumatra tsunami, occurred on December 26, 2004, when a magnitude 9.15 earthquake (third-largest recorded on a seismograph), known as the Sumatra–Andaman Islands earthquake, occurred just north of Simeulue Island, 100 miles west of Sumatra (box figs. 1A and 1B). The earthquake was caused by the complex movement of the Indian-Australian plate beneath the Eurasian Plate (fig. 6.1 and box fig. 1B). It resulted from oblique movement on the fault (an average of 216.5 feet of vertical slip and up to 118 feet of lateral slip), along the estimated 124-mile-wide by 806-mile-long dipping fault plane. The earthquake lasted close to 10 minutes, which is much longer than most major earthquakes, which generally last no more than a few seconds. It displaced an estimated 7 cubic miles of water, triggering devastating tsunami waves along the entire length of the rupture, which is the longest known rupture to have been caused by an earthquake. The tsunami was not a single devastating wave but was several waves, with the third wave the most powerful and destructive. The third wave occurred up to an hour and a half after the first wave. The resulting tsunami propagated across the Indian Ocean and inundated or destroyed the coasts of Indonesia, Malaysia, Thailand, Myanmar, India and Sri Lanka (box fig. 2). The effects were felt as far away as

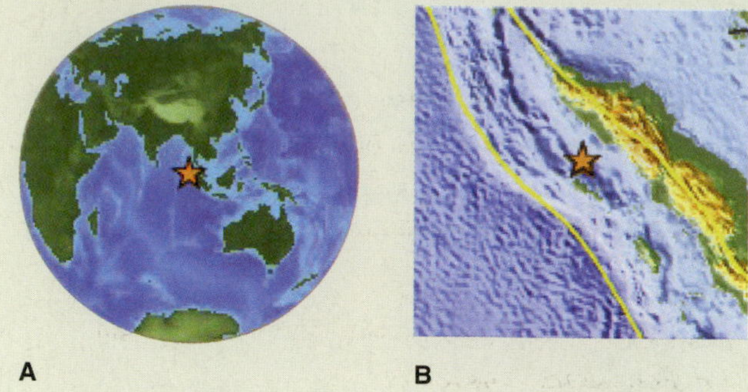
A B

Figure 1
(A) Site of the Sumatra Andaman Islands earthquake.
(B) Site map of the Sumatra Andaman Islands.
U.S.G.S. National Earthquake Information Center.

A

B

Figure 2
(A) Banda Aceh shore before the Sumatra tsunami. (B) Banda Aceh shore after the Sumatra tsunami.
© Digital Globe Photos.

the Maldive Islands and the coast of Somalia. These effects occurred within 0.2–8 hours and were the result of waves up to 106 feet high that killed nearly 300,000 people. The tsunami waves traveled around the globe and were measured in the Pacific Ocean and many other places by tide gauges thousands of miles from the source. The before-and-after aerial photographs of Banda Aceh shore (box fig. 2) reveal the enormous erosive and destructive power of a significant tsunami such as the Sumatra tsunami.

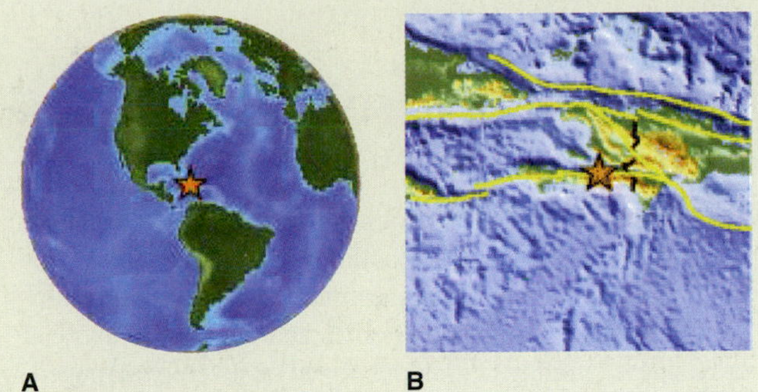

Figure 4

(A) Site of Hispanola (Haiti) earthquake. (B) Site map of focus of the Haiti earthquake.

U.S.G.S. National Earthquake Information Center.

2010 Earthquakes in Haiti and Chile

The recent earthquakes in Haiti and Chile allow the compariosn of two major earthquakes (box figure 3). Both earthquakes were generated on plate boundaries but with quite different motions involve.

Motion between the two plates occurs on the Septentrional-Oriente and the Enriquillo-Plantain Garden transform fault systems in the north and south respectively. The Caribbean Plate is moving eastward with respect to the North American Plate at about 2 cm per year. Since the 18th century plate motion has loaded the fault at a rate of 0.7 cm per year to produce a slip deficit of about 175 cm based on GPS data. Port-Au-Prince is located on the seismically active Enriquillo fault and historically there have been several destructive earthquakes that have impacted Haiti (1751 and 1770) and the Dominican Republic (1842 and 1946).

The epicenter of the earthquake was about 25 km (16 miles) to the west of Port-Au-Prince along the fault at a depth of 13 km. The rupture was estimated to be 65 km long and over 1000 km² in area with a mean slip of about 170 cm. The fault size and slip were estimated from seismic stations around the world.

On the Modified Mercalli Scale, the intensity was VIII or IX as recorded in Port-au-Prince. Over 230,000 people were killed, another 300,000 were injured, and an estimated 1,000,000 were left homeless following the quake.

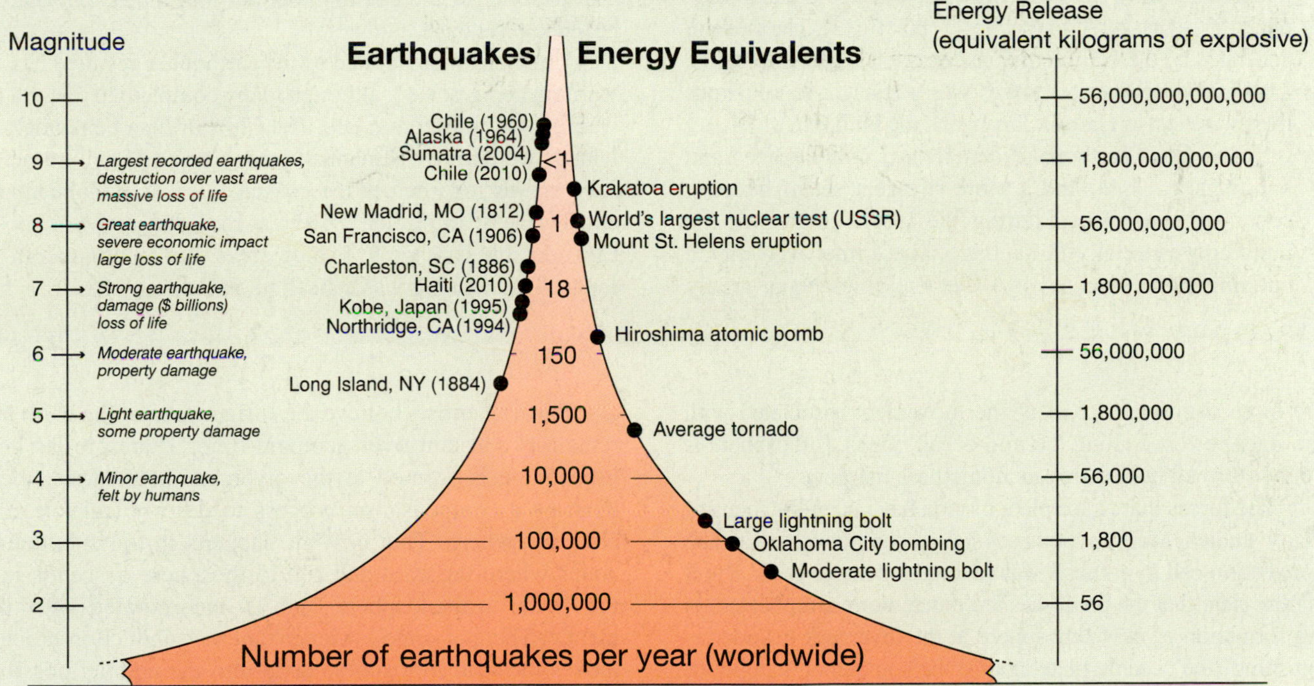

Figure 3

Earthquake energy equivalents. Number and magnitude of earthquakes per year (worldwide).

(continued)

BOX 6.1

Recent Destructive Major Earthquakes and Tsunamis *(Continued)*

The infrastructure of the country was almost completely destroyed—hospitals, port facilities, roads, airports, communication systems, etc. The estimates of damage to buildings included over 200,000 homes and 25,000 commercial or business buildings either destroyed or severely damaged.

Following the earthquake, a tsunami warning was posted but it was cancelled after a short time as the motion of the fault was such that no tsunami was generated. It was later reported that a small fishing village was hit by a wave that probably was generated by a submarine landslide triggered by the earthquake.

In contrast to the Haitian earthquake, the earthquake that impacted Chile on February 27, 2010, took place in an area of subduction between the Nazca Plate to the west and the South American Plate to the east. The two plates are converging at about 7 m per century. The epicenter was offshore from the Maule province, Chile, at a depth of 35 km. The fault rupture was estimated to be 100 km in width and 500 km in length parallel to the Chilean Coast (box figs. 5A and B). The magnitude of the earthquake was 8.8 M_w (box fig. 3), the result of west directed thrust faulting that displaced the city of Concepcion 3.04 meters to the west.

In 1960 an earthquake of magnitude 9.5 M_w was generated along this same subduction zone. This seismic event had the greatest energy release and calculated earthquake magnitude ever measured in the world (box fig. 3). The tsunami generated by the 1960 earthquake was felt across the Pacific. Destruction of seacoast towns was a result, and numerous lives were lost in Hawaii, Japan, and the Philippines.

The 1960 earthquake demolished many homes and buildings in Chile, and a more stringent set of building codes were emplaced during the reconstruction of the buildings, especially in Santiago. These improvements in building codes resulted in fewer buildings being totally destroyed but did not save all of the buildings from damage. In cities such as Concepcion there was major damage, and in the smaller cities and towns there was widespread destruction of all types of buildings. An intensity level of VIII was reported from two towns south of Concepcion. The intensity reported from Santiago was VII.

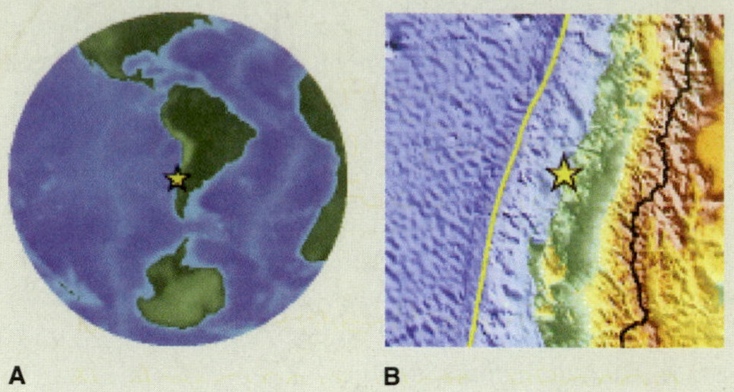

Figure 5

(A) Site of Chilean earthquake. (B) Site map showing focus of Chilean earthquake.

U.S.G.S. National Earthquake Information Center.

It is estimated that over 500,000 homes were damaged across the country. Close to 500 people were killed. The majority of the country was put into a blackout that lasted several days.

The tsunami generated by the earthquake resulted in a warning being posted, although many complained that the warning was posted too late. The Chilean Juan Femandez Islands were hit by tsunami waves; a wave of amplitude 2.34 m was recorded at the coastal town of Talcahuano and a wave of amplitude 2.6 m was recorded at Valpariso. Considerable damage to fishing areas was reported from Japan. The warning was cancelled on February 28.

are more common as part of the movement associated with the fracture zones along the mid-ocean ridges. Lithosphere is neither formed nor destroyed along these margins.

The forces that cause plate motion have been the focus of many studies. Many hypotheses have been put forward. The **convection cell** hypothesis was favored for many years. But it is now clear that the processes are much more complicated. Is the formation of new lithosphere at the divergent boundaries "pushing" the oceanic plate against the continental margin, or is the subduction of the heavier oceanic lithosphere under the continent "pulling" the oceanic plate? Do the overriding plates get pulled toward the subduction zone? Is the convection due to heating from within the earth, or does the subduction process induce it? All of these hypotheses are still under study.

Earth scientists believe the surface area of the earth has remained constant over geologic time. Therefore, as new lithosphere is formed at the divergent boundaries, there must be a compensation process to destroy (recycle) the crust elsewhere. This is what happens in the subduction zones where the cool and rigid lithosphere descends into the hotter asthenosphere and is incorporated into the mantle. There is some evidence that the subduction process moves the cold lithosphere deep into the mantle, possibly as deep as the mantle–core boundary, rather than restricting the subduction to the asthenosphere and the upper part of the mesosphere.

The rate at which the plates move is also of great interest. Methods used to determine the velocity of the plate

Sea-floor Spreading in the South Atlantic and Eastern Pacific Oceans

Background

A basic premise of plate tectonics is that the crustal plates have moved with respect to each other over geologic time and, in fact, are moving today. The rates of movement of crustal plates can be determined by using data from the plate margins along the mid-ocean ridges, where the amount of movement can be measured.

To measure the movement of two adjacent crustal plates along the margins of a divergent plate boundary, you must know two things: (1) two points on adjacent diverging plates that were once at the same geographic coordinates but have since moved away from each other over a known distance and (2) the time required for the two points to move from their original coincident position to their present positions. If the two points can be identified and plotted on a map, the distance between them can be measured by use of the map scale. Determining the age in actual years of the two points involves the earth's magnetic field. It is therefore necessary to review this subject in order to understand how it relates to the movement of crustal plates.

The Earth's Magnetic Field

The earth is encompassed by a magnetic field. The source of this magnetic field is in the liquid metal outer core. The field generated is analogous to the lines of force produced by a bar magnet with a north pole at one end and a south pole at the other. Imagine an immense bar magnet passing through the center of the earth with its north and south poles located near the North and South poles of the earth's axis of rotation (fig. 6.7A). A magnetic compass placed in this field would align itself parallel to the lines of magnetic force. The direction of this force is shown by the arrows in figure 6.7B, which point from the south magnetic pole toward the north magnetic pole. This condition is called **normal polarity.**

A magnetic compass does not point to the north geographic pole (true north), however, because the magnetic poles are not coincident with the geographic poles. The geographic poles define the earth's axis of rotation and remain fixed with respect to the equator over geologic time. The magnetic poles, however, shift over time with respect to the geographic poles. A plot of the magnetic poles during historical time shows that they tend to stay in close proximity to the geographic poles, so on a geologic time scale, it is assumed that the magnetic poles and the geographic poles have remained within about 10 degrees of each other.

Reversals of the Magnetic Poles

Lava flows may contain minerals with magnetic properties (domains) that align themselves parallel to the lines of force in the earth's magnetic field when the lava solidifies and the temperature falls below 580°C (the Curie Point). These magnetic minerals are like minute magnetic compasses frozen in the rock. Other rocks, such as sandstone, also contain magnetic minerals that become aligned with the existing magnetic field when they sink to the lake bottom or sea-floor at their site of deposition.

One of the amazing features of the earth's magnetic field is that its polarity has reversed itself many times over geologic time. That is, the north and south magnetic poles abruptly changed places so that a magnetic compass would point to the south magnetic pole during periods of **reversed polarity** (fig. 6.7C).

The study of magnetism in ancient rocks is called **paleomagnetism.** The paleomagnetic features of rocks studied over the entire world have provided the basis for a detailed chronology of times of normal and reversed polarities during the last 170 million years. The rock layers from which this chronology has been assembled have been dated by radioactive means, thereby providing an absolute time scale that identifies the times when the periods of normal and reversed polarities occurred.

Both normal and reversed polarities are called **magnetic anomalies** or simply **anomalies.** A chronology of magnetic anomalies is given in figure 6.8. Three elements are contained in this chronology: (1) a time scale in millions of years before the present (Ma Age in figure 6.8), (2) the periods of normal (black) and reversed (white) polarities, and (3) the conventional identification numbers that have been assigned to each anomaly. These identification numbers are arranged in chronological order with the youngest anomaly designated by number 1 and successively older anomalies by numbers 2, 2a, 3, and on up to anomaly 33. (Figure 6.8 is a shortened version of a chronology that extends to anomaly 37, which is about 170 million years old.)

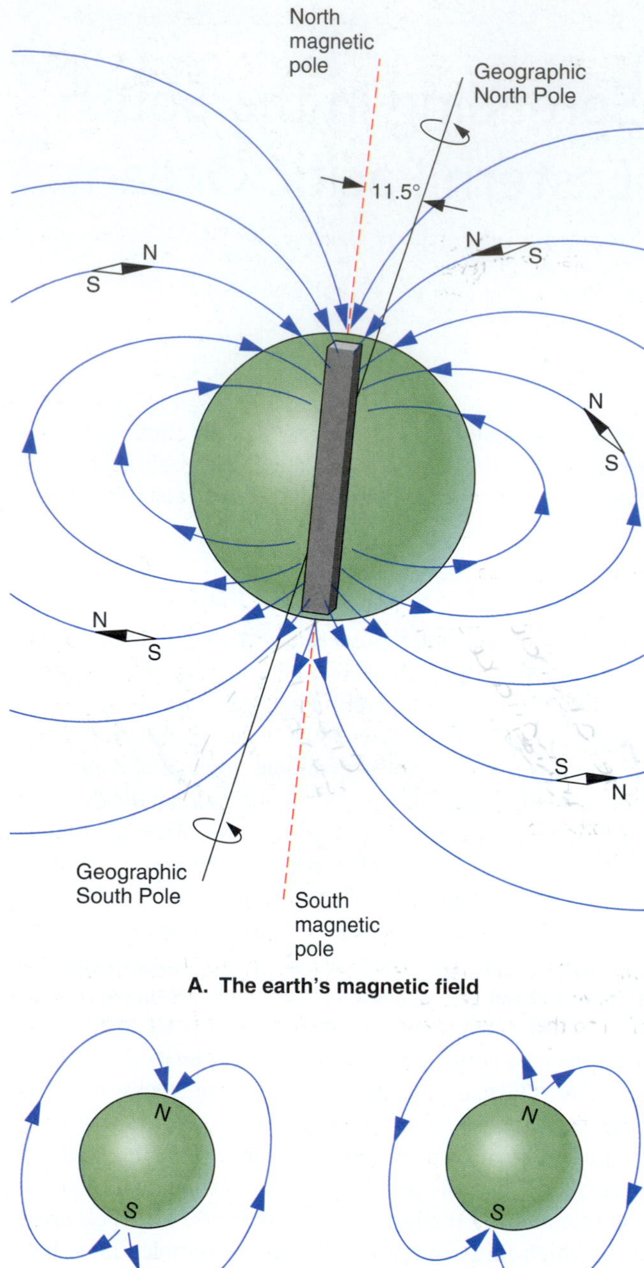

Figure 6.7

The earth's magnetic field. (A) The north magnetic pole and the north geographic pole are not coincident, and over geologic time they have been not much further apart than they are today. (B) Normal polarity characterizes the earth's magnetic field when the direction of the magnetic lines of force is from the south magnetic pole toward the north magnetic pole. (C) Reversed polarity occurs when the direction of the magnetic lines of force is toward the south magnetic pole. The earth's magnetic field has reversed many times during the history of the earth.

The time scale of figure 6.8 will be used later in an exercise, so it will be useful to become familiar with it. For example, the anomaly with the identification number 6 is a positive anomaly that is 20 million years old. Notice that the identification number assigned to an anomaly bears no relationship to that anomaly's absolute age. In other words, the series of identification numbers represents a **relative chronology,** and the series of dates in millions of years is an **absolute chronology.**

Magnetic Anomalies

An anomaly is a departure from the normal scheme of things. With respect to rocks that contain magnetic minerals, a **magnetic anomaly** is a magnetic reading that is greater or less than the normal strength of the magnetic field where the rock occurs. To illustrate, consider a series of bands of lava beds (basalt) lying on the sea-floor along both sides of an active spreading ridge between two crustal plates. An instrument that measures the intensity of the magnetic field is called a **magnetometer.** When one is towed on a long cable behind a ship headed along a course across the trend of the lava beds, the strength of the earth's magnetic field is recorded continuously on board the ship, and a satellite navigational system simultaneously records the ship's position.

When the seaborne magnetometer passes over basalts that are normally polarized, the strength of the earth's magnetic field is slightly intensified because the ancient magnetism in the normally polarized rocks adds a small component to the normally polarized earth's field. In this case, a **positive anomaly** is recorded. If the rocks over which the magnetometer is passing were formed during a time of reversed polarity, the strength of the earth's field is slightly reduced, and a **negative anomaly** is recorded.

Figure 6.9A shows a hypothetical record of magnetic field strengths in relation to basalts on either side of a mid-ocean ridge. High points on the curved line indicate a normal or positive anomaly, and low points indicate a reversed or negative anomaly. The pattern of negative and positive anomalies is repeated on either side of the spreading ridge; that is, the magnetic curve on one side of the ridge is a mirror image of the curve on the opposite side. The spreading ridge crest is thus flanked by stripes of alternating positive and negative anomalies **that increase in age with distance from each side of the spreading ridge** (fig. 6.9A and B).

One can deduce from this relationship that the tectonic plates are moving away from the divergent plate boundary. If the **absolute age** of the various anomalies on either side of the ridge is known and the distance between anomalies of the same age is measured, the **spreading rate** of the two adjacent plates can be determined. Using such information, the previous relative positions of two continents, such as South America and Africa, can be determined for various times in the geologic past.

In Exercise 28, you will determine the spreading rates on segments of the East Pacific Rise and the Mid-Atlantic Ridge, and in Exercise 29, you will determine the relative positions of South America and Africa at a specific time in the geologic past.

EXERCISE 29

Name

Section Date

Restoration of the South Atlantic Coastline 50 Million Years Before the Present

Given the evidence of spreading along the Mid-Atlantic Ridge, it can be deduced that the Africa and South America plates are moving away from each other, carrying the *continents* of Africa and South America with them. By using the pattern of magnetic lineations shown on figure 6.12, it is possible to reverse the spreading process and restore the positions of the African and South American coastlines to a time when a particular set of magnetic lineations was being formed on the Mid-Atlantic Ridge.

For the purposes of this exercise, we will use reversal number 21, which, according to the magnetic lineation time scale of figure 6.8, was formed 49.6 (or roughly 50) million years ago. Proceed as follows.

1. On figure 6.12, draw a red line over each of the magnetic lineations of reversal number 21 on the South American side of the Mid-Atlantic Ridge. Connect the segments of the number 21 reversal with a red line drawn along the fracture zones against which they terminate. Start with the point where reversal 21 touches the Ascension F.Z. Follow reversal 21 with your red pencil southward until it reaches the Bode Verda F.Z., then along the Bode Verde F.Z. westward to the northern end of the next fracture zone. Continue until you have reached the southernmost fracture zone on the map.
2. Attach a piece of tracing paper over figure 6.12 with tape or paper clips and repeat the process described previously for reversal 21 on the African side of the Mid-Atlantic Ridge. Draw this line in red pencil on the tracing paper.
3. With the tracing paper still in place, trace with black pencil the coastlines of Africa and South America. Also, trace in black pencil on the tracing paper the boundaries of figure 6.12 and the 20° South latitude line.
4. Detach the tracing paper and slide it toward South America until the red line on the tracing paper matches the red line on figure 6.12. When the two lines are matched as closely as possible, hold the tracing paper in place and trace the coastline of South America in red pencil on the tracing paper. Trace also the 20° South line on the tracing paper.
5. The map you have constructed on the tracing paper shows the Mid-Atlantic Ridge as it existed when magnetic reversal 21 was being formed. Your tracing paper also shows the *relative* positions of segments of the coastlines of Africa in black pencil and South America in red pencil as they were approximately 50 million years ago. This reconstruction is based on the assumption that the *continents* of Africa and South America were fixed to their respective plates during the spreading process over the past 50 million years. The *continents* moved with respect to each other because the *tectonic plates* to which they were attached moved as spreading continued along the Mid-Atlantic Ridge. What is the evidence that the movement of the two plates was not strictly in an east–west direction?
6. Was the earth's magnetic field normal or reversed at the time represented by your map on the tracing paper?

Normal.

Volcanic Islands and Hot Spots

The Hawaiian Islands consist of a northwest-trending chain of volcanic islands. Only the largest of these, Hawaii, has active volcanoes, whose eruptive history you are already familiar with from a previous exercise in this manual. The other islands in the chain are also volcanic in origin, but they ceased erupting some millions of years ago and have since been severely eroded by wave action and intense surface runoff. The Hawaiian Islands are, in fact, part of a longer chain of extinct volcanic islands that compose the Emperor Seamount and the Hawaiian Ridge (fig. 6.13 and fig. 6.14).

It has been postulated that the alignment of the extinct volcanoes forming the Emperor Seamount and the Hawaiian Ridge was formed as follows. A **hot spot,** whose latitude and longitude have remained relatively fixed over many millions of years, lies in the asthenosphere beneath the island of Hawaii. This hot spot is the source of heat that produces the magma that feeds the volcanoes on Hawaii. The oceanic lithosphere has been moving northwest over this spot in a more or less straight line, and as the ocean floor is carried over the hot spot by the conveyor action of the lithosphere, a new volcano is born. Evidence in support of this hypothesis lies in the absolute dates of the old lava flows along the Hawaiian Ridge and Emperor Seamount (fig. 6.13 and fig. 6.14). These are successively older the farther they are from the island of Hawaii.

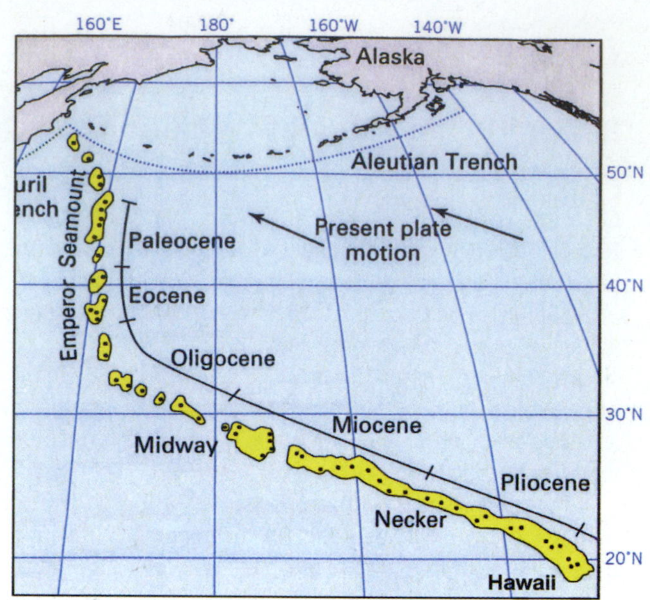

Figure 6.13

The Hawaiian Ridge and Emperor Seamount showing numerical ages and relative movement over time.

Modified from Strahler and Strahler, 2006.

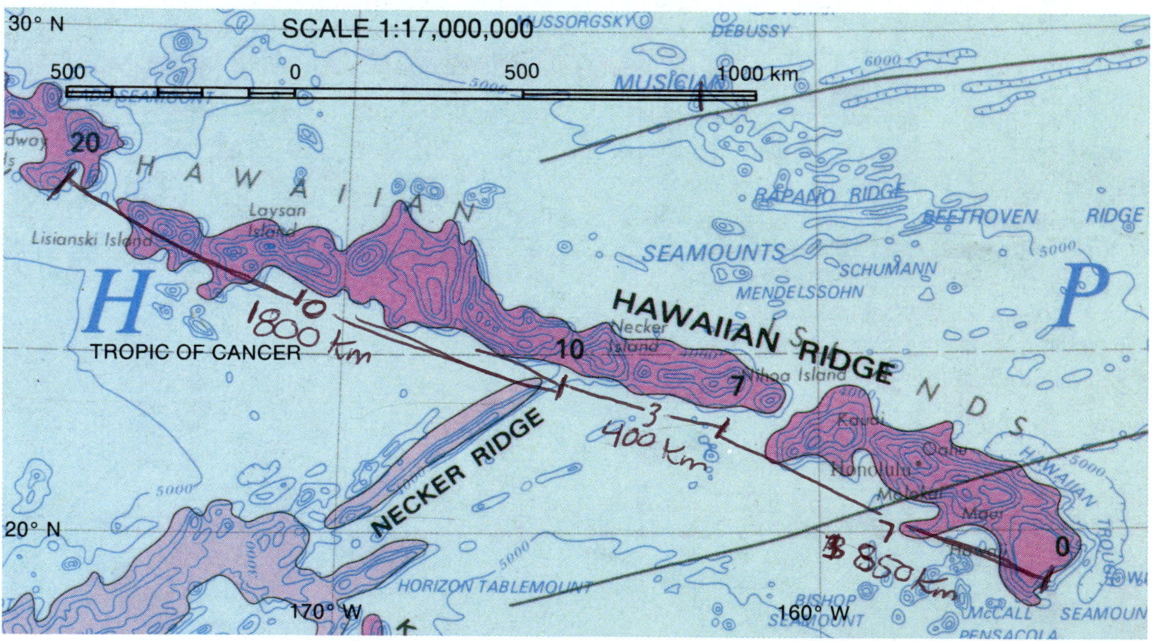

Figure 6.14

Map of the Hawaiian Ridge in the Pacific Ocean. Contour intervals, 1,000 m.

EXERCISE 30A

Name

Section Date

Movement of the Volcanoes in the Hawaiian Ridge over the Hawaiian Hot Spot

1. The map of figure 6.14 shows part of the Hawaiian Ridge and the absolute dates of lava along it printed in bold black numbers that represent millions of years before the present. Hawaii contains an active volcano, Kilauea, so the lava from it is zero years old. The lava on Nihoa Island is 7 million years old. Thus, according to the hot spot hypothesis, Nihoa Island was once an active volcano standing where Hawaii stands today.
   (a) Determine by simple arithmetic the rate of movement of the oceanic lithosphere over the Hawaiian hot spot. Figure the rate in centimeters per year using the distances from Hawaii to each of the three dated lavas on the map of figure 6.14. Make your distance measurements from the center of the zero on Hawaii to the center of each of the boldface numbers on the ridge.

   (b) Do the rates of movement based on the three dates indicate that the movement has been constant or variable?

   Variable

   (c) What was the rate of movement during the formation of the Emperor Seamount?

   800 km per 25 mill. years.

2. What might have caused the change of direction of movement as shown in figure 6.13?

   Volcanic movement disrupted the plate movement.

Volcanic Islands and Atolls

An **atoll** is an oceanic island that in map view appears as a narrow strip of land with low relief that forms a closed loop. Inside the loop is a shallow lagoon. The loop itself may contain gaps that allow access by ships from the surrounding ocean to the lagoon. An atoll is made chiefly of **coral reef,** but it also contains assorted other marine organisms such as algae and mollusks. Some of the coral may be weathered and eroded by wave action to form **coral sand.**

Corals are marine animals that thrive in warm, shallow waters of the world's oceans, most notably in the equatorial regions of the Pacific Ocean. Reef-forming corals grow in colonies that attach themselves to the shallow sea-floor in subtropical and tropical climatic zones. Corals live in water that is no more than about 50 m deep, is relatively free of sediments, is penetrated by sunlight, and has an abundant food supply of small marine organisms.

In 1842, Charles Darwin proposed a theory for the origin of atolls. He based this theory on his observations of the tropical islands during the voyage of the *Beagle* through the equatorial waters of French Polynesia. Darwin recognized three stages in the evolution of atolls in the islands around Tahiti. These stages are illustrated schematically in figure 6.15.

Stage I consists of a newly formed volcanic island surrounded by a **fringing coral reef.**

Stage II is reached after the volcanic peak has been eroded by surface runoff and wave action and has partly subsided beneath the sea due to the contraction of the cooling volcanic rock and isostatic adjustment. As the island subsides, the corals of the fringing reef die because the water becomes too deep for their survival. However, the remaining mass of dead coral forms a platform on which new corals establish themselves continuously. This process allows the upward growth of the reef to maintain pace with the sinking of the island. The fringing reef now becomes a **barrier reef,** and a shallow lagoon develops between it and the shore of the volcanic island. The low-lying surface of the barrier reef also may be colonized by vegetation.

Stage III is reached when the eroded remnants of the volcanic peak disappear beneath the sea through continued surface erosion and subsidence of the sea-floor. The coral reefs continue their upward growth and ultimately encircle the enclosed lagoon.

Darwin's theory was challenged by others on the grounds that no mechanism was known that could cause a volcanic island in the middle of the ocean to sink. These critics proposed instead that the evolution of an atoll was due to a rising sea level, not a sinking island. With the advent of plate tectonics, however, Darwin has been vindicated. Volcanic islands are formed by submarine eruptions over hot spots and are carried on the oceanic lithosphere to deeper water by plate movement.

The rising sea level proponents of atoll evolution were not totally wrong, however. Sea level has fallen and risen during past geologic time in response to the waxing and waning of Pleistocene ice sheets, and these sea-level fluctuations account for some *dead* coral reefs that now lie above sea level on the flanks of some volcanic islands.

The modern theory of atoll formation, therefore, calls for a subsiding volcanic island on which sea-level fluctuations over geologic time have been superimposed.

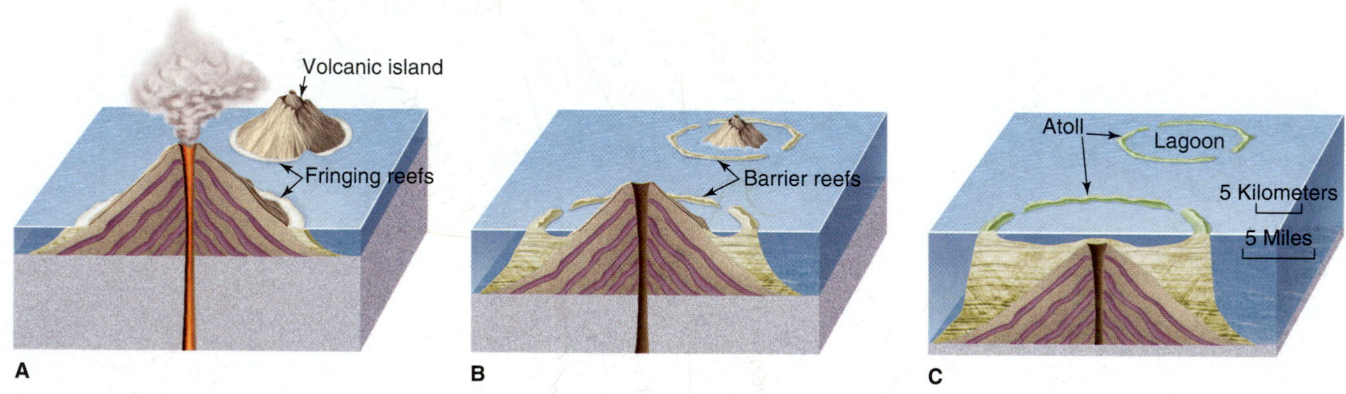

Figure 6.15

Types of coral–algal reefs. (A) Fringing reefs are attached directly to the island. (B) Barrier reefs are separated from the island by a lagoon. (C) Atolls are circular reefs with central lagoons. Charles Darwin proposed that the sequence of fringing, barrier, and atoll reefs forms by the progressive subsidence of a central volcano, accompanied by the rapid upward growth of corals and algae.

EXERCISE 30B

Name ______________________

Section ______________________ Date ______________________

Islands in French Polynesia of the South Pacific Ocean

1. Figures 6.16, 6.17, and 6.18 represent the three stages in atoll evolution as proposed by Darwin. What are the main identifying characteristics of each that can be observed in the photographs?

   16 – Volcanic Island

   17 – Barrier Reefs

   18 – Atoll / lagoon

2. What is the cause of the gaps in the fringing reef of figure 6.16?

   water being released from the island via cave ECT.

3. What kind of rock would you expect to be encountered by a drill that penetrated the coral and sediments in the center of the lagoon in figure 6.18?

   Barrier Reefs.

4. Assume that another hole was drilled somewhere on the atoll itself in figure 6.18.
   (a) Would the thickness of the coral penetrated by the drill be greater or less than in the center of the lagoon?

   greater.

   (b) Sketch a cross section through the center of the atoll in figure 6.18. Exaggerate the vertical scale and show the following:
   1. The foundered volcanic island remnant.
   2. Sea level.
   3. Coral reef material above and below sea level.
   4. The location of the two drill holes: A, through the lagoon; B, through the atoll itself.

Figure 6.16
Mooréa Island in the Society Islands of French Polynesia in the equatorial Pacific Ocean, an example of Stage I in the evolution of an atoll.
With the permission of and copyrighted by Erwin Christian.

Figure 6.17

Bora Bora Island in the Society Islands of French Polynesia in the equatorial Pacific Ocean, an example of Stage II in the evolution of an atoll.

With the permission of and copyrighted by Erwin Christian.

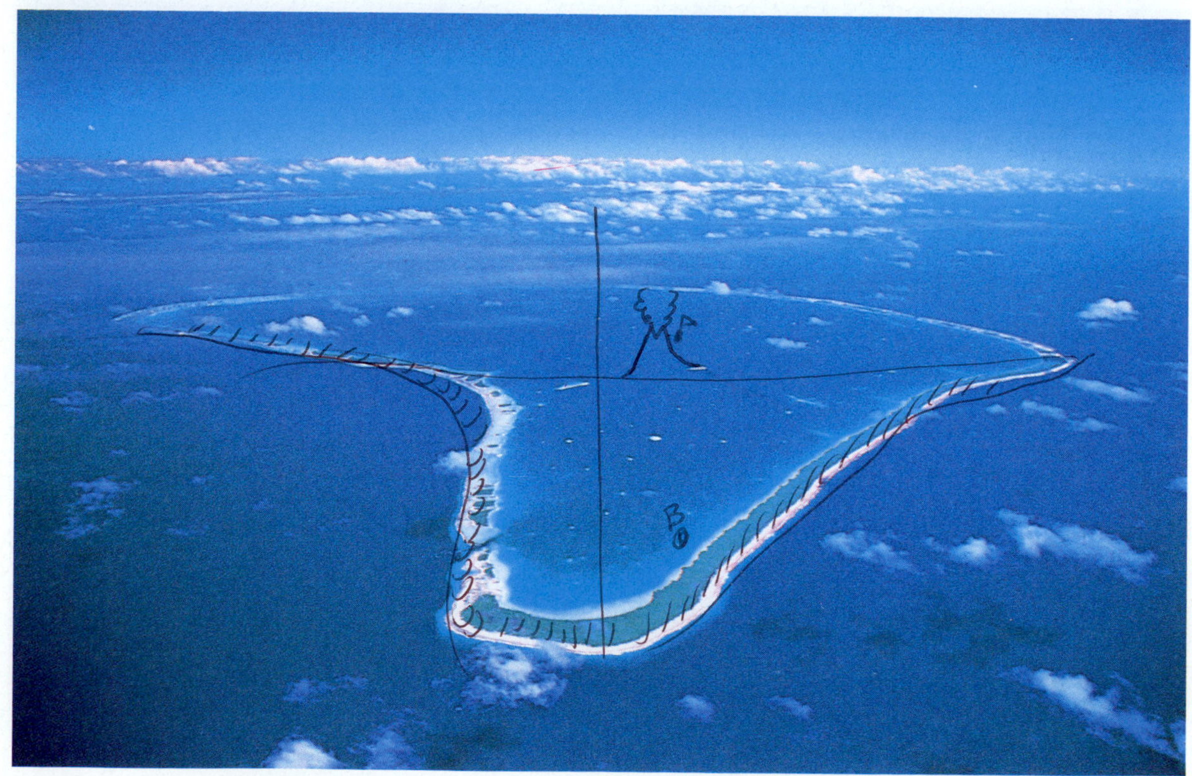

Figure 6.18

Aratica Island in the central Tuamotas of French Polynesia, an example of Stage III in the evolution of an atoll.

With the permission of and copyrighted by Erwin Christian.

APPENDIX A

International Geologic Time Scale

INTERNATIONAL STRATIGRAPHIC CHART

International Commission on Stratigraphy

Eonothem Eon	Erathem Era	System Period	Series Epoch	Stage Age	Age Ma	GSSP
Phanerozoic	Cenozoic	Quaternary	Holocene		0.0117	GSSP
			Pleistocene	Upper	0.126	
				"Ionian"	0.781	
				Calabrian	1.806	GSSP
				Gelasian	2.588	GSSP
		Neogene	Pliocene	Piacenzian	3.600	GSSP
				Zanclean	5.332	GSSP
			Miocene	Messinian	7.246	GSSP
				Tortonian	11.608	GSSP
				Serravallian	13.82	GSSP
				Langhian	15.97	
				Burdigalian	20.43	
				Aquitanian	23.03	GSSP
		Paleogene	Oligocene	Chattian	28.4 ±0.1	
				Rupelian	33.9 ±0.1	GSSP
			Eocene	Priabonian	37.2 ±0.1	
				Bartonian	40.4 ±0.2	
				Lutetian	48.6 ±0.2	
				Ypresian	55.8 ±0.2	GSSP
			Paleocene	Thanetian	58.7 ±0.2	GSSP
				Selandian	~ 61.1	GSSP
				Danian	65.5 ±0.3	GSSP
	Mesozoic	Cretaceous	Upper	Maastrichtian	70.6 ±0.6	GSSP
				Campanian	83.5 ±0.7	
				Santonian	85.8 ±0.7	
				Coniacian	~ 88.6	
				Turonian	93.6 ±0.8	GSSP
				Cenomanian	99.6 ±0.9	GSSP
			Lower	Albian	112.0 ±1.0	
				Aptian	125.0 ±1.0	
				Barremian	130.0 ±1.5	
				Hauterivian	~ 133.9	
				Valanginian	140.2 ±3.0	
				Berriasian	145.5 ±4.0	

Eonothem Eon	Erathem Era	System Period	Series Epoch	Stage Age	Age Ma	GSSP
					145.5 ±4.0	
Phanerozoic	Mesozoic	Jurassic	Upper	Tithonian	150.8 ±4.0	
				Kimmeridgian	~ 155.6	
				Oxfordian	161.2 ±4.0	
			Middle	Callovian	164.7 ±4.0	
				Bathonian	167.7 ±3.5	GSSP
				Bajocian	171.6 ±3.0	GSSP
				Aalenian	175.6 ±2.0	GSSP
			Lower	Toarcian	183.0 ±1.5	
				Pliensbachian	189.6 ±1.5	GSSP
				Sinemurian	196.5 ±1.0	GSSP
				Hettangian	199.6 ±0.6	GSSP
		Triassic	Upper	Rhaetian	203.6 ±1.5	
				Norian	216.5 ±2.0	
				Carnian	~ 228.7	GSSP
			Middle	Ladinian	237.0 ±2.0	GSSP
				Anisian	~ 245.9	
			Lower	Olenekian	~ 249.5	
				Induan	251.0 ±0.4	GSSP
	Paleozoic	Permian	Lopingian	Changhsingian	253.8 ±0.7	GSSP
				Wuchiapingian	260.4 ±0.7	GSSP
			Guadalupian	Capitanian	265.8 ±0.7	GSSP
				Wordian	268.0 ±0.7	GSSP
				Roadian	270.6 ±0.7	GSSP
			Cisuralian	Kungurian	275.6 ±0.7	
				Artinskian	284.4 ±0.7	
				Sakmarian	294.6 ±0.8	
				Asselian	299.0 ±0.8	GSSP
		Carboniferous: Pennsylvanian	Upper	Gzhelian	303.4 ±0.9	
				Kasimovian	307.2 ±1.0	
			Middle	Moscovian	311.7 ±1.1	
			Lower	Bashkirian	318.1 ±1.3	GSSP
		Carboniferous: Mississippian	Upper	Serpukhovian	328.3 ±1.6	
			Middle	Visean	345.3 ±2.1	GSSP
			Lower	Tournaisian	359.2 ±2.5	GSSP

Eonothem Eon	Erathem Era	System Period	Series Epoch	Stage Age	Age Ma	GSSP
					359.2 ±2.5	
Phanerozoic	Paleozoic	Devonian	Upper	Famennian	374.5 ±2.6	GSSP
				Frasnian	385.3 ±2.6	GSSP
			Middle	Givetian	391.8 ±2.7	GSSP
				Eifelian	397.5 ±2.7	GSSP
			Lower	Emsian	407.0 ±2.8	GSSP
				Pragian	411.2 ±2.8	GSSP
				Lochkovian	416.0 ±2.8	GSSP
		Silurian	Pridoli		418.7 ±2.7	GSSP
			Ludlow	Ludfordian	421.3 ±2.6	GSSP
				Gorstian	422.9 ±2.5	GSSP
			Wenlock	Homerian	426.2 ±2.4	GSSP
				Sheinwoodian	428.2 ±2.3	GSSP
			Llandovery	Telychian	436.0 ±1.9	GSSP
				Aeronian	439.0 ±1.8	GSSP
				Rhuddanian	443.7 ±1.5	GSSP
		Ordovician	Upper	Hirnantian	445.6 ±1.5	GSSP
				Katian	455.8 ±1.6	GSSP
				Sandbian	460.9 ±1.6	GSSP
			Middle	Darriwilian	468.1 ±1.6	GSSP
				Dapingian	471.8 ±1.6	GSSP
			Lower	Floian	478.6 ±1.7	GSSP
				Tremadocian	488.3 ±1.7	GSSP
		Cambrian	Furongian	*Stage 10*	~ 492 *	
				Stage 9	~ 496 *	
				Paibian	~ 499	GSSP
			Series 3	Guzhangian	~ 503	GSSP
				Drumian	~ 506.5	GSSP
				Stage 5	~ 510 *	
			Series 2	*Stage 4*	~ 515 *	
				Stage 3	~ 521 *	
			Terreneuvian	*Stage 2*	~ 528 *	
				Fortunian	542.0 ±1.0	GSSP

This chart was drafted by Gabi Ogg. Intra Cambrian unit ages with * are informal, and awaiting ratified definitions.

Eonothem Eon	Erathem Era	System Period	Age Ma	GSSP GSSA
			542	
Precambrian: Proterozoic	Neoproterozoic	Ediacaran	~635	GSSP
		Cryogenian	850	GSSA
		Tonian	1000	GSSA
	Mesoproterozoic	Stenian	1200	GSSA
		Ectasian	1400	GSSA
		Calymmian	1600	GSSA
	Paleoproterozoic	Statherian	1800	GSSA
		Orosirian	2050	GSSA
		Rhyacian	2300	GSSA
		Siderian	2500	GSSA
Precambrian: Archean	Neoarchean		2800	GSSA
	Mesoarchean		3200	GSSA
	Paleoarchean		3600	GSSA
	Eoarchean		4000	
Precambrian	*Hadean (informal)*		~4600	

Subdivisions of the global geologic record are formally defined by their lower boundary. Each unit of the Phanerozoic (~542 Ma to Present) and the base of Ediacaran are defined by a basal Global Boundary Stratotype Section and Point (GSSP), whereas Precambrian units are formally subdivided by absolute age (Global Standard Stratigraphic Age, GSSA). Details of each GSSP are posted on the ICS website (*www.stratigraphy.org*).

Numerical ages of the unit boundaries in the Phanerozoic are subject to revision. Some stages within the Cambrian will be formally named upon international agreement on their GSSP limits. Most sub-Series boundaries (e.g., Middle and Upper Aptian) are not formally defined.

Colors are according to the Commission for the Geological Map of the World (*www.cgmw.org*).

The listed numerical ages are from 'A Geologic Time Scale 2004', by F.M. Gradstein, J.G. Ogg, A.G. Smith, et al. (2004; Cambridge University Press) and "The Concise Geologic Time Scale" by J.G. Ogg, G. Ogg and F.M. Gradstein (2008).

Sept. 2010

APPENDIX B

Standard Symbols Used in Topographic Maps Published by the U.S.G.S.

CONTROL DATA AND MONUMENTS

Aerial photograph roll and frame number*	3-20
Horizontal control	
Third order or better, permanent mark	Neace Neace
With third order or better elevation	BM 45.1 Pike BM 45.1
Checked spot elevation	19.5
Coincident with section corner	Cactus Cactus
Unmonumented*	
Vertical control	
Third order or better, with tablet	BM × 16.3
Third order or better, recoverable mark	× 120.0
Bench mark at found section corner	BM 18.6
Spot elevation	× 5.3
Boundary monument	
With tablet	BM 21.6 BM 71
Without tablet	171.3
With number and elevation	67 301.1
U.S. mineral or location monument	

CONTOURS

Topographic	
Intermediate	
Index	
Supplementary	
Depression	
Cut; fill	
Bathymetric	
Intermediate	
Index	
Primary	
Index Primary	
Supplementary	

BOUNDARIES

National	
State or territorial	
County or equivalent	
Civil township or equivalent	
Incorporated city or equivalent	
Park, reservation, or monument	
Small park	

*Provisional Edition maps only

Provisional Edition maps were established to expedite completion of the remaining large scale topographic quadrangles of the conterminous United States. They contain essentially the same level of information as the standard series maps. This series can be easily recongnized by the title "Provisional Edition" in the lower right hand corner.

LAND SURVEY SYSTEMS

U.S. Public Land Survey System	
Township or range line	
Location doubtful	
Section line	
Location doubtful	
Found section corner; found closing corner	
Witness corner; meander corner	WC MC
Other land surveys	
Township or range line	
Section line	
Land grant or mining claim; monument	
Fence line	

SURFACE FEATURES

Levee	Levee
Sand or mud area, dunes, or shifting sand	Sand
Intricate surface area	Strip mine
Gravel beach or glacial moraine	Gravel
Tailings pond	Tailings Pond

MINES AND CAVES

Quarry or open pit mine	
Gravel, sand, clay, or borrow pit	
Mine tunnel or cave entrance	
Prospect; mine shaft	
Mine dump	Mine dump
Tailings	Tailings

VEGETATION

Woods	
Scrub	
Orchard	
Vineyard	
Mangrove	Mangrove

GLACIERS AND PERMANENT SNOWFIELDS

Contours and limits	
Form lines	

MARINE SHORELINE

Topographic maps	
Approximate mean high water	
Indefinite or unsurveyed	
Topographic-bathymetric maps	
Mean high water	
Apparent (edge of vegetation)	

Figure Appendix

Standard symbols used on topographic maps published by the U.S.G.S.

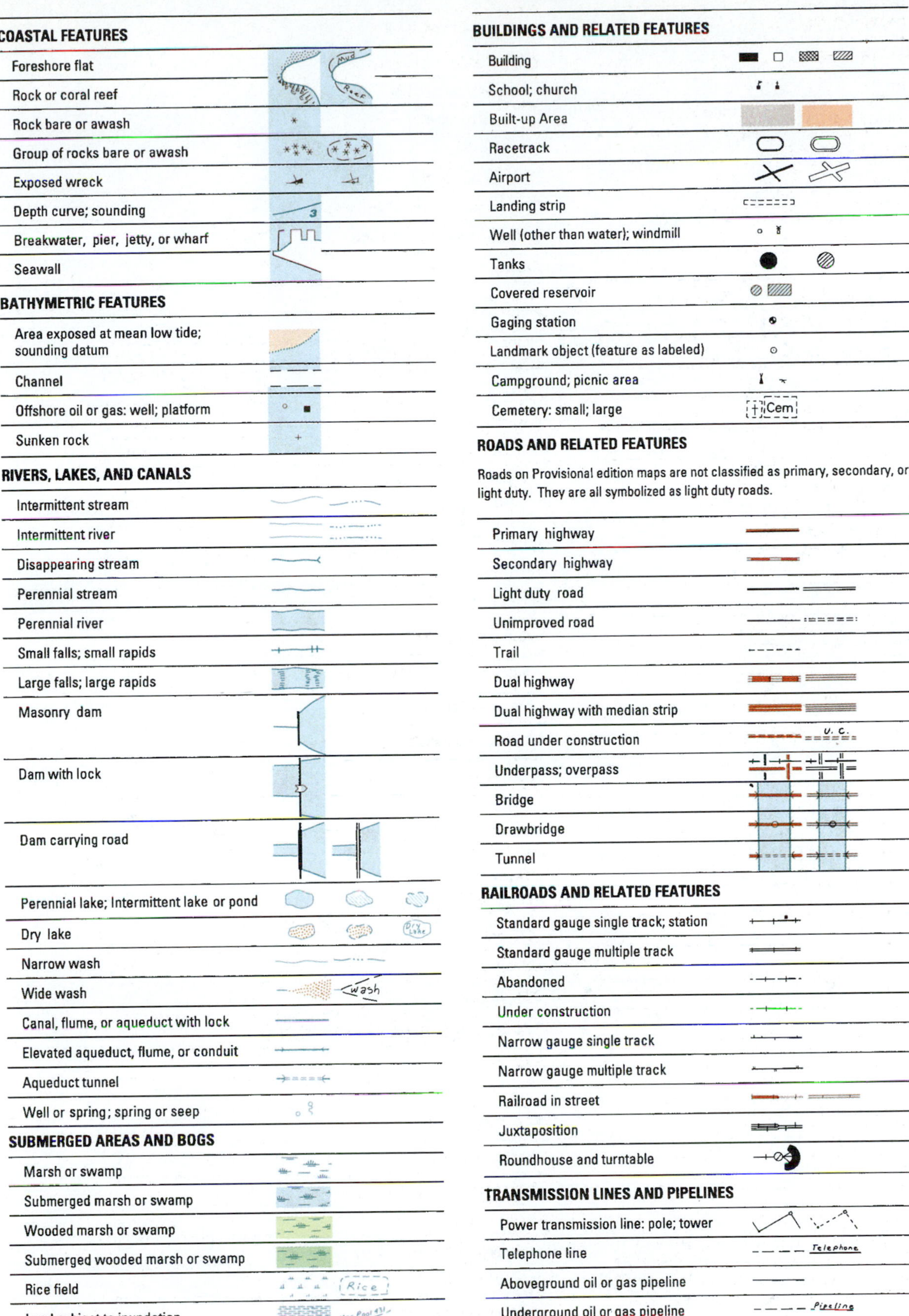

Figure Appendix (Continued)

Standard symbols used on topographic maps published by the U.S.G.S.

GLOSSARY

The numbers following each definition refer to the page numbers in the text where the term is "first" used.

A

Aa A lava flow, usually basaltic in composition, that has a jagged, rubbly surface. 189

Ablation zone That area on a glacier where snow and ice are lost on an annual basis by melting, wind erosion, sublimation, and calving (processes of ablation). 141

Absolute ages (absolute chronology) Sequential order of geologic events (a series of dates, usually in millions of years), based on radiometric measurements. 56, 260

Accessory mineral A mineral not important in the classification of a rock, but may be used as a descriptor. 25

Accumulation See Zone of Accumulation.

Active dune A sand dune that migrates in a downwind direction or is constantly changing in shape in response to multiple wind directions at different times. 163

Active fault A fault along which movement has occurred sparadically during historical time. 228

Active volcano A volcano that is in an eruptive phase, has erupted in the recent past, or is likely to erupt in the future. 187

Alluvial fan A fan-shaped depositional feature produced where the gradient of a stream changes from steep to shallow as the stream emerges from a mountainous terrain. 107

Alpine glacier A river of ice confined to a mountainous valley. 141

Amorphous (1) A rock, mineral, or substance without structure. (2) A very fine-grained or glassy rock. 5, 36

Amygdaloidal An adjective describing an igneous rock containing amygdules. 25

Amygdule A vesicle in an igneous rock filled with secondary mineral matter by groundwater. 25

Angular unconformity An unconformity in which younger sediments rest upon the eroded surface of titled or folded older rocks. 61

Anticline A fold, generally convex upward, whose core contains the stratigraphically older rocks. 201

Aphanitic The texture of an igneous rock composed of microscopic crystals. 25

Arête A sharp, narrow, rugged ridge (divide) between two parallel glacial valleys. 142

Assimilation The process in which a magma melts country rock and assimilates the newly molten material. 23

Asthenosphere The weak, ductile layer of the earth's upper mantle (beneath the lithosphere); the layer on which the lithospheric plates move. 248

Atoll A roughly circular narrow strip of continuous or discontinuous coral reef with low relief; encloses a lagoon and is bounded on the outside by the deep water of the open ocean. 264

Attitude The orientation of a structural element (line or plane) relative to a horizontal plane. Attitudes of planes are expressed in terms of strike and dip, attitude of lines are expressed by plunge and bearing (compass direction). 200

Aureole The zone of contact metamorphism in the host rock surrounding an intrusion. 45

Avulsion The process by which a river suddenly leaves its old bed and forms a new bed. 108

Axial plane An imaginary plane that divides a fold into halves as symmetrically as possible. 201

Axis of symmetry An imaginary line about which a crystal may be rotated, during which there may be two, three, four, or six repetitions of an identical face. 5

B

Bajada An alluvial plain formed at the base of a mountain by the coalescing of several alluvial fans. 107

Barchan A crescent-shaped sand dune in plan view with the highest part in the center and the tips (horns) pointing downwind. 162

Barrier island A long, narrow island of sand parallel to the shoreline and built by wave action. 170

Barrier reef A reef separated from the shoreline by a lagoon. 268

Base level The lowest level to which a stream can erode its bed; usually sea level. 106

Batholith A large mass of intrusive igneous rock that crops out over an area of more than 100 square kilometers. 24

Beach The wave-washed sediment along a coast, extending throughout the surf zone to a cliff or zone of permanent vegetation. 170

Bedding plane A horizontal surface along which a sedimentary rock breaks or separates; represents changes in the depositional history. 35, 199

Beds A sedimentary unit composed of layers of sedimentary rock. 35

Bench mark A relatively precisely located point marked by a brass plate with a cross, elevation, and coordinates permanently fixed on the ground. 79

Bird-foot delta A delta with many distributaries. 171

Block diagram A perspective drawing in which the information on a geologic map and geologic cross section are combined. 201

Blowout A shallow depression on the land surface caused by the removal of sand or smaller particles by wind erosion. 162

Bowen's reaction series A schematic description of the order in which minerals crystallize and react during the cooling and progressive crystallization of a magma. 22

Breccia A rock composed of angular fragments of broken rock cemented together. 40

Budget year of a glacier A 12-month period of winter snow accumulation and summer melting. 141

C

Calcareous An adjective used for a rock containing calcium carbonate; usually refers to a sedimentary rock in which a carbonate mineral is present but is not a major constituent. 40

Calving A process whereby masses of glacier ice become detached from a glacier and terminate in deep water to form icebergs. 142

Capacity The total sediment load a stream can carry at a given discharge. 106

Cirque A steep-sided amphitheater-like feature carved into a mountain at the head of a glacial valley and formed mainly by glacial abrasion and plucking and by frost wedging. 142

Clay mineral A group of platy hydrous aluminosilicate minerals that have layered atomic structures that formed from the weathering and hydration of other silicate minerals. 36, 40

Clay-size A particle 4 microns or less in diameter. 36, 40

Cleavage (1) The tendency of a mineral to break along certain preferred crystallographic planes. (2) The tendency of a rock to separate into platelike fragments along certain planes; usually the result of preferred orientation of the minerals in the rock. 2, 3

Color The quality of a mineral or rock with respect to light reflected by it, usually determined visually by the measurement of hue, saturation, and brightness of the reflected light. 2

Columnar section A vertical display of the sequence of rock strata in a geologic cross section from oldest at the bottom to youngest at the top. 59

Competence The ability of a current of water or wind to transport detritus, measured as the diameter of the largest particle transported. 106

Composite volcano (stratovolcano) A conical volcano with steep sides composed of interbedded layers of viscous lava and pyroclastic material. 187

Confining (static) pressure The pressure (equal in all directions) on deeply buried rocks resulting from the weight of the overlying rocks. 45

Confluence The point at which a tributary joins the main stream. 118

Conglomerate A coarse-grained rock composed of rounded to subangular clasts set in a fine-grained matrix; the lithified equivalent of gravel. 40

Contact (thermal) metamorphism The metamorphism of the country rock adjacent to an intrusion of igneous rock; metamorphism in which temperature is the dominant factor. 45

Continental crust That portion of the crust (outer lithosphere) that is mainly granitic in composition, varies from 20–70 km thick, and which underlies the continents. 248

Continental glacier An ice sheet that covers an area of continental proportions and is not confined to a single valley and spreads outward in all directions under the influence of its own weight. 141, 152

Contour interval The difference in elevation of any two adjacent contour lines. 78

Contour line A line on a topographic map connecting points of equal elevation. 78

Core The central part of the earth below about 2,900 km. Composed mainly of iron and nickel. The outer core is liquid, the inner core solid. 248

Country rock The rock mass intruded by a magma. 23

Creep The slow downslope movement of rock and regolith under gravitational forces. 162

Crevasse A deep, gaping, vertical fissure in the upper 40 to 50 meters of a glacier. 148

Cross-bedding Sediments laid down at an angle to the main sedimentary layering. 37

Cross section See Geologic cross section.

Crust The outermost layer of the solid earth (outer lithosphere). 248

Crystal A homogeneous solid body having a regularly repeating atomic arrangement that is outwardly expressed by plane faces that have a definite geometric relationship to one another. 5

Crystal form In a definite geometric relationship, the assemblage of faces that constitute the exterior surface of a crystal. 2, 5

Crystal habit The crystal form commonly taken by a given mineral. 5

Crystal symmetry The geometric relationship between crystal faces. 5

D

Deflation The picking up and removal by wind of loose, dry, surface material. 162

Delta A body of sediment deposited by a river where it enters an ocean or lake. 171

Detritus Fragments of minerals, rocks, and skeletal remains of dead organisms. 35

Diagenesis The changes, both physical and chemical, that occur in a sediment after deposition and during compaction and lithification as the sediment is transformed into a sedimentary rock. 34

Diaphaneity The ability of a thin slice of a mineral to transmit light. 5

Dike A tabular igneous intrusion whose contacts cut across the trend of the country rock. 24

Dip The vertical angle in degrees measured downward between a horizontal plane and an inclined plane perpendicular to the strike. 200

Disconformity An unconformity in which the bedding planes above and below a break (visible and irregular or uneven erosion surface of appreciable relief) are essentially parallel, indicating a considerable interval of erosion or nondeposition. 59

Dolostone A carbonate sedimentary rock consisting of more than 50% dolomite. 40

Dormant volcano An active volcano during periods of quiescence. 187

Double refraction The splitting of light into two components when it passes through certain crystalline substances. 7

Drainage system A stream together with its tributaries. 106

Drowned river mouth (estuary) A long embayment formed by the encroachment of the sea into the mouth of a river. 171

Drumlin An elongate and streamlined hill of till whose long axis is parallel to the direction of ice movement. 152

E

Effluent stream A stream fed by groundwater because its channel lies below the water table. 124

End moraine The accumulation of till at the terminus of a glacier. 142

Epicenter The point on the earth's surface that lies directly above the focus (hypocenter) of an earthquake. 230

Equator A *great circle* formed by a plane through the center of the earth (globe) and equidistant from the poles that divides the earth (globe) into a northern and southern hemisphere. It is the line from which latitude is measured north and south. 72

Equilibrium line The line on a glacier that separates the accumulation and ablation zones for a given budget year. 141

Esker An ice-contact deposit in the form of a continuous, low ridge of stratified sand and gravel, often sinuous, formed beneath the glacier surface by a sediment-laden stream flowing through an ice tunnel. 152

Essential mineral A mineral necessary to classify a rock. 25

Estuary A semi-enclosed coastal body of water that has a free connection with the open sea and within which seawater is measurably diluted with freshwater from land drainage. 171

Eustatic A change in sea level that affects all the oceans. 171

Extinct volcano A volcano in which all volcanic activity has ceased permanently. 187

F

False color image An image from remote sensors combined with other wavelengths in the processing of the imagery data resulting in a colored image in which the true colors are replaced by other colors. 71, 96

Fault A fracture or break in the earth's crust along which differential movement of the rock masses has occurred. 199, 228

Fault plane The plane that best approximates the fracture surface of a fault. 199, 228

Fault scarp The surface expression of the fault plane. 228

Fault trace The intersection of the fault plane with the ground surface. 228

Felsic An adjective used to describe a light-colored igneous rock that is poor in iron and magnesium and contains abundant quartz and feldspars. 23

Ferruginous An adjective used for a sedimentary rock that is cemented with iron oxide. 42

Floodplain A flat erosional river valley floor on either side of a stream channel; inundated during floods and containing silt and sand carried out and deposited from the main channel during floods. 107

Flow line (1) The path a water molecule follows from the time it enters the zone of saturation until it reaches a lake or stream, where it becomes surface water. (2) The path a particle takes from the point where it is incorporated into a glacier until it ceases movement. 124

Focus (hypocenter) The place where rupture occurs (movement begins) on a fault plane during an earthquake. 230

Fold A portion of strata that is bent (an anticline or syncline) or that connects two horizontal or parallel portions of strata of different levels (a monocline). 199

Foliation A metamorphic texture in which the mineral constituents are oriented in a parallel or subparallel arrangement. 45

Footwall The surface of the block of rock below an inclined fault. 228

Fracture (1) Breakage that forms a surface with no relationship to the internal structure of a mineral. (2) A crack or joint in bedrock. 4

Fringing coral reef A coral reef attached to or bordering a landmass. 268

G

Geologic age The age of a geologic event or feature referred to the geologic time scale and expressed in terms of years (absolute age) or of comparison with the immediate surroundings (relative age). An age dateable by geologic methods. 56

Geologic column A composite diagram that shows in a single column the subdivisions of part or all of geologic time or the sequence of stratigraphic units of a given locality or region (the oldest at the bottom and the youngest at the top) so arranged as to indicate their relations to the subdivisions of geologic time and their relative positions to each other. 55, 56

Geologic cross section A diagram or drawing that shows geologic features transected by a given plane. 59, 201

Geomorphology The association of geologic agents with the origin of various landforms. 105

Geothermal gradient The rate of increase of temperature with depth within the earth. 22

Glacier A mass of flowing land ice (derived from snowfall) that moves because of its own weight. 141

Graben A wedge-shaped, downthrown block of rock bounded on its sides by normal faults. 229, 230

Graded bedding Gradual vertical shift from coarse to fine clastic material in the same bed; resulting from deposition by a waning current. 37

Gradient The vertical drop of a stream over a given horizontal distance. 106

Groundmass The fine-grained matrix of an igneous rock. 25

H

Hand specimen A specimen of mineral or rock that can be held in the hand for study. 1

Hanging valley A valley formed where a tributary glacier once joined the main glacier, and a waterfall cascades from it to the floor of the main valley. 142

Hanging wall The surface of the block of rock above an inclined fault. 228

Hardness Resistance of the surface of a mineral to abrasion. 3

Headland Part of the coast that juts out into a lake or ocean. 170

Heavy minerals Minerals having a specific gravity greater than 2.85, and commonly found as minor constituents or accessory minerals of a rock. 40

Heft To estimate the weight of an object by lifting it. 5

Hinge line Axis of a fold along which the curvature is the greatest. 202

Horn (1) A sharp, pyramid-shaped mountain peak near the heads of valley glaciers or glaciated valleys. (2) The ends of a barchan or parabolic dune. 142

Hornfels A fine-grained nonfoliated rock usually formed by contact metamorphism. 46, 48

Horst An upthrown block of rock bounded on its sides by normal faults. 230

Hot spot A deep-seated zone of intense heat whose geographic coordinates remain fixed for several million years. 187, 266

I

Ice Age A time of extensive glacial activity, usually referring to the Pleistocene epoch. 152

Iceberg A mass of glacier ice calved into a lake or ocean. 142

Ice sheet A very large area of perennial and continuous ice cover with considerable thickness and an area greater than 50,000 sq. km (ice caps are less than 50,000 sq. mi.). An ice sheet flows outward under the force of gravity and is not constrained by topography. A continental glacier. 141, 152

Igneous rock A rock formed by the cooling and crystallization of molten material within or at the surface of the earth. 21

Inactive dune A sand dune that has become stabilized by the growth of vegetational cover to the extent that its migration ceases. 163

Inactive fault A fault in which no movement has occurred during historical times. 228

Inclusion (xenolith) A fragment of country rock surrounded by igneous rock. 23

Index fossil A fossil that has wide geographic distribution but narrow geologic time range. 62

Index mineral A mineral that forms or is stable over a limited range of temperature and pressure conditions and whose first appearance marks the outer limits of a specific zone of metamorphism. 46

Influent stream A stream that lies above the water table and in which groundwater flows away from the stream channel. 124

Intermittent stream A stream that flows only during certain times of the year when rainfall is sufficient to supply surface runoff directly to it. 124

Interpolate To estimate values of a function between two known values. 81

K

Kame A knob, hummock, or conical hill composed of coarse stratified drift formed as a delta at the front of a glacier by meltwater streams. 152

Karst topography A terrain marked by many sinks and caverns and usually lacking a surface stream. 131

Kettle In an end moraine, a depression formed by the melting of an underlying large block of ice left behind by a receding glacier. 152

L

Laccolith A lenticular pluton, many times wider than it is thick, intruded parallel to the layering of the intruded rock, above which the layers of intruded rock have been bent upward to form a dome. 24

Lagoon A shallow-water body lying between the main shoreline and a barrier island or inshore from an enclosing reef. 170

Lateral moraine An accumulation of till along the sides of an Alpine glacier. 142

Lava fountain Lava spouting from a fissure in a rift zone as a result of gas and fluid pressure that builds up in the crust below. 189

Law of constancy of interfacial angles The angle between similar crystal faces of a mineral is constant. 5

Law of crosscutting relationships A rock is younger than any rock across which it cuts. 56

Law of faunal assemblages Similar assemblages of fossils indicate similar geologic ages for the rocks that contain them. 56

Law of faunal succession Each geologic formation has a different total aspect of life (fossils) from that in the formation above it and below it. 56

Law of inclusions (xenoliths) Fragments of rocks in a sedimentary or igneous unit are older than the host rock containing them. 56

Law of original horizontality Sediments deposited in water are laid down in strata that are horizontal or nearly horizontal and are parallel or nearly parallel to the earth's surface. 56

Law of original (lateral) continuity Sedimentary rock units extend laterally in all directions until they thin or pinch out at their margins due to nondeposition or abutting against the edge of the basin of deposition. 56

Law of superposition In any undisturbed sequence of sedimentary rocks, the layer at the bottom of the sequence is older than the layer at the top of the sequence. 56

Law of unconformities Rock units above an unconformity are younger than those below an unconformity. 56

Leeward The downwind side of a sand dune. 162

Limestone A carbonate sedimentary rock consisting of more than 50% calcite. Limestones include calcarenite, chalk, micrite, coquina, and travertine. 40

Lithification The combination of processes that convert a sediment into a sedimentary rock. 21, 34

Lithosphere The rigid outer 100 km of the solid earth consisting of the rigid upper mantle and crust of the earth. 248

Longitudinal profile A line showing a stream's slope, drawn along the length of the stream as if it were viewed from the side. 106

Longitudinal dune (seif) A long, linear symmetrical dune with its long axis parallel to the wind direction. 162

Longshore drift The movement of particles obliquely up the slope of a beach; caused by the swash, and directly down this slope, by the backwash. 170

Luster The appearance of a fresh mineral surface in reflected light. 2

M

Macroscopic (megascopic) A feature of a mineral or rock that can be distinguished without the aid of magnification. 1

Mafic An adjective used to describe a dark-colored, silica-poor igneous rock with a high magnesium and iron content and composed chiefly of iron- and magnesium-rich minerals. 23

Magma Molten material generated within the earth and consisting of a complex solution of silicates plus water, dissolved gases, and any suspended crystals. 22

Magnetic anomaly A magnetic reading that is greater or less than the normal strength of the magnetic field where the rock occurs. 259, 260

Magnetometer An instrument that measures the intensity of the earth's magnetic field. 260

Mantle The largest portion of the solid earth, separating crust above from the core below. It is mainly ultramafic in composition. 248

Map scale The ratio between linear distance on a map and the corresponding distance on the surface being mapped. 78

Meander A looplike bend of a stream channel; develops as the stream erodes the outer bank of a curve and deposits sediment against the inner bank. 107

Meander line A surveyed line, usually of irregular course, that defines the highwater line to mark the sinuosity of the bank or shoreline. 120

Meandering course The circuitous course of a stream flowing across its floodplain. 107

Medial moraine A linear moraine formed by the joining of two lateral moraines as the two glaciers flow together. 142

Mesosphere The portion of the mantle between the bottom of the asthenosphere and the core-mantle boundary. 248

Metamorphic facies The pressure and temperature stability fields for metamorphic rocks as determined by mineral assemblages. 45

Metamorphic grade The intensity of metamorphism in a given rock; the maximum temperature and pressure attained during metamorphism. 45, 46

Metamorphic rock A rock resulting from the change of a preexisting rock as a result of the effects of heat, pressure, chemical action, or combinations of these. 21, 45

Metasomatism The process of solution and deposition by which a new mineral or minerals of partly or wholly different chemical composition may grow from externally supplied fluids and elements in the body of an old mineral or mineral aggregate forming a metasomatic rock. 45

Microscopic A feature of a mineral or rock that can be distinguished only with the aid of magnification. 1

Mohs scale of hardness Ten common minerals arranged in order of their increasing hardness from 1 to 10. 3

Monocline A one-limb flexure in which the strata have a uniform direction of strike but a variable angle of dip. 201

N

Natural levee A narrow ridge of flood-deposited sediment found on either side of a stream channel, which thins away from the channel. 107

Nonconformity An unconformity developed between sedimentary rocks and older plutonic igneous or metamorphic rocks that had been exposed to erosion before the overlying sediments covered them. 61

Noncrystalline A textural term in which the rock is amorphous or consists of very finely divided material deposited by chemical precipitation. 36

Nonfoliated No preferred orientation of mineral grains in a metamorphic rock. 46

Normal fault A fault in which the hanging wall has moved downward relative to the footwall. 228

O

Oceanic crust The crust (outer lithosphere) beneath the oceans consisting of a 7–15-kilometer-thick layer of mainly basalts and thin sedimentary rocks and overlying sediments. 248

Oolitic A sedimentary texture formed by spheroidal particles (oolites) less than 2 mm in diameter. 36

Outwash plain A deposit of sand and gravel formed beyond the front of a glacier by streams flowing from the glacier terminus. 152

Oxbow lake A crescent-shaped shallow lake formed where a meander loop is cut off from the main stream and its ends plugged with sediments. 107

P

Pahoehoe A smooth surface, "ropy" lava usually basaltic in composition. 189

Paleomagnetism The study of natural remanent magnetism in ancient rocks, recording the direction of the magnetic poles at some time in the past. 259

Parabolic dune A deeply curved dune in a region of abundant sand. The horns point upwind and are often anchored by vegetation. 162

Pediment A gently sloping erosional surface cut into the solid rock of a mountain range and covered with a thin, patchy veneer of coarse colluvium that slopes away from the base of a highland in an arid or semiarid environment. 107

Pegmatite A very coarse-grained igneous rock of any composition but most frequently granitic. 26

Phaneritic The texture of an igneous rock composed of macroscopic minerals. 24

Plan view As viewed from above. 85

Plate tectonics A theory that the earth's surface is divided into a few large, thick, rigid plates that are slowly moving and changing size with respect to one another. 247

Playa A flat undrained desert basin that contains intermittent lakes. 107

Playa lake A lake formed on the flat floor of a desert basin having interior drainage, usually formed after a heavy rain. 107

Plunging fold A fold with an inclined axis. 205

Pluton A body of igneous rock, regardless of shape or size, that crystallized underground. 24

Porphyritic An igneous rock in which phenocrysts (macroscopic minerals) are embedded in a fine-grained matrix. 25

Porphyroblast A metamorphic mineral crystal that is much larger than the matrix in which it grew. 48

Primary minerals Minerals that crystallized from a cooling magma. 25

Prograding shoreline A shoreline that moves seaward by deltaic growth. 172

Protolith The rock from which a metamorphic rock was formed. Example: a quartz sandstone is the protolith of a quartzite. 45

Pyroclastic Accumulations of material ejected from explosive-type volcanoes. 25, 187

R

Recessional moraine Successive end moraines lying beyond the snout of a retreating glacier. 142

Recumbent fold An overturned fold in which the axial plane and both limbs are essentially horizontal. 202

Relative chronology (relative ages) Sequential order of geologic events based on relative position, fossil content, and cross-cutting relationships. 56, 260

Remote sensing A process whereby the image of a feature is recorded by a camera or other devices that are not in direct contact with it and reproduced in one form or another as a "picture" of the feature. 71

Reverse fault A fault in which the hanging wall has moved up relative to the footwall. 228

Ripple marks The parallel ridges and troughs left by running water or wind on a former sedimentary surface. 37

Rock A naturally formed, lithified aggregate of one or more minerals or lithified organic matter. 21

Rock cycle The sequence of events in which rocks are created, destroyed, altered, and reformed by geologic processes. 21

S

Saltation The movement of sediment by bouncing along the bottom surface. Applies to transport both by water and by wind. 162

Sand dune A ridge or mound of sand deposited by wind. 162

Saturated zone Beneath the surface of the earth, the zone that is saturated with water. 124

Secondary mineral A mineral formed later than the rock enclosing it, usually at the expense of an earlier-formed primary mineral. 25

Section A rectangular block of land 1 mile long by 1 mile wide. 76

Sedimentary rock A rock formed from precipitation from a solution, by accumulation of organic materials, or by sedimentation and cementation of sediments derived from preexisting rocks and transported to a site of deposition by water, wind, or ice. 21

Serrate divide The divide between the headward regions of oppositely sloping glacial valleys. 142

Shield volcano A broad, gently sloping dome built of thousands of highly fluid (low viscosity) lava flows generally basaltic in composition. 187

Sill A tabular-shaped igneous rock mass that lies parallel to the layering of the intruded rock. 24

Sink (sink hole) A surface depression in limestone caused by the collapse of the roof of an underground dissolution cavity. 131

Slip face The steep leeward slope of a sand dune formed from loose, cascading sand that generally keeps the slope at the angle of repose (about 34 degrees). 162

Sole marks Marks left at the top of a soft sediment by bottom-dwelling organisms. 37

Specific gravity A number stating the ratio of a mineral's weight to the weight of an equal volume of pure water. 5

Spit An elongate ridge of sand or gravel extending from shore into a body of open water; deposited by longshore currents. 170

Stack An isolated rocky pillar that is the remnant of a retreating wave-cut cliff. 170

Stock An irregular body of intrusive igneous rock, smaller than a batholith, that cuts across the layering of the intruded rock. 24

Strata Plural of stratum. A stratigraphic unit composed of a number of beds or layers of sedimentary rock. 34

Streak The color of a mineral's powder usually obtained by rubbing the mineral on a streak plate (unglazed porcelain). 5

Strike A horizontal line in a plane of the bedding or fault surface expressed as a compass direction from true north. 200

Strike-slip fault A fault with relative displacement along the fault plane in a horizontal direction parallel to the strike of the fault plane. 230

Subduction The process of one lithospheric plate descending beneath another. 249

Syncline A fold, generally concave upward, whose core contains the stratigraphically younger rocks. 201

T

Tarn A small lake within a cirque. 142

Tenacity A mineral's resistance to being broken or bent. 5

Tephra A collective term assigned to all sizes of airborne volcanic ejecta. 189

Texture The appearance of a rock; results from the size, shape, and arrangement of the mineral grains or crystals. 24, 35, 46

Thalweg A line connecting the deepest parts of a stream. 118

Thrust fault A reverse fault in which the fault plane dips less than 45 degrees over most of its extent. 230

Tidal inlet A break or passageway through a bar or barrier island; allows water to flow alternatively landward with rising tide and seaward with falling tide. 170

Till The nonsorted debris carried and deposited directly by a glacier. 142

Topographic map A map on which elevations are shown by means of contour lines. 78

Topography The relief and form of the land surface. 78

Township An area containing 36 sections. 76

Transform fault A strike-slip fault that offsets spreading ridges. 230

Transverse dune A linear ridge of sand with a gently sloping windward side and a steep lee face with its long dimension oriented perpendicular to the prevailing wind direction. 162

Tuff The consolidated ash and pyroclastic fragments generated from a volcanic eruption. 27

U

Unconformity A surface that represents a break in the geologic record, such as an interruption of deposition of sediments or a break between eroded igneous or metamorphic and overlying sedimentary rocks. 59

U-shaped valley A glacially eroded valley with a characteristic U-shaped cross section. 142

V

Vertical exaggeration The ratio between the horizontal scale and the vertical scale (Caution: units must be the same). Used to emphasize the relief shown on a cross section, model, etc. 85

Vesicle A small cavity in an igneous rock, usually extrusive, that was formerly occupied by a bubble of escaping gas originally held in solution under high pressure while the parent magma was deep underground. 25

Vesicular A texture characterized by the presence of vesicles. 25

Wacke A sandstone that contains a variety of poorly sorted mineral and rock fragments and has a considerable amount of clay and fine silt as matrix. A "dirty" sandstone. 40

Water table The top of the zone of groundwater saturation. 124

Wave-cut cliff A cliff formed by wave refraction against headlands. 170

Wave-cut platform A gently sloping rock surface lying below sea level and extending seaward from the base of a wave-cut cliff. 171

Windward The side of a dune ridge facing the wind. 162

Xenolith See inclusion.

Z

Zone of accumulation The upper reaches of a glacier over which more snow is deposited than is melted each year. 141

CREDITS

Illustrations and Text

Part I:

1.5, Copyright © 1974 McGraw-Hill Book Co. (UK) Ltd. From Cox, Price & Harte: An Introduction to the Practical Study of Crystals, Minerals and Rocks. **Table 1.2,** Data from *Dana's New Minerology,* 8th Edition, Richard V. Gaines, H. Catherine, W. Skinner, Eugene E. Foord, Bryan Mason and Abraham Rosenzweig (New York, NY: John Wiley and Sons, Inc., 1997. Reprinted with permission. **Table 1.4,** From *Mineralogy,* 2nd Edition by L. G. Berry, Brian Mason and R. V. Dietrich (New York, NY: WH Freeman, 1983).

Part II:

Table 2.1, Courtesy of Geological Society of America.

Part III:

3.2, Data from United States Geological Survey. **3.8,** Data from United States Geological Survey. **3.9,** Courtesy U.S.G.S. National Mapping Program Pamphlet, Map Scales 1981. **3.12,** Data from United States Geological Survey.

Part IV:

Page: 108, Van Zandt, Franklin K. 1976. Boundaries of the United States and the Several States. U.S.G.S. Professional Paper 909, p. 4. **Exercise 13E,** Van Zandt, Franklin K. 1976. Boundries of the United States and the Several States. U.S.G.S. Professional Paper 909, p. 4. **Table 4.2,** Source: Courtesy of Andrew G. Fountain, *U.S. Geological Survey,* Tacoma, Washington. **4.25,** Source: Simplified from The Geology of the Rhoda Quadrangle, USGS Map GQ-219 (1963). **4.30,** Source: Based on U.S. Geological Survey Professional Paper 715-a, 1971. **4.52,** After Regional Geomorphology of the United States by W. D. Thornbury. Copyright © 1956 John Wiley & Sons. **Box 4.1, Figure 1,** Reprinted with permission of J. Thomas McGlothlin IV. **Box 4.1, Figure 2,** Reprinted with permission of J. Thomas McGlothlin IV. **4.53,** After Elements of Geology, 3rd Edition by J. H. Zumberge and C. A. Nelson. Copyright © 1972 John Wiley & Sons. This material is used with permission of John Wiley & Sons, Inc. **4.60,** Based on data published by the U.S. Army Corps of Engineers, Detroit, MI 48231. **4.63,** Source: From R. I. Tilling, et al., 1987. Eruptions of Hawaiian Volcanoes: Past, Present and Future. *U.S. Geological Survey.* **4.65,** Source: From R. W. Decker et al., eds., 1987, Volcanism in Hawaii. *U.S.G.S. Professional Paper 1350* **4.66A, 188,** Source: Modified from R. I. Tilling, et al., 1987. Eruptions of Hawaiian Volcanoes: Past, Present and Future. *U.S. Geological Survey.* **4.67,** Source: From R. I. Tilling, et al., 1987. Eruptions of Hawaiian Volcanoes: Past, Present and Future. *U.S. Geological Survey.* **4.68,** From *J. P. Lockwood and P. W. Lipman, 1987,* Holocene Eruption History of Mauna Los Volcano, *Chapter 18 in R. W. Decker et al., editors,* Volcanism in Hawaii, U.S. *Geological Survey Professional Paper 1350.*

Part V:

5.1, From Elements of Physical Geology by J. H. Zumberge and C. A. Nelson. Copyright © 1976 John Wiley & Sons, Inc. This material is used by permission of John Wiley & Sons. **5.33,** *Source: R. H. Campbell, 1976. "Active faults in the Los Angeles-Ventura area of Southern California." ERTS-1: A New Window on Our Planet, U.S. Geological Survey Professional Paper 929, pp. 113–116.* **5.35,** *Source: R. H. Campbell, 1976. "Active faults in the Los Angeles-Ventura area of Southern California." ERTS-1: A New Window on Our Planet, U.S. Geological Survey Professional Paper 929, pp. 113–116.* **5.37,** Source: Based on data from Charles G. Sammis, University of Southern California. **5.38,** Source: Based on data from Charles G. Sammis, University of Southern California. **5.39,** Data from United States Geological Survey.

Part VI:

6.2, Adapted from Brian J. Skinner and Stephen C. Porter, 1989, Physical Geology: John Wiley & Sons, Inc. **Box 6.1, Figure 2,** Data from United States Geological Survey. **6.3,** Sources: M. Nafitoksoz, "The Subduction of the Lithosphere," Scientific American, Nov. 1975; and Peter J. Wylie et al., "Interactions Among Magmas and Rocks in Subduction Zone Regions: Experimental Studies from Slab to Mantle Crust: in European Journal of Mineralogy, vol. 1, pp. 165–179. **6.4,** Source: Computerized Digital Image and Data Base available from the National Geophysical Data Center, National Oceanic and Atmospheric Administration, U.S. Department of Commerce, Code E/GC3, Boulder, Colorado 80303. **6.5,** Compiled from the Plate Tectonic Map of the Circum-Pacific Region, 1982; and the Preliminary Tectonostratigraphic Terrane Map of the Circum-Pacific Region, copyrighted and published by the American Association of Petroleum Geologists, Post Office Box 979, Tulsa, Oklahoma 74101. Reprinted by permission of the AAPG whose permission is required for further use. **6.8,** From Magnetic Lineations of the World's Ocean Basins. Copyright © 1985 the American Association of Petroleum Geologists, Post Office 979, Tulsa, Oklahoma 74101. Reprinted by permission of the AAPG whose permission is required for further use. **6.10,** From Magnetic Lineations of the World's Ocean Basins. Copyright © 1985 the American Association of Petroleum Geologists, Post Office 979, Tulsa, Oklahoma 74101. Reprinted by permission of the AAPG whose permission is required for further use. **6.12,** From Magnetic Lineations of the World's Ocean Basins. Copyright © 1985 the American Association of Petroleum Geologists, Post Office 979, Tulsa, Oklahoma 74101. Reprinted by permission of the AAPG whose permission is required for further use. **6.14,** From the Preliminary Tectonostratigraphic Map of the Circum-Pacific Region. Copyright © 1985 the American Association of Petroleum Geologists, Post Office Box 979, Tulsa Oklahoma 74101. Reprinted by permission of AAPG whose permission is required for further use.

Appendix A:

Page: 271, www.Stratigraphy.org

Appendix B:

Pages: 273–275, Data from United States Geological Society.

INDEX

Note: Figures and tables, indicated by *f* and *t* respectively, are noted only when they occur on pages outside the related text discussion.

A

B

C

D

E

F

G

H

I

L

M

N

O

P

Q

R

S

T

U

Z